DIY GUNS

RECOIL Magazine's Guide to Homebuilt Suppressors, 80 Percent Lowers, Rifle Mods And More!

COMPILED BY
THE EDITORS OF
RECOIL MAGAZINE

Published by

RECOIL

Recoil Books, an imprint of Caribou Media Group, LLC
5600 W. Grande Market Drive, Suite 100
Appleton, WI 54913
recoilweb.com | www.gundigest.com

To order books or other products call 920.471.4522 ext.104
or visit us online at www.gundigeststore.com

ISBN-13: 978-1-951115-56-2

Edited and designed by Recoil Staff

Printed in the United States of America

10 9 8 7 6 5 4 3 2

TABLE OF CONTENTS

PROLOGUE

Building your own firearms is as American as an apple pie wrapped in a Gadsen flag. We have a long and proud tradition of independence, and this extends to creating the tools of freedom without let or hindrance from Big Brother, a fact sometimes lost on those who worship at the alter of the state. DIY firearms have become latest cause celebre of the anti-gun movement, and their timing coincides with that of increased censorship from tech giants who fancy themselves gatekeepers of all that is good and proper. This unholy combination means that all the work we at RECOIL have done over the past decade to bring quality information to our readership risks being undone. Although we regularly publish articles online regarding DIY firearms, gunsmithing etc., there's no guarantee any of it will remain available.

Looking over the horizon, the one thing we can do to ensure an end-run around digital censorship, is to roll the presses. Despite their power, Facebook and Google still can't control what we print, and should the information in these pages disappear from the online realm, at least it can still be found on your bookshelves. It's profoundly sad that we even have to consider the possibility, but historical precedent leaves us with little choice in the matter. In the event of a digital Kristallnacht, at least we can say we've done our part to preserve a small element of freedom, and save knowledge that might otherwise be suppressed.

That's the depressing aspect of this book, and we shan't dwell on it any longer. The more positive side is that home gunsmithing, tinkering with and building firearms is just plain fun, and educational to boot. If you want to have the deepest level of understanding of how your firearms work, there's no better way than to build one from scratch. In doing so, you'll learn how critical components interact, where the potential failure points are, and how to mitigate problems as they occur — it's also an opportunity to build a firearm exactly how you want it. DIY gun projects are a great way to introduce new shooters to the world of firearms, so make sure you have a spare spot at your workbench to share with a friend or relative. Passing along practical information is one way we can all help preserve our Second Amendment rights, and ensure the next generation benefits from the rights we've all enjoyed.

Iain Harrison
EDITOR-IN-CHIEF

CH. 1

Gun Builds

DISCLAIMER:
It's legal for you to build your own firearms. Convicted felons or other prohibited persons, however, cannot. There may also be other applicable laws where you live, so double check the regulations in your jurisdiction. The following article is for educational and entertainment purposes only. CMG West, its parent entities, and subsidiaries shall be held harmless in the event any entity acts on information provided in this article.

THERE ARE INHERENT RISKS AT EVERY TURN WHEN IT COMES TO FIREARMS, AND LIKE ANYTHING WITH A MODICUM OF DANGER, THEY'RE UNFORGIVING OF HUMAN ERROR.

BUILDING JOE BIDEN'S BLASTER

Making Your Own Double-Barreled Shotgun at Home

By Mike Searson

A homemade shotgun is one of the few firearms you can build without using any actual gun parts; the JBB-12 is a volley fire 12-gauge AOW (Any Other Weapon).

Earlier this year, we built a 410-gauge shotgun from a kit and a lot of pipe fittings for our sister magazine RECOIL OFFGRID. One thing about building that particular piece was that even the parts kit that was provided by Pop-A-410 could easily be sourced at a home improvement store. So this time, we set about building one relying purely on Home Depot's inventory.

Some people who consider the firearm lifestyle a mere hobby couldn't get over themselves decrying how dangerous and irresponsible these types of builds are for a shooter. There are inherent risks at every turn when it comes to firearms, and like anything with a modicum of danger, they're unforgiving of human error.

With that in mind, we set out to build a somewhat different version in 12-gauge.

BUILDING

It's legal for gun owners to build their own firearms, so long as they can legally own and possess what they're building — convicted felons or other prohibited persons cannot. There may be additional restrictions where you live, so be sure to double check.

Schedule 40 black iron pipe is sold in a number of sizes. For 12-gauge, you'll need ¾- and 1-inch diameter pipes, plus a 1-inch cap to thread on the end of the 1-inch pipe that'll act as your receiver.

It gets tricky here, as not all pipes are created equal, and the sizes aren't close to being uniform. The reason for this is that plumbing isn't a modern invention, and most standards are really just guidelines. Furthermore, pipe threads will certainly not be concentric to the bore.

When selecting a ¾-inch pipe, try to bring a few different types of shotgun shells with you to see if they'll fit. Some will, but most won't. If you check several pipes on the same shelf, you'll often find inside diameter variances between all of them. Ideally, the front of the rim should sit flush, and the shell itself should have very little, if any, play at all.

The next step is to check that the ¾-inch pipe will fit inside the 1-inch pipe. If it drops in with a tight fit, even better. Most likely, you'll need to hog out the inside with a sanding bit. Ideally, you want enough clearance and fit so that the ¾-inch pipe freely moves back and forth through the 1-inch pipe.

Ensure that the cap easily threads down on the 1-inch pipe. A proper seal is critical for this next step. This was our stumbling block, so we turned to good friends Tyler Norona and Jarrett Bushey at Illumined Arms to cut and rethread the pipe ends.

Back at the shop, drill and tap the center of the cap to install the firing pin. Centering will be the hardest part, because you'll find that even the cap is woefully out of spec when compared to most gun parts. You can mark the primer of your shotgun shell, load it in the ¾-inch pipe, and see where the marking agent hits the inside of the cap. This is where you need to drill and will become your breech face.

Once you have a hole all the way through, you can start your tap from the outside. Go slow and use cutting oil; back out every so often to keep it clean. We used an 8-40 tap

because that was the size of the set screws we had on hand.

Thread the set screw into the threads you created. You just need the tip to barely protrude. Too deep and you run the risk of piercing a primer. Once the tip of the firing pin is set, hit it for about 15 seconds with a butane torch to harden it.

Lastly, we attached a ¾-inch tee to the threaded end of the ¾-inch pipe, making a forend grip to more easily slam-fire the gun.

These five parts comprise a basic slam-fire gun. Aside from preparing the cap, the only real tools you need are a pipe wrench to ensure everything is snug, tools for cutting the pipe, and a file or Dremel to size the insides if needed.

We christened our creation the JBB-12.

INSPIRATION

Former Vice President Joe Biden once remarked that all anyone needs for self-defense is a shotgun:

"I said, 'Jill, if there's ever a problem, just walk out on the balcony here, walk out and put that double-barrel shotgun and fire two blasts outside the house,'" Biden said.

"You don't need an AR-15 — it's harder to aim," he added, "it's harder to use, and in fact you don't need 30 rounds to protect yourself. Buy a shotgun! Buy a shotgun!"

We decided to build this firearm just like he advised.

It's a bit more of a challenging build as you need to connect the two outer pipes and ensure that the barrels can move back and forth. This means welding. Once we spaced the chambered receivers, we did the same with the barrels, using ¾-inch pipe tees to add a foregrip made of shorter pieces of pipe. Fitting the pipes was tough, and we used a liberal amount of Lucas Oil Gun Grease to keep the action smooth. As it was a bit heavy and cumbersome, we mounted a piece of pipe to act as a pistol grip.

This was intended to mostly be fired from the hip. You could add a stock, but firing two 12-gauges simultaneously, inches away from your face, didn't seem like a good idea.

Construction is simple, using four pieces of pipe, two pipe caps, two tees, two set screws and some scrap pieces of steel. The only element of complexity is in lining up the welds properly to ensure consistent ignition.

YOU ALSO MIGHT WANT TO SAY A HAIL MARY THAT YOU DON'T BLOW YOURSELF UP IN THE PROCESS.

FIRING THE JBB-12

To fire the JBB-12, you load the chambers, take hold of the forward grip, aim, and slam the barrels rearward. The impact of the firing pins against the primers of the shells causes ignition and fires the projectile.

You also might want to say a Hail Mary that you don't blow yourself up in the process.

Our first try was experimental to lessen the chances of grave injury or death. We emptied the shot out of a pair of mini shells about 1.75 inches long. Only one barrel successfully discharged, so we tightened down the set screw on the faulty chamber and slammed it again. It went bang. Next, we moved up to live ammo, again in the form of mini shells, then eventually a few 2¾-inch loads from bird shot to #4 buck shot. Both barrels fired simultaneously 90 percent of the time. You can somewhat-intentionally fire a single barrel instead of two by altering your angle when you slam.

We had built a fully functional double-barrel, slam-fire gun and have to admit the recoil was much milder than anticipated, due to the weight of the piece and the thickness of the pipe. It was much quicter than a typical double-barrel, but hip firing may have played a role there.

NFA

The barrels measured 19.5 inches, and the overall length was 20 inches. While the barrel length was certainly legal if you were building a shotgun, the overall length was below 26 inches. Had we placed a stock on the JBB-12 or added an extra 6 inches to the barrels, this would be a typical Title 1 firearm.

You can make the frond easier to rack by wrapping the pipe sections in burlap (as shown), parachute cord, wire, or rubber from a bicycle inner tube.

She may not be pretty, but the JBB-12 is effective and can be either a fun "I made this" project, or a weapon of last resort when it's the only thing you can lay hands on.

THIS IS A GUN FOR A TIME WE HOPE NEVER COMES - WHEN YOUR OWN GUNS ARE CONFISCATED OR DESTROYED, AND YOU HAVE NO CHANCE OF BUYING ANOTHER ONE.

However, there was no need for a stock on this project. In this case, the end result was to register it as an AOW (Any Other Weapon).

This is a catch-all classification for firearms that don't have a traditional design, such as pen guns, glove guns, and so forth. If your pipe gun is extremely simple, with no grips, stocks, or triggers, looking like one piece of pipe inside another with a cap, that'll be an AOW. The classification is also applied to firearms with one rifled and one smoothbore barrel under the 18- or 26-inch minimum requirements, like the old Marble's Game Getter. Likewise, if you have a smoothbore firearm that has never had a stock installed and doesn't meet both length requirements, an AOW tax stamp is the order of the day.

It costs $200 to build an AOW, just like any other NFA item on a Form 1. However, the transfer tax is only $5, so there's a benefit when you transfer it to another entity.

ALTERNATIVES AND LOOSE ROUNDS

This was yet another fun project. It was definitely less expensive than building the slam-fire 410-gauge, and you can walk away with the satisfaction that you built a firearm on your own. Moreover, putting it together may give you a better sense of how firearms work.

This is a gun for a time we hope never comes — when your own guns are confiscated or destroyed, and you have no chance of buying another one. Had we built a single-shot version, our cost would've been about $45. This one came in at about $95.

If you just want a sterile firearm, there are plenty out there in the form of shotguns and rimfire rifles built prior to 1968 that have no serial numbers. They can be had for not much more than what was invested in this project.

Look for names like Hopkins & Allen, New England Firearms, Harrington & Richardson, Iver Johnson, Savage/Stevens, Sears, Western Auto, and the like. You'll end up with something safe and definitely a more accurate shooter.

The home-built slam fire gun is more of a tool to help you understand how a firearm works in most simplistic terms, providing a basic blueprint for building something like this when you have no other choice. Beyond the fun factor, this is a firearm of last resort.

It may also be your only alternative should a certain political party get their way in all branches of government and decide to shred the um, uh, you know the thing. **R**

Below: The JBB-12 breaks down into two components for ease in transportation, with each section measuring 20 inches and 14 inches, respectively.

Bottom: Two shots are better than one sometimes, but make sure your shells will fit in the chambers as you build.

THE CONDO LOWER

Finishing an 80 Percent Arms AR-15 Lower in Your Apartment

By Dave Merrill

There was a time that finishing a so-called 80-percent lower receiver required a milling machine and the requisite machining skills to operate it. These early 80-percent lowers were advertised as "paper weights" and definitely worked better as bookends than anything resembling an actual, functioning AR-15 receiver.

While what comprises an "80-percent" receiver is subject to interpretation, and therefore legally ambiguous; the current crop of what the BATFE considers 80 percent is much different than in those early years.

Advances in materials science made polymer AR-15 lowers more problematic to the BATFE, and official determination letters regularly contradicted one another. A lower with polymer in different colors to denote what needed to be removed became an issue, and some stores selling these receivers were raided.

And while there are several manufacturers of all-the-same-color polymer 80-percent lowers currently on the market, nothing beats a genuine 7075 aluminum type III hard-anodized lower within that weight class. Well, titanium exists, but not only are they unavailable, you'll pay for it in terms of materials and tooling for exceedingly little benefit.

The current crop of 80-percent lowers that the BATFE finds acceptable incorporate some significant machining — the magazine well is cut, as are takedown pin holes, the bolt catch slot, pistol grip, takedown detent holes, and critically, threading for standard receiver extensions. The machining that needs to be performed typically consists of the trigger pocket and the selector and fire control group (FCG) holes.

There are many incomplete lowers and jigs on the market, but the 80 Percent Arms kits are definitely in the upper echelon in terms of quality and ease of use.

80 Percent Arms sells lowers, tooling, jigs, and the actual tools to complete your lower. And not only do they not require a milling machine and an engineering degree, you can get everything shipped to your door to finish a receiver in your apartment living room if you so desire. (While I wouldn't say building guns in the living room led to my divorce, building guns in my living room certainly didn't help my cause.) Actually, you do need one tool they don't sell: that cordless drill that's stashed under your sink right now waiting for you to hang pictures again. Also, a vise, readily available on Amazon or at your local store. No need for anything crazy here, just something decent.

And while 80 Percent Arms sells incomplete receivers, their jig arrangement will fit most of what you'll find on the market; while we're sure they'd prefer that you buy their lowers, you still have ample options.

To be clear, up until our current (and previous) panics, you could easily buy a complete lower for less money than this combo — but that isn't the point. There's something incredibly special about making a completely legal rifle with zero markings on it, sans the fire selector (be sure to check your local laws, as this isn't the case

80 PERCENT ARMS
ALL-YOU-NEED KIT (JIG, ROUTER, BITS, AND LOWER)

MSRP: $460

URL: 80PERCENTARMS.COM

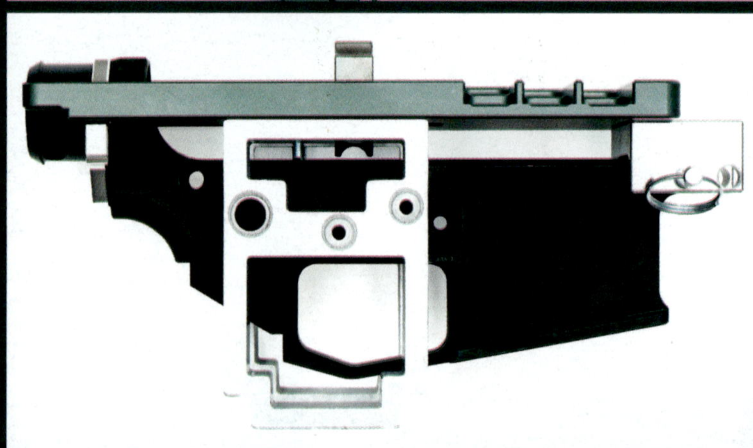

First-time assembly will look daunting—just take your time.

80 Percent Arms put a lot of work into clear labeling and markings.

everywhere). You're doing something totally legal — that you've been told is a bad thing — and have a fun weekend project to boot!

Moreover, this definitely isn't a huge project. You're not building a barn here; our first lower took about 70 minutes after setup and our next was completed within the hour.

80 Percent Arms sells not only AR-15 80-percent lowers, but also 7.62 and 9mm lowers that all work with the same jig and tools.

To complete your first lower, our suggestion is to first watch the videos on their webpage or YouTube and then to follow-on with the included written directions as you do the work. There are three sets of router spacers and a quick reference guide built-in the jig itself that make for an easy day.

We have some lessons learned we'd like to particularly emphasize for those going through this process for the first time. In no particular order:

TAKE YOUR TIME

When in doubt, stop! Included with the jig is a guide to physically attach to your router; it doesn't matter if you're using the 80 Percent Arms Freedom Router or one from Harbor Freight. Ensure all tools are setup correctly prior to cutting a damned thing. An ounce of prevention is worth a pound of cure here.

USE A TON OF CUTTING OIL

Cutting oil is cheap (we got a quart for about $6 locally). Don't use motor oil, which is intended to ensure parts don't touch each other — you want the opposite. A bottle of cheap vegetable oil would actually be better than motor oil, but we don't recommend it. Buy some cutting oil and apply a liberal amount before every cutting or drilling operation.

CHIP BRUSH

Have either a Dollar Tree paint-brush, air compressor, or both on-hand to clear out aluminum chips as you work. Not only do you want them out of the way, you don't want chips clogging your movement track for each step.

SLOW AND DELIBERATE

Remember, we're making a functional firearm, not an actual paperweight. The slower you go, the smoother your cuts will be. Though the included milling bit is capable of cutting more metal than the guide shows, it'll severely reduce the lifespan of your bits. Better to take slightly more time than to ruin your hardware. It's particularly important to go slow when the cutting bit hits corners.

UNPLUG TOOLS DURING TRANSITIONS

This is one of those tips that's easily ignored but is pivotally important: If you're swapping guides, unplug the router first. If you're check-

The only thing not included is the vise.

Left and below: Cutting oil and loose chips will quickly make for sludge; a chip brush is required, and an air compressor a bonus.

Be sure to fully clean out the entire trigger pocket before moving to your next pass.

ing screws for tightness, unplug the router first. If you're admiring your handiwork, unplug the router first. Respecting the raw power and cutting ability of your tools is never wasted. Be deliberate.

MORE LIGHT, MORE BETTER

There's probably no such thing as too much light in a work area. A dank, dark, dirty basement isn't the place to check your work. If you think there's already too much light, you're probably about halfway there.

WAIT FOR THE ROUTER TO STOP ROTATING

Want some weird ugly? Get ahead of yourself. Even while the router is rotating down, it'll still cut. Check out this ugly we got from our impatience:

Take your time. Don't rush. It'll be fine.

CHECK SCREW TIGHTNESS

Regularly check if your hardware fasteners are tight before proceeding to the next step. Though technically you could do this before every step, we found it beneficial to check when replacing router guides or if something gets weird.

FOLLOW EACH PASS

Following just the perimeter of each guide will result in an unmilled section right in the middle of the trigger pocket. Chase each pass with the router with a slow, clockwise pass to ensure all metal that was supposed to be cut is cut.

SHIELD WORK AREA

There will be aluminum bits absolutely everywhere. This is a given, but you don't want residual parts and pieces to fly everywhere — a couple of strategic gutted garbage bags will help quell the chips, and your possible divorce.

Check out the gouge we got on the back of the trigger pocket for not being patient.

PRETTY DAMN QUICK

Installing the Teal Blue Bravo Ambidextrous Release

By Dave Merrill

A proper receiver block is *always* preferred, but we're keeping the specialized tools and equipment to a minimum.

Ambidextrous bolt releases that aren't add-a-bracket pieces like the Magpul BAD lever mostly lay in the realm of higher-end AR-15s like Knights Armament and Grey Ghost Precision. The Teal Blue Bravo PDQ allows you to add a seamless bolt release to your regular AR-15 lower. Though the PDQ has been around for a few years, the fact it requires permanent modification makes many people shy away. But we figured since we were grinding on lowers in this issue, might as well grind just a little more.

To use the PDQ on your rifle, you need to extend the channel for the bolt release and remove some material from the magazine fence. You can use a milling machine if you have access to one, but we decided on the everyman Dremel option to see exactly how hard it would be.

As we followed the instructions TBB provided, we picked up some tips and suggestions to make the process easier.

SKIP THE EMERY WHEEL

The instructions say to use a #425 emery wheel to extend the channel for the bolt release all the way to the right side of the receiver. We ran into two issues with this. Firstly, the emery wheel wears down incredibly fast, taking two of them to go through a standard lower. Secondly, the diameter of the #425 emery wheel decreases as it wears, and the radius is too small to cut the channel squarely with a standard Dremel.

Our solution was to stack multiple, larger-diameter cutting wheels until they're 1/8-inch thick. The extra space allows you to perform a nice cut without going off-kilter.

USE EXISTING BOLT CATCH AS A GUIDE

To install the PDQ, the channel needs to be uniform across the entire lower. A set of needle files is extremely helpful here. You can use your existing bolt catch to check fit. Once it slides in and out, you're A-OK.

After shaving down a bit of the fence with your sanding drum around the magazine fence, give it a polish and a touch of Alumablack.

Congratulations, you've made an advanced lower for $60 and around an hour of your time.

TEAL BLUE BRAVO PDQ

MSRP: $60

URL: TEALBLUEBRAVOLLC.COM

DIY BOLT GUN BUILD

ARs can be screwed together with a vise and a couple of wrenches. Here's what it takes to go a little deeper into the art of gunsmithing

By Keith Wood

As a kid, I spent a lot of Saturdays standing around a grimy tool and die shop in a South Florida industrial park. My dad's friend was a Serbian machinist who had escaped the communist oppression in Yugoslavia and applied his considerable skill to making precision parts for the aviation and marine industries on his manual equipment. Watching Sean, whose thick accent was barely decipherable, turn hunks of metal into functional treasures was a huge influence — I've spent the three decades since with a strong desire to make things with my own hands, particularly out of metal. With a keen interest in firearms since the age of 3, I've steadily applied my desire to improve and apply shop skills into building my own guns. This is the latest of those builds, a 6.5x47L built with long-range precision in mind.

I've been a huge fan of the short 6.5mm's since before it was cool, before Dave Emary gave the world the 6.5 Creedmoor, and when everyone still thought the .308 was a "long-range" round because the military used it. This rig was definitely going to be a 6.5, but the question of "which one?" caused some heartburn. I'm in the process of building hunting rifles in 6.5-06 and 6.5 Creedmoor, and with more .260s than I can justify in the safe those were off the table. In the end, nostalgia won out: My first published article was a review of a rifle chambered in 6.5x47L, and I didn't own a rifle in that cartridge, so I decided to build myself one to scratch that itch. It's not like it's a terrible choice for such a rifle, either.

Custom actions are fantastic choices for a rifle build since the best of them come pretty close to perfect right out of the box. That said, there's nothing wrong with a Remington 700 once the factory tolerances have been addressed, especially if you already have an action on hand. A short-action 700 sitting in the safe was perfect to use for this project. Greg Tannel at Gre-Tan rifles trues actions so well and at such reasonable prices that sending him the work was a no-brainer.

Greg single-point re-machined the action threads, trued the receiver face, bolt face, bolt lugs, and lug seats, and double-sleeved the bolt body with stainless steel to keep the bolt from jumping during ignition. He altered the action to accept his own double-pinned 0.250-inch recoil lug, bushed the firing pin hole, and installed his own lightweight firing pin and spring, turned to fit. All of this work ensures that the relevant surfaces of the action are square and true and, like the foundation on a building, provides a suitable and stable platform for the barrel.

This rifle wasn't designed to be a lightweight, but given the components involved, it could end up being excessively heavy. I wanted to cut weight where I could without compromising performance, so I

1, 3. Once the barrel is dialed into the lathe's customized four-jaw chuck, a facing cut is made to square-up the breech end.

2. No barrel is perfectly straight. Using a pair of 0.0001-inch test indicators and a close-fitting range rod, the author was able to ensure that the rifle's bore ran perfectly true over the length of the chamber.

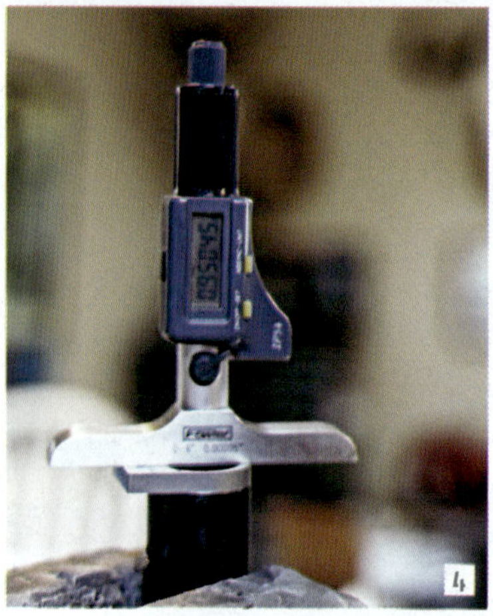

4. Using a depth micrometer to measure the distance to the bolt face and locking lugs, the proper dimensions of the barrel tenon are determined.

5. With the barrel spinning at 450 rpm, the tenon is cut in a series of passes using a carbide turning tool.

chose a 26-inch bull profile carbon-fiber–wrapped barrel from Proof Research with a 1:8-inch twist. What's great about Proof's barrels is that they offer the handling and shooting characteristics of an all-steel heavy barrel without all of the weight. All other things being equal, using a Proof barrel cut 4 pounds compared to an all-steel barrel with the same profile.

McMillan's A5 stock is a great design for the type of shooting that this rifle would be used for, but I was impatient and didn't want to wait for a custom stock made to my component specs. Grayboe is a McMillan-affiliated company (run by Ryan McMillan) that offers solid fiberglass stocks in finished configurations at very competitive prices. Greyboe's Renegade is an A-5 clone that comes as a drop-in stock complete with aluminum pillars, a 1-inch recoil pad, sling swivels, and paint. This stock wasn't designed to accommodate the exact components planned for the build, but since it was constructed using solid fiberglass, I could machine it to fit my needs. Best of all, the stocks are only $349.

With the majority of the components in place, it was time to start making chips fly. Much is made of CNC machining these days, and that technology is amazing, but CNC really shines when you want to make several of something. For one-off projects, manual machines can be just as precise (or even more so) when used properly. My personal lathe is a Chicom Grizzly that I've adapted to better suit my needs — there are far better tools on the market, but since I derive zero revenue from building guns, a larger investment wasn't warranted.

When it first arrived, it vibrated so badly that it just about hopped its way out the door when run at high rpm. Over time and with plenty of tinkering, I've gotten it pretty well dialed-in. As in just about anything, the manner in which the equipment is used is far more important than the price tag. I learned how to fit and chamber barrels from Robert Gradous who is sort of a quirky genius when it comes to building rifles. Using his methods, I've learned to turn out some real shooters. This article isn't an absolute step-by-step how-to article, but will give you a flavor of what goes into building a custom rifle.

1

2

3

4

5

1, 2. A full-profile threading tool is used to cut the 16TPI barrel threads. As the threads are deepened 0.001-inch at a time, the action is test-fit onto the barrel.

3-5. The bolt nose on a Remington 700 protrudes past the locking lugs and requires that a recess be cut into the barrel face. A piloted tool from Manson Reamers does quick work of cutting the proper-dimensioned clearance.

6, 7. A piloted chambering reamer secured in a Manson Precision Floating Reamer Holder is used to cut the 6.5x47L chamber, 0.050 inch at a time. The lathe is running at its slowest rpm and lots of cutting oil is used — this is no time to be in a rush.

8. To ensure that the chamber is cut to the proper depth, a headspace gauge is used to track the progress of the cuts.

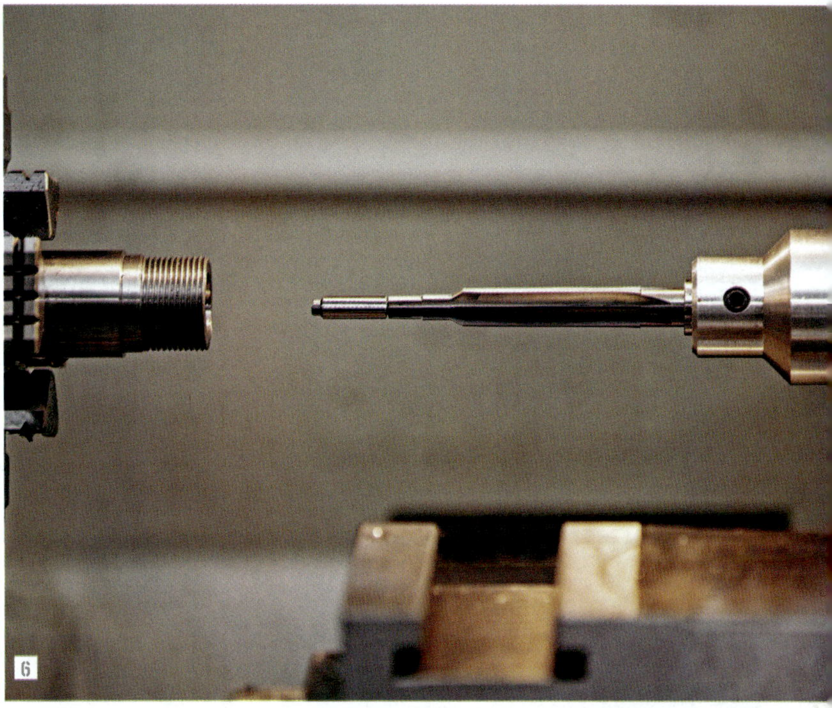

6

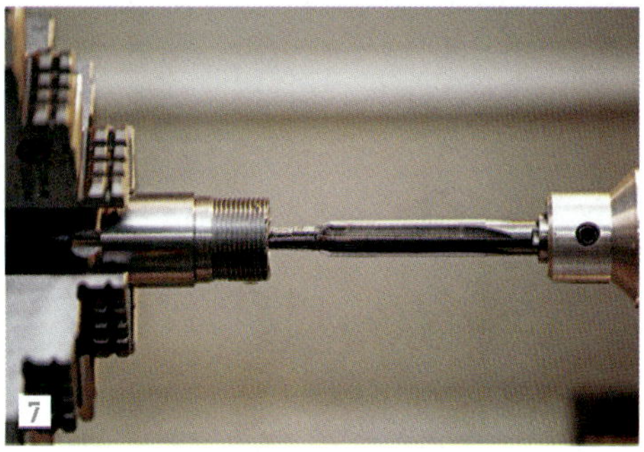

7

8

Some guys spin a barrel into a collet chuck on a lathe until it doesn't visibly wobble and chamber rifles that shoot pretty well. That may be fast and it may work for them, but it's not for me. If I'm bothering to build my own rifles, I'll do so to the absolute best of my ability. This starts with ensuring that the chamber end of the barrel is running as true as possible with the cutting, threading, and reaming tools that we'll use to fit it.

Building on the techniques taught by Gradous, I use a gimbal system described in William Hambly-Clark's excellent book, *Centerfire Rifle Accuracy.* It doesn't matter what brand of custom barrel is chosen or how much you pay for it; the bore won't be perfectly straight. A bore may wander by 0.050-inch over its length, so dialing in both ends of a barrel is a fool's errand. Using the gimbals to ensure that the chamber area of the bore is axially aligned with the center of the lathe's chuck, we can ensure that the bullet enters the rifling as perfectly as is possible.

How do we accomplish this? By fitting a range rod into the bore using a removable pilot bushing, ensuring a snug fit. The rod allows us to use two test indicators to ensure that the bore is running true over the length of the chamber rather than at a single point. This is a tedious process than can take well over an hour to get right. My goal is to have both indicators running at less than 0.0001 inch (that's one

1. If the bolt closes on the "go" headspace gauge and doesn't close on the "no go" gauge, the chamber is in spec.

2. A barrel wrench is used to torque the action onto the barrel — this often leads to 0.002 inch or so of "thread crush" on the headspace measurement.

3. The stock came pillar-bedded, but the author re-bedded the rifle for a perfect fit. An end mill was used to create some locking recesses for the bedding epoxy.

ten-thousandths of an inch) over the 3 or so inches that we'll ream. Once dialed in, we begin by cutting the barrel tenon to the proper dimensions, determined by measuring the action with a micrometer. These cuts are pretty basic lathe operations, but we want to take light cuts so as not to stress the setup that we spent so much time perfecting.

With the tenon cut, it's time for the most nerve-wracking part of the build: cutting the threads. Threading is a fast and potentially tricky procedure, especially when you don't do it on a daily basis. Make a mistake at this stage, and your $900 barrel may very well end up in the trash. Remington actions are threaded at 16 threads-per-inch; I prefer to use full-profile threading inserts, especially once an action has been single-point recut.

After triple-checking the setup and with plenty of lube on the steel, the phone is turned off, the door is locked, and the process begins. Pass by pass, the threads grow deeper and deeper until it's time to test-fit the action. Once the action starts onto the threads, .001-inch passes are made until a perfect rattle-free fit is achieved. Every Grizzly lathe I've encountered leaves a wavy finish on the threads that's visible under close scrutiny but this effect is purely cosmetic; the barrel to receiver fit on this rifle is about perfect.

The bolt nose on a Remington 700 protrudes beyond the locking lugs, and a recess must be cut into the barrel to accommodate it along with suitable clearance. This can be done relatively simply with a boring bar, but Manson Precision makes a handy piloted tool that reams this recess rather quickly. Once this relief is cut and test-fit to ensure that the bolt isn't contacting the barrel, we can turn our attention to reaming the chamber. In theory, if we did our dialing-in correctly, we could run our chamber reamer into the bore using a fixed mount and encounter no issues.

Since nothing about my lathe is 100-percent perfect, I use a piloted reamer and a floating reamer holder (also from Manson Precision) to keep the reamer on track with the bore. I use lots of lube, my lathe's slowest rpm, keep my passes short and slow, and clean the reamer thoroughly after each pass. Using a 0.001-inch dial indicator rigged to a homemade stop on my lathe's tailstock, it can hit the headspace measurement precisely, keeping in mind that this measurement will "crush" by 0.001 to 0.003 inch when we torque the barrel onto the action. A set of "go" and "no go" gauges confirm that our headspace is within spec.

With the barreled action ready to test fire, the stock was next. The Renegade's fore end is molded to accommodate a Sendero barrel contour, so a ¾-inch ball end mill in the Bridgeport opened it up to fit the 1.200-inch Proof. I also opted for bottom metal from Seekins Precision, requiring the Badger Ordnance-sized inlet to be recut. Using Marine-Tex epoxy, I pillar

1 & 2. Since the Grayboe stock was made to fit a Sendero barrel contour, the author used a ¾-inch carbide ball end mill to open the stock's barrel channel.

3. Modeling clay and various thicknesses of tape are used to prepare the action for glass-bedding.

2

1

3

bedded the barreled action into the stock as well as any voids made by my substituted components. The barrel was free-floated to leave 0.010-inch clearance except for a small pad left to support the shank just ahead of the recoil lug.

The exposed stainless steel sections of the barrel along with the recoil lug were bead blasted, and the receiver went off to Glenrock Blue in Wyoming to be caustic blued. Finishes such as Cerakote add dimension to the action that wouldn't be compatible with the tight bedding that I worked hard to create; besides, bluing is a highly underrated finish if you take decent care of it. The stock went to Custom Gun Coatings for a Flecktarn camo paintjob using Sherwin-Williams Polane epoxy, one of the most durable paints on the market.

1. A carbide cutter on the milling machine is a fast and precise way to remove excess Marine-Tex epoxy from the stock.

2. The action was sent to Glenrock Blue in Wyoming for a satin black finish — the bolt sleeves are made from stainless steel so they didn't take on any finish.

Once all of the parts came back, I mounted a Vortex Razor HD Gen II 4.5-27x56mm in a set of Seekins Precision 34mm rings and a Nightforce base. With a 50-round brick of Lapua factory 139-grain ammo, it was time to see whether my work had paid off. This gun is no lightweight. With the addition of Vortex's flagship monster mounted on it, it weighs a full 13 pounds, 5 ounces. The result of that mass is a very forgiving rifle with nearly zero felt recoil — ideal for calling one's own hits or misses on steel.

Factory loads made little 3/8's triangles at 100 yards and 123-grain Scenars and Varget showed even more promise. Yes, the lighter bullets give up some BC at long range, but they're fine for most of the shooting I'll be doing for now. With Lapua cases and Redding bushing dies, I'd cobbled together some loads that sounded good in theory. I settled on 38 grains of powder and experimented with three different primers, with the bullets seated 0.015 inch from the lands.

It's often surprising what a difference a humble little primer can make: the loads using CCI 450s shot 0.61 inch, BR-4s 0.44 inch, and #41s put two shots into 0.13 inch until I yanked the trigger on the third round and threw it ¼ inch to the left. Velocities were all in the 2,950 fps range. I headed home and loaded a half-dozen rounds with the #41 primers, hoping to get consistent groups under 0.25 inch. Though from a practical standpoint, a gun that shoots 0.5 MOA consistently is a keeper.

A few days later I was back at the range on a hot and humid but windless afternoon. On a gun this heavy, trigger control is everything — I focused on getting three clean breaks as carefully as possible. Through the 27x of magnification I could tell that the first two rounds were in one oblong hole — better not mess up now. Two pounds of sear never felt so heavy until, finally, the third shot broke. The calipers say it's 0.253 inch — it's a keeper.

A gun that shoots ¼ inch at 100 yards is nothing to write home about these days, but when you wear a suit and tie to work and can build such a rifle yourself, it's a different story. Since I had some of the parts already on-hand and others were donated by their makers, I probably have less than $1,200 in the rifle. That said, there's nothing economical about building your own rifles — the cost of setting up my shop would've paid for a lifetime of gunsmithing bills. That's not the point, though; not only has building my own guns allowed me to fulfill whatever firearm wish I can come up with, it's given me a better understanding and appreciation of the factory and custom guns that I evaluate as a writer. There's a lot of bullsh*t in the gun business, and this process has helped me to sniff it out a mile away. **R**

DIY 6.5X47L

ACTION:
Short-action Remington 700 (Brownells)
› $459

MACHINE WORK:
Action blueprinting by Gre-Tan Rifles
› $450

BARREL:
26-inch Proof Research 6.5 Bull profile, 1:8-inch twist
› $900

STOCK:
Grayboe Renegade
› $349

BOTTOM METAL:
Seekins Precision AICS
› $35

MAGAZINE:
Magpul PMAG AICS
› $208

TRIGGER:
Timney Calvin Elite
› $227

SCOPE RINGS:
Seekins Precision 34mm
› $149

SCOPE MOUNT:
Nightforce
› $57

OPTIC:
Vortex Razor HD Gen II 4.5-27x56mm
› $3,399

BLUING (GLENROCK BLUE):
› $50

STOCK PAINTING (CUSTOM GUN COATINGS):
› $175

WEIGHT:
13 lb. 5 oz. (w/scope)

OVERALL LENGTH:
45½ inches

URLS:
www.brownells.com
www.customguncoatings.com
www.glenrockblue.com
www.grayboe.com
www.gretanrifles.com
www.magpul.com
www.mansonprecision.com
www.nightforceoptics.com
www.proofresearch.com
www.seekinsprecision.com
www.timneytriggers.com
www.vortexoptics.com

THE GHOST GLOCK

Completing an 80% Glock Lower

By David Merrill

No doubt you've seen 80-percent AR-15 lowers. They've taken many forms over the years from simple forgings to billet pieces in various states of completion. If the Bureau of Alcohol, Tobacco, Firearms, and Explosives (BATFE) has a special calculation as to the difference between an 80-percent lower and an 81- or 82-percent lower, they haven't shared it with us.

By and large, what the BATFE considers a firearm seems mostly to be a subjective measure. There've been incidents in the past where a lower was deemed kosher by the BATFE — until it wasn't. A notable example of this in the AR-15 world was the EP Armory 80-percent lower that utilized two different colors of polymer: one that needed to be drilled and milled out, and the other that didn't. Another was an "AR flat" that consisted of slabs that could be drilled, tapped, and screwed together (a weldable version is still allowed at the time of this writing).

It was only a matter of time before we saw 80-percent Glock lowers for sale. There've been some scratch-made Glock receivers built in the past, with materials ranging from sheet steel to converted airsoft lowers (the latter was harder than you'd think). Polymer 80 first wet its feet with AR-15 lowers and more recently moved into the world of Glock.

There've been a few third-party complete Glock receivers, all with varying levels of success and failure, but Polymer 80 has produced the first commercially available 80-percent: the Spectre. Polymer 80 started with a standard 9mm/40 frame, suitable for constructing third-generation Glocks (17/17L/22/34/35) in a variety of colors. Likely by the time of publication, they'll have a Glock 19-sized option available.

For $160 you get the incomplete frame, a disposable jig, mill and drill bits, and a proprietary locking block. All the small parts and pieces you'll have to provide yourself, as well as the slide and barrel assembly. A quick jaunt to Gunbroker and other online auctions shows complete Glock receivers for as low as $165, so some of you

WARNING!
When handling firearms, always observe safety rules and the precautions set forth in the firearm's owner's guide. Be certain that your firearm is unloaded and made safe before proceeding with this DIY.

may be asking "why bother?" Why not? Not every DIY project one undertakes has to be done solely for cost savings; the pure pleasure of working with your hands is enough for many. Still, there may be those worried about prospective bans, and perhaps more who simply like the idea of having a legal, homemade gun for personal use (subject to your state laws) — like the Ghost Guns California State Senator Kevin de Leon is so worried about.

The parts you need to complete this build include the following:

- Complete slide assembly, including barrel and recoil spring
- Trigger and trigger bar
- Rear housing assembly, including trigger spring, ejector in appropriate caliber, and connector
- Magazine catch and spring
- Slide stop lever
- Slide locking lever and spring
- Pins (trigger, locking block, and rear housing assembly)

Going through our box o' parts resulted most of what was needed, and the rest was obtained for under $20. The magazine catch spring, one of the simplest parts, was the hardest to source. For those without a box o' parts, some aftermarket shops have started selling receiver completion packages. Lone Wolf Distributors currently has them listed for $79.

For complete slides, Polymer 80 sells their own, and auction sites like GunBroker are a safe bet if you want OEM. Some resellers seem to do little more than separate the slides and receivers and sell them separately. Of course, a ton of aftermarket slides and barrels are out there. We decided it would be best to go with an OEM slide to minimize possible issues with fitment.

The Spectre frame isn't just a facsimile of the OEM model. It has a larger gripping area, no finger grooves, an integral beavertail and magwell, and a full Picatinny rail. It's also available in a variety of colors. Just giving the kit a glance over, it doesn't seem like there's actually too much to do — mill off the front bits that stick out, drill three holes, and machine the rail in the back. The rail is what we expected to give the most trouble, in no small part, because while a traditional Glock

Though it might look like a lot of work to the uninitiated, there's not too much required to bring this project to completion.

Complete slides are readily available online and can ship directly to your door.

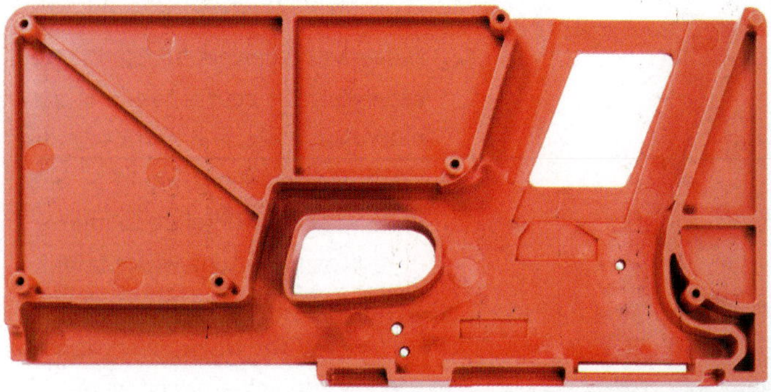

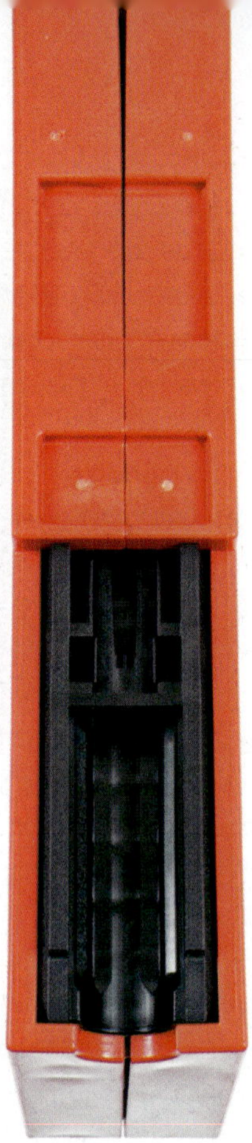

handgun frame is mostly polymer, the rear rails themselves are metal.

The Polymer 80 webpage has links to video instructions, which we dutifully watched for 30 seconds before deciding to just dive right in. Let's talk about how we did, where we messed up, and how to avoid the same pitfalls yourself.

While we have access to a milling machine, in order to make this more of a "home build" we decided to use some more common tools. An old 12-inch tabletop drill press, Dremel tool (of course), a drill press vice, and some clamps. The idea was to put the end mill bit into the drill press, secure the jig in the vice, raise the drill press table itself to achieve proper height, and move the vice itself around to perform the milling. The jig wasn't a perfect fit, so the clamps ensured it fit tightly around the incomplete frame as we went.

We're talking plastic here, so you don't have to mill this area down in 40 passes like you're working something like hardened steel. Still, a few slow passes makes for much cleaner work. We did bobble it a couple times and you can see the scars on the jig itself. Oh well, it's single use anyway.

You can see why a clamp is recommended to keep the jig tight around the frame.

Go slow if you want a job clean enough to show off to your friends.

When drilling the holes, ensure you're perfectly lined up — it's probably best to drill each side separately. One of the holes we drilled through without being perfectly level, making for a larger headache later. Then came milling the rear rails. While you'll be cleaning up this area later during fitting, this isn't something you want to mess up. Go nice and slow.

And that's it! Well, almost. The actual machining operations are fairly simple, but parts fitment can be more time consuming. Start

A flat file is perfect for final fitting.

with a stripped slide to check for fit. Use a flat file to make modifications as required, but continue checking for fit often — you can always take off material, but adding it on is a different story.

Next, you'll want to install your small parts. The included locking block is the first to go in. It'll absolutely be tight, but don't remove any material. Line it up and tap it in. From there you might think you can install your trigger, but you'd be wrong. The holes themselves are undersized, and there are reports by some of frame cracking. Some builders have used larger bits, but we used a hand reamer until the pins went in without an undue amount of force.

After your pins are installed, bust out your Loctite or other thread locker, and secure the included machine screws to ensure your locking block stays in place. Your slide locking lever and spring go in next.

And check fitting. Again. Start again with a bare slide. Then, an assembled slide without barrel. And finally, with the barrel. Fitting in steps allows you to isolate and diagnose any potential issues with the parts and your machining. Don't go nuts and take off too much at once.

Though it took a bit of time, we felt fairly confident it would be easier to do another. And with fewer tools. So we did. Though we didn't have time to grow a scraggly beard and find a proper cave to work in, we settled for not showering for a day and doing everything on a dirty basement floor. You gotta set the mood. Glocks are like AKs in that no matter how dicked up they are, there's always some assh*le who (drunkenly) says, "I bet it still shoots good!" Let's test that.

Another Spectre was obtained, this time in FDE. That desert color also helped with the mood. For tools we went with a Dremel (naturally), cordless drill, and a clamp. In a pinch, you could probably just use the cordless drill. Sanding drums on the Dremel made quick, albeit dirty, work of the front area that has to be milled out. The

Use the best tools you can, but you don't necessarily need the best tools.

Drilling and Dremeling, and not much more.

The beavertail is a welcomed touch.

rear rail and the holes were done freehand. To fit the rails a stripped slide and hunk of wood were used to rough it out. Total time? Twelve minutes — with most of that being swapping out sanding drums and snapping photos.

Assembled, she's ugly as hell. But, do they shoot?

The first time shooting any home build gives us pause. Yes, people have done this before, but we can't help feeling like we should be using a vice, a string, and a ballistic shield. I settled for extending my arm and turning my head to the side — all that was missing was a hearty "Hold my beer, and watch this!" We wish we could report that both pistols were phenomenal right out of the box ... err basement? Ultimately it took a couple range trips along with some tinkering before they became more reliable. The black Spectre suffered from ejection issues and the FDE would fail to return to battery.

With a generous amount of lubrication and rounds sent downrange, the

POLYMER 80
80% PISTOL FRAME V.1

PRICE: $160

URL: www.polymer80.com

FDE Glock self-corrected. Apparently a rushed Dremel job doesn't make for the smoothest of actions. Huh. The issue with the black Spectre was a bit more nuanced. There was some play in the rear trigger housing, likely due to mucking up the pin hole a bit, Ultimately we ended up shimming it in place to restrict movement, though there are still some phantom malfunctions to track down.

All told, the Spectre was a fun afternoon project with some more work on the back end at the range. Don't be like us — watch the official videos and use the best tools that you have available. But even a Dremel can probably do some excellent work if you have a steady hand and aren't racing the clock. In terms of receiver completion, this is a very easy project. With the frames being polymer, you can also burn and stipple to your heart's content, or until you mess something up so badly you have to scrap the whole thing.

The longevity of the polymer rails at the rear of the receiver is our main concern, and the only way to see if it holds up is to continue to shoot the hell out of it. R

ROLL YOUR OWN

Building Your Own Lightweight AR-15

By Steven Kuo

If you've read through the past five pages on the OIP rifle, you've seen what three grand (or more) will buy from a custom rifle builder. It's a hefty price tag, but you get a purpose-built weapon that's well-sorted, perfectly tuned, and fully optimized. But what does it take to roll your own lightweight AR-15? It's the golden age of the AR platform, with a mind-boggling array of components available to suit every imaginable use and taste. So, utilizing freely available components and with an eye on your pocketbook, what can you pull off in your own workshop?

DEATH BY A THOUSAND CUTS

Bentwood Gunsmithing achieved extreme weight savings in their OIP carbine by lightening every part of the rifle they could. The problem for home builders is the expense of acquiring the lightest available alternative for every single part — and even then Bentwood invested in additional customization and gunsmithing work. We sought to put together our own lightweight build for less scratch, though we'll tell you upfront that it's still not cheap...nor is it the prettiest rifle you've ever seen.

The heaviest parts in an AR-15 are typically the barrel, handguard, upper and lower receiver, and buttstock. We targeted those components first for hefty weight reductions and justified some splurging here to avoid compromises in function. Fortunately, we found a very reasonably priced pencil barrel from Faxon, the same manufacturer as used by Bentwood. While their 14.5-inch version was very light at 17.5 ounces, and its mid-length gas system tempting, as home builders we opted for their 18.8-ounce, mid-length, 16-inch barrel. This meant we avoided pinning and welding the muzzle device, preserved serviceability, and let us take advantage of the wee Taccom compensator since we didn't need to extend overall length to 16 inches. Compare this to M4 profile barrels, which can range from 1.75 to just over 2 pounds.

For the handguard, we chose one of our favorites, BCM's popular KMR, made of aluminum-magnesium alloy with a full top rail and KeyMod all around. It's a great combination of functionality and reduced weight, but it's a bit pricey. We selected a 13-inch variant as a compromise between versatility and weight — 7.5 ounces including the barrel nut, as compared to cheese-grater Picatinny handguards that can clock in at a pound and beyond for 12-inch versions. Opting for the 10-inch KMR would save you 1.1 ounces, and upgrading to the 15-inch adds 0.7 ounces. For those watching their budget, a 12-inch carbon-fiber handguard from AP Customs provides a 30-percent cost savings with similar weight and one less inch in length.

There are a couple lightweight aluminum AR receiver sets, including the Battle Arms Development BAD556-LW (similar to the OIP) and the 2A Arms Balios, featured in Issue 16. However, they're both quite expensive at $550 and $520, respectively. To save some cash, we went with X-7 Fusion polymer receivers from Kaiser Shooting Product, which cost less than half the price of the lightweight aluminum sets at $240 for the set. While reading the words *polymer* and *AR receiver* placed next to each other, you might have some reservations — as did we. But Kaiser's offering is unique in its use of aluminum inserts at key locations: the trigger and hammer pin holes, the receiver extension threads, and the pivot and takedown pin holes. Fears were further allayed when we learned Kaiser stands behind the X-7 with a lifetime warranty. At a combined weight of 11.1 ounces, they're quite a bit lighter than typical forged receivers, which are almost a pound.

Buttstock assemblies are heavier than you might expect. A complete setup from BCM (carbine receiver extension, end plate, castle nut, and BCMGunfighter stock) weighs almost 13 ounces. We'll admit our buttstock solution was the most WECSOG (Wile E. Coyote School of Gunsmithing) aspect of the whole build. The Kaiser lower relocates the channel for the takedown pin detent and spring to a vertical orientation so the pistol grip retains it rather than the end plate or buttstock. Not needing to worry about retaining it with the stock assembly, we screwed on a rifle-length receiver extension and affixed the buttplate from a standard A2 stock we had lying around — for a total weight of just 5 ounces. Want something a little less ghetto? Ace's ultralight stock reinforces the buffer tube at its weakest point and adds a more secure buttplate, but at a weight penalty of an additional couple of ounces.

We used an aluminum bolt carrier from Whiskey Arms, about half the weight of a BCM Mil-spec unit from which we scavenged a bolt, firing pin, and other bolt carrier group (BCG) parts. Additional reciprocating weight was saved courtesy of the Taccom adjustable recoil system in place of a standard buffer. With all these feathery race gun parts, we needed an adjustable gas block to dial down the gas. SLR Rifleworks sets the standard for robust and easily adjustable gas blocks, and have served us well on other builds — we installed an older 0.625-inch set screw model we had on hand. SLR now makes a titanium version that would have been even lighter, but $60 more.

Titanium variants of pins and other small parts are available on the market. But they're expensive and with a total weight of just 1.3 ounces for the components to be replaced, we saved our money and opted for an affordable lower parts kit from Strike Industries. Also from Strike came our safety selector with a wide selection of light, polymer levers. Final touches came in the form of the Alexander Arms Tactical Trigger (which we chose because of its skeletonized trigger and hammer), and an old-school A2 pistol grip, to save weight and money since we all probably have a drawer full of old A2 grips that we discarded in favor of heavier but nicer aftermarket grips. We're trigger snobs, so we opted for a nice aftermarket trigger, but you could save yet more money by using a stock trigger or an ALG.

We noticed after a range session that the ejection port cover cracked. We're not sure how it happened, but it still functions.

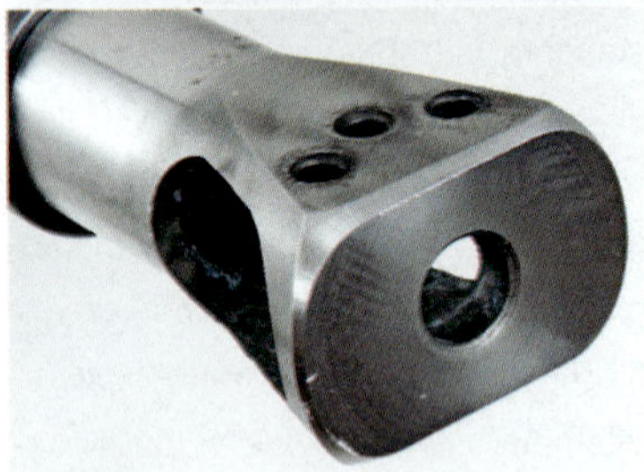

The miniscule Taccom compensator can be tuned with set screws — we ended up opening all of them.

The Proctor sling requires no hardware and easily loops onto the rifle.

Strike Industries provides 10 different selector levers to choose from. Note the outlines of the aluminum inserts in the Kaiser X-7 Fusion polymer receiver.

The Trijicon RMR06 and Alamo Four Star mount are extremely light and compact.

IN THE WORKSHOP

With the help of a solid vise and good armorer tools (see page 144), our build was straightforward to accomplish in a garage workshop. In particular, the Geissele reaction rod (or equivalent) was mandatory in order to safely install the barrel nut on the polymer upper. We must say that seeing a few polymer shavings appear during installation of the barrel extension raised an eyebrow, but all was well in the end.

We discovered that the lugs on the Kaiser upper are a bit oversized, making it impossible to install on most of the aluminum lowers that we had. Out of about seven or eight different lowers, we could only seat it on a Colt — with the help of a hammer. On its matching polymer lower, the upper fit as snug as a bug. The Kaiser lower, on the other hand, had no problems accommodating various aluminum uppers.

The fire control pocket in the lower was rather tight. After squeezing the Alexander Arms trigger into the lower, it was stiff at first, but loosened over time. The single-stage trigger pull started at 4.5 pounds after the initial build, falling with use to a 4-pound pull.

After all was said and done, the rifle weighed in at exactly 4 pounds. Not too shabby. The total MSRP was a princely $1,600, but with some patience and smart discount shopping you can collect your parts over time for less.

We installed a Trijicon RMR on an Alamo Four Star mount, adding just 3.7 ounces to the rifle. Bobro's new Lowrider backup iron sights are incredibly thin and light, and a slim Proctor sling (designed by RECOIL contributor Frank Proctor) rounded out the minimalist theme. The final package was 4 pounds, 7 ounces.

AT THE RANGE

We eagerly took our handiwork out to the range, first dialing in the gas block for our lightened BCG and buffer. Six clicks later, the rifle was reliably ejecting, feeding, and locking back on empty. Then we moved on to our bench work.

With Freedom Munitions 55-grain FMJ ammo traveling at 3,228 fps, our best group with a 10x scope was a hair under 1 MOA, but most other groups spread out to 1.5 to 2 MOA. Surplus Ammo 55-grain FMJ turned in 2 MOA groups, and our 69-grain Gorilla SMKs posted 1.5 MOA groups.

When it comes to recoil impulse, 16-inch mid-length gas systems are at a bit of a disadvantage when compared to mid-length 14.5 inch or rifle-length 18-inch barrels. As expected, while the rifle was pleasantly flat and soft shooting (especially for such a light gun), it didn't quite measure up to the OIP in this regard.

The rifle's ergonomics were great, snapping on target quickly and transitioning easily. The RMR is light and tiny, but its small field of view demands a disciplined cheek weld. The A2 pistol grip's finger groove proved to be annoying and will probably meet a Dremel in the near future. The skeletonized buttplate was surprisingly comfortable, but not particularly well suited for shooting from the bench. With just a screw securing it to the receiver extension, when struck or yanked in a counterclockwise direction it would come loose. Some epoxy solved that problem, but we wouldn't feel confident if we needed to mortar the weapon at some point to clear a malfunction.

With the aluminum Whiskey Arms BCG, we ran the gun sloppy wet and didn't have a single problem. If you prefer the peace of mind of a steel BCG, the AP Customs Rhino low mass bolt carrier is only about 2.4 ounces heavier and about the same price. Cheaper alternatives are out there, but they aren't as light. Overall, we were quite pleased with this build. Assembling a rifle to your exact specifications and preferences is a lot of fun. You needn't take it to the airy extreme shown here, but you can apply similar logic to customize your own builds. Give it a try — you'll be glad you did! R

PART	MAKE & MODEL	PRICE	WEIGHT (OZ)
Lower and upper receiver	Kaiser Shooting Products X-7 Fusion	$240	11.1
Barrel	Faxon Firearms 16-inch Pencil	$175	18.8
Handguard	BCM KMR 13	$300	7.5
Muzzle device	Taccom .625-inch ULW Compensator	$55	1.1
BCG	Whiskey Arms Aluminum Lightweight Bolt Carrier	$200	3.9
BCG	BCM bolt parts	$103	2.1
Gas block	SLR Rifleworks Sentry 6 set screw adjustable gas block (steel)	$120	1.2
Gas tube	Faxon Firearms mid-length gas tube	$16	0.7
Charging handle	BCM Gunfighter Mod 5	$48	1.2
Lower parts	Strike Industries Lower Parts Kit	$35	1.3
Safety selector	Strike Industries Ultimate Selector Switch	$40	0.3
Fire control group	Alexander Arms Tactical Trigger	$204	2
Receiver extension	BCM rifle-length receiver extension	$20	3.3
Buffer	Taccom Adjustable Rifle System	$37	1.9
Buffer spring	Rifle action spring	$5	2.1
Buttstock	Buttplate only from A2 stock		1.7
Pistol grip	A2 pistol grip	$3	2.9
RIFLE-ONLY		**$1,601**	**64**
Optic	Trijicon RMR06	$708	1.1
Optic mount	Alamo Four Star RMR mount	$119	2.4
BUIS	Bobro Lowrider Sights	$211	1.6
Sling	Proctor Sling	$29	2
TOTAL		**$2,668**	**71**

WILE E. COYOTE LIGHTWEIGHT

CALIBER: 5.56mm

MAGAZINE CAPACITY: 30 rounds

BARREL LENGTH: 16 inches

OVERALL LENGTH: 35 inches

WEIGHT: 4 pounds (rifle only)

ACCESSORIES:

Trijicon RMR06 $708
Alamo Four Star RMR mount $119
Proctor Sling $29
Bobro Lowrider Sights $211

MSRP: $1,601

PRICE AS FEATURED: $2,668

THE 1,000-YARD AR-15

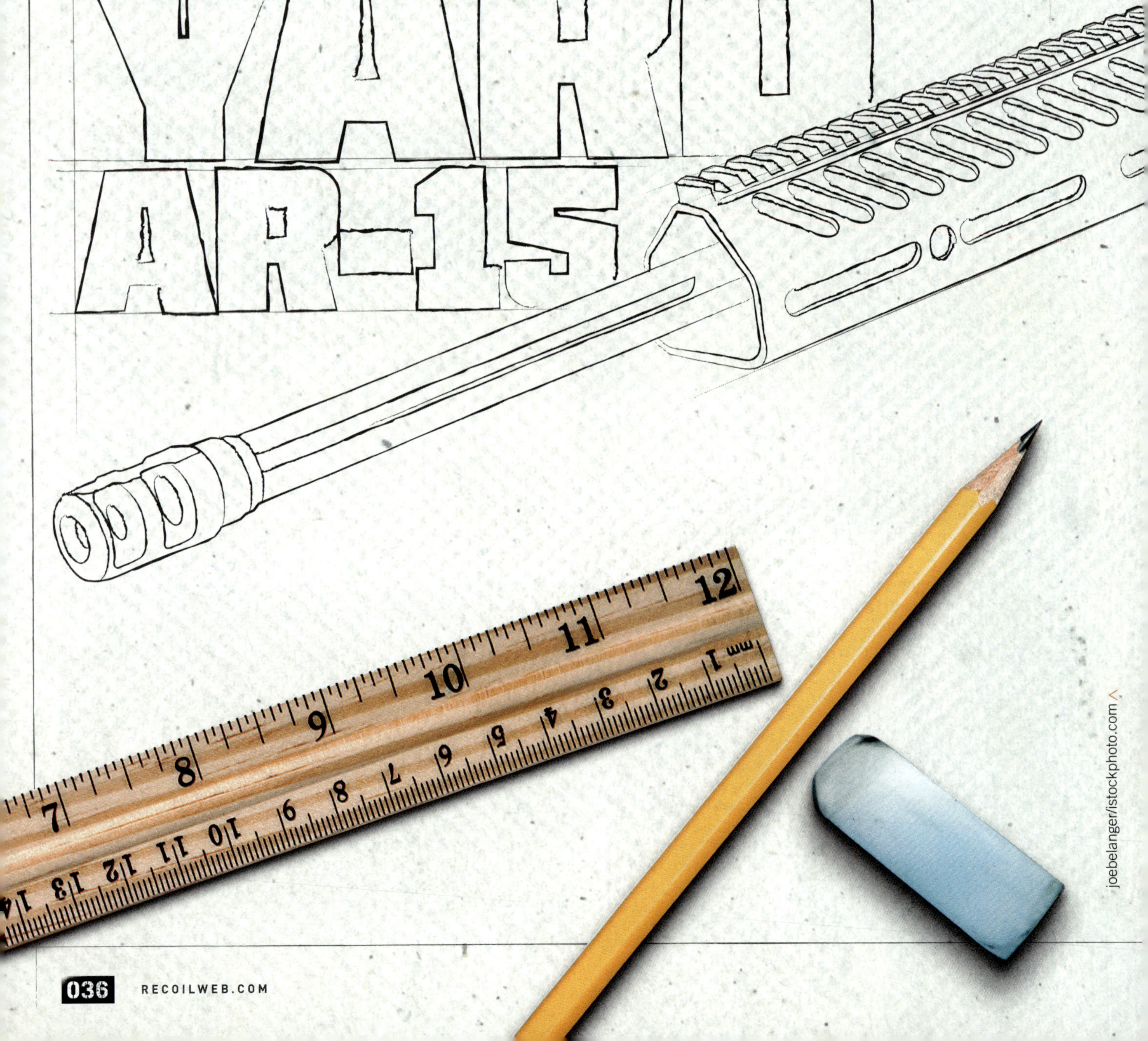

joebelanger/istockphoto.com

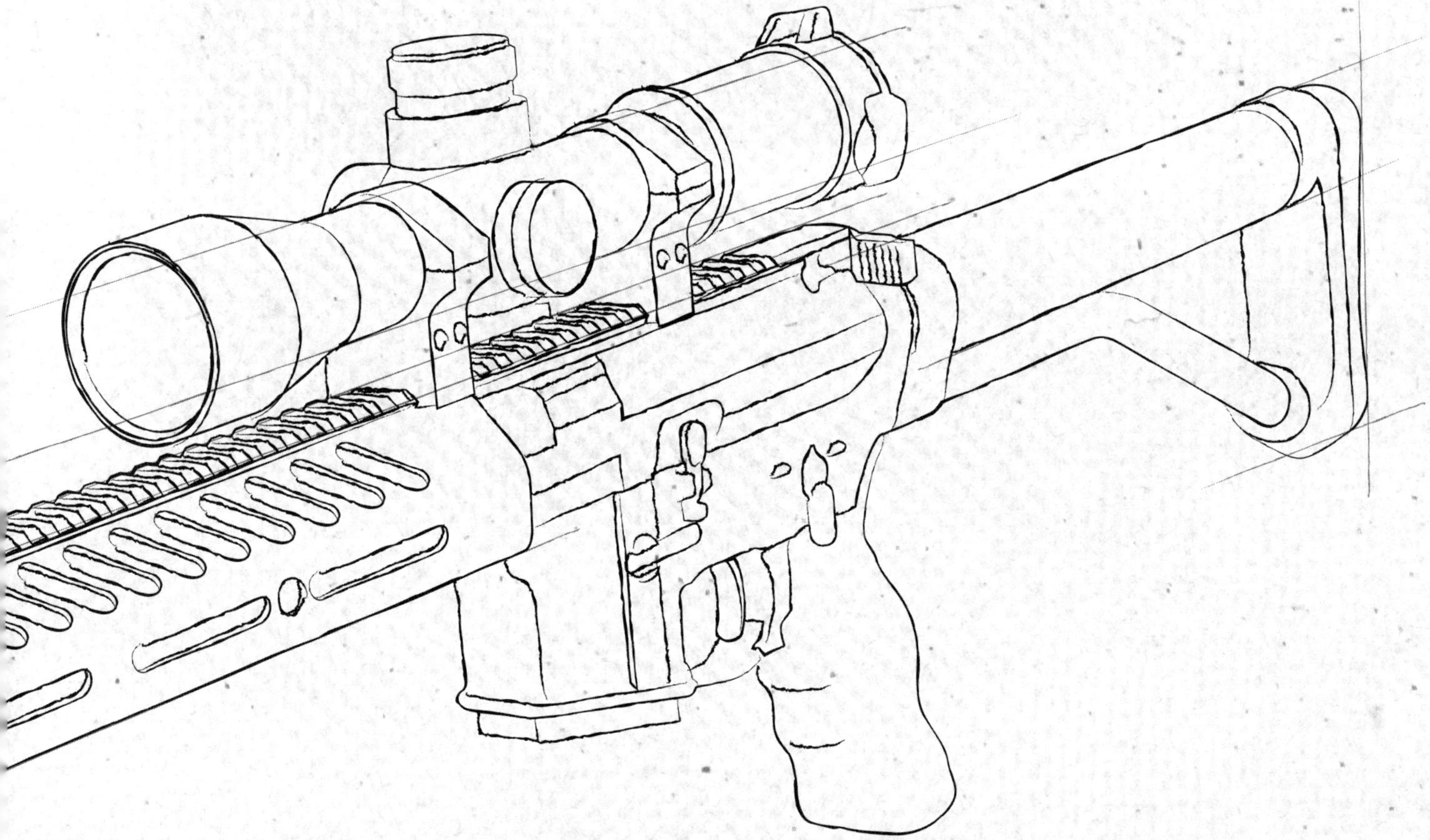

Going the Distance With a 6mm Mousegun

By Iain Harrison | Photos by Kenda Lenseigne

After six decades of development, it's safe to say that the AR-15 is one of the most tinkered-with firearms designs ever. It's available in more flavors than can be found in a supermarket soup aisle, but due to Stoner's original being developed around the .223 Remington cartridge, there are very clear limits as to how far the recipe can be pushed. The key parameters are those of magazine length and width, and bolt head diameter — go outside of these boundaries and you wind up with either a single-shot rifle or a face full of steel and brass.

The armed forces are not going to give up the 5.56mm round any time soon; despite its shortcomings in range and lethality, there's an entire logistics chain built around it. As they say, amateurs talk tactics, and professionals talk logistics — and the .22 caliber is good enough for most tasks to resist any siren calls for its replacement. Civilians don't have to put up with "good enough," though. As shooters, we benefit from a cottage industry composed of smart, talented people who A) don't have to conform to Big Green's procurement protocols, and B) are driven to build a better mousetrap.

So what does a shooter do when he's dissatisfied with the .223's ballistics? Say we want to effectively engage targets out past 1,000 yards with an AR-15; what options are available? Well, the first is to handload the .223 and boost its performance a little. In doing so, our goal should be to keep our bullet flying at supersonic velocities for as long as conditions allow (see sidebar). The formula for stretching the envelope is no secret; just launch the sleekest bullet available at the highest velocity possible.

Compared to, say, a 55-grain M193 bullet, heavyweights from Nosler, Berger, and Barnes (among others) boast very high ballistic coefficients (BC), meaning that although they start out slower, they lose velocity less quickly than lightweight projectiles. Unfortunately, the very characteristic that makes them fly so sweetly (i.e, their length) also means we quickly run into problems when we try to stuff them into a magazine. They either take up room in the case that could be used for powder, thus limiting their muzzle velocity, or else

they stick out so far that they have to be loaded one at a time. Although the latter strategy works well for NRA High Power shooters, most of us prefer to use that awfully convenient spring-loaded box to feed fresh rounds whenever we press the trigger.

While it's just about possible to get a .223 to reach the magic (or arbitrary and meaningless, depending on your point of view) 1,000-yard line, it's very difficult to achieve with rounds that fit into a magazine, especially at sea level or lower temperatures. Even if everything goes as planned, the bullet arrives with so little energy remaining that it has a tough time doing much damage. And we want damage.

THROWING TOMAHAWKS

We also wanted to stick with the AR-15, so stepping up to the bigger AR-10 and its wide array of longer, fatter cartridges was a non-starter. Looking at the commercially available offerings, the most successful long-range round is the 6.5mm Grendel. While we're big fans of the 6.8 SPC, it doesn't offer a high enough BC to make it a viable 1,000-yard cartridge, with the slickest bullets available in this loading (the 115-grain versions) hitting the transonic mark only halfway there. And .300 Blackout aficionados, get back in your box.

Were we rational, objective human beings, then the most logical thing to do would be to call up Alexander Arms, have them ship us one of their 6.5mm uppers and call it a day. But we're not, so we didn't. And here's why.

Despite having proven itself in the 10 years since its introduction, Bill Alexander himself would be the first to admit that his brainchild is running at the ragged edge of the AR-15's bolt geometry. Sheared lugs are a much more common occurrence than in .223 chambered guns, so we started casting around for something that would give a bit more margin for error for those times when pressures spike, such as when the gun is dirty and hot. A smaller diameter case puts more steel in the bolt head — and while the 6.8 case head is almost the same size (.421 versus .441 inches), that 20,000 delta could make a decent cushion.

Although 6.5mm-diameter projectiles have been very successful in long-range competition, the most aerodynamic, commonly available 140-grain 6.5mm bullets are too long to use in the Grendel case, so 123-grain projectiles are the ones usually encountered. These turn in a very respectable G1 BC of around .510, but we can exceed their performance with a lighter-weight .243 bullet, while leaving more room for powder as well as giving lower recoil. Although neither round is

›BUILD LIST

Because the goal of this build was accuracy at 1,000 yards and beyond, we picked quality components for the upper.

UPPER RECEIVER: Seekins Precision IRMT-3 / $299
RAIL SYSTEM: Seekins SP3R 12-inch / $225
BOLT CARRIER GROUP: Rubber City Armory 6.8 SPC / $200
CHARGING HANDLE: Strike Industries ARCH-EL/ $30
BARREL: Bartlein 22-inch 5R, chambered and profiled by Keystone Accuracy / $550
GAS BLOCK: SLR Rifleworks Sentry 7 Adjustable / $119
MUZZLE BRAKE: Badger Ordnance Micro / $100
SCOPE: Leupold Mark 6 3-18x44mm (H58 reticle) / $3,250
RINGS: Talley 34mm Tactical / $211
TOTAL COST: $4,984

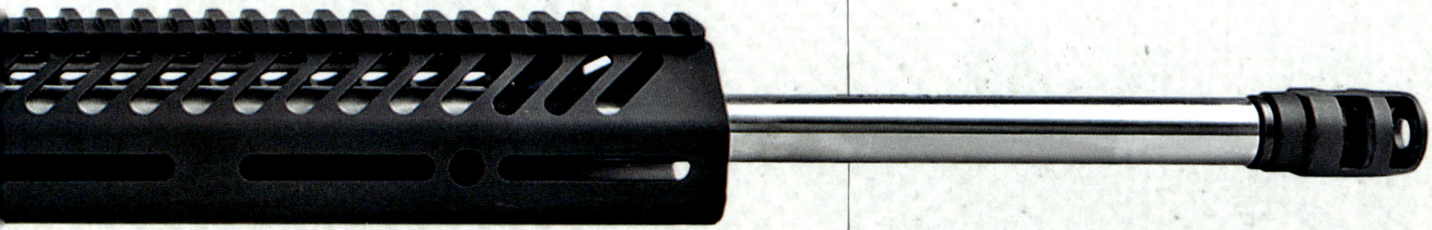

punishing on the shoulder, there are big advantages to being able to spot trace and strike through the scope.

WILDCATTING THE AR

There are several very good, long-established 6mm wildcat (i.e. non-factory) cartridges that fit inside the AR-15 envelope. After all, shooters are an innovative crowd and started tinkering with the .223 round almost as soon as it was launched. Simply expanding the case neck to 6mm produced the 6x45mm, occasionally offered as a factory load by Black Hills. Based on the 6.8 case, the 6mm Hagar was used to good effect by Carl Bernosky to win at Camp Perry several years in a row and has the advantage of Hornady supplying factory brass (but no factory ammo) for it. Designed by John Scandale at Keystone Accuracy, the .240 Tomahawk is currently the big kahuna with the most internal volume, so despite potential feeding problems caused by its very steep 40-degree shoulder, we ordered up a barrel and dies.

Once you stray outside of the 5.56mm lane, the support of big ammo and gun companies, along with their dedicated R&D departments, dwindles rapidly, so sticking to the holy tenets of load development is critical. If you already handload, then you're probably already aware of them, but for the potential newbie, here's the scoop.

1. Treat any load you find on the Internet as highly suspect and verify from another credible source before even considering it. Couch commandos are well known for pulling load data out of their flabby arses.
2. Published load data may not perform exactly the same in your firearm. Chambers are cut to different tolerances and dimensions (just ask anyone who loads for 6.8), and powder varies from lot to lot.
3. Just because two bullets from different manufacturers weigh the same doesn't mean they'll give identical pressures.
4. When developing loads, go slowly and change only one variable at a time.
5. Start out 10 percent below your chosen load and work up, checking for pressure signs on your cases as you go. If you don't know what overpressure signs are, be sure your medical insurance is up to date.

GLOSSARY OF TERMS

Ballistic Coefficient: A measure of how air resistance affects a bullet in flight, when compared to a standard projectile. Higher numbers mean the bullet cuts through the air better, and as a result retains more of its velocity for longer. There are currently two standard projectile models in common use for modeling rifle bullet trajectories.

BTHP: Boat tail, hollow point — a match bullet typically made to exacting specs.

Fire forming: Using the action of firing a cartridge in order to expand the case to fit the chamber. This method is usually employed when making wildcat cases from standard brass, where the finished case is bigger than the original.

G1 BC: The G1 BC (ballistic coefficient) model is based on a projectile that looks like a 19th century artillery shell. Which is unsurprising when you consider that's exactly why it was created in the first place. If your rifle bullets look like, well, rifle bullets, then you might consider using the G7 BC, which actually looks like a match bullet, complete with boat model.

Trace: Bullets displace air as they fly downrange and under the right conditions, it's possible to see the "wake" that they create on their way to the target. Even if the bullet strike can't be observed, oftentimes trace is visible and this allows the shooter to correct if needed.

Wildcat: A cartridge that has not been standardized by the Sporting Arms and Ammunition Manufacturers Association. Many popular factory cartridges start out as wildcats before being added to the catalog, e.g. .300 Blackout, .260 Rem.

The six steps needed to create a finished .240 Tomahawk round. L-R, loaded 6.8SPC round, 6.8 empty case, false shoulder formed in sizing die, loaded round ready for fire forming, fire formed case, loaded .240 Tomahawk round.

In order to form cases for the .240 Tomahawk, 6.8 SPC brass is first run through the sizing die to produce a false shoulder, just south of the case neck. This allows assembly of a round for fire forming, the objective of which is to blow the case out to its final profile, ready to reload with a full powder charge. This step was accomplished with 28 grains of Accurate 2520 under a Nosler 90-grain Ballistic Tip. Normally, we don't get too wrapped up in accuracy testing at this stage, as our end goal is just to get some fully formed brass. But when the very first five-shot group came in at 0.7 MOA, we figured there might just be something to this little round.

Due to the number of steps involved in making them, empty cases are precious — so a case catcher was used to avoid having to search the crevices of our range bag. Tumbling and processing gave shiny, new-looking hulls that we then used to work up high-accuracy loads with IMR 8208 XBR powder and Hornady 105-grain BTHPs. Settling on a final recipe of CCI small rifle primers, 28 grains of 8208, and an overall length of 2.275 inches, we saddled up the Bronco and rolled out to Cowtown's west range for a date with some distant steel. The observant among y'all will have noticed that OAL is longer than the good book would have us believe is the maximum for the .223. Using stainless steel magazines allows us to cheat the bullet out a little more than usual — and every thousandth of an inch counts when case volume is at a premium.

Dialing in on the 300-yard line, spotter and shooter worked their way out through the 500-, 600-, 750-, and 850-yard torso-sized gongs. At each distance it was possible to see both a vapor trail and impacts through the scope, making instant corrections via the Leupold's Horus reticle as the DOPE (data on previous engagements) book gained its first entries. Deciding to go for broke, we picked out an empty 25-pound propane tank on a far hill and called for the spotter to range it. Verifying his call with a Bushnell ARC 1 Mile rangefinder, the triggerman came up 13 mils and held slightly left for wind. With the first round impacting a hair low and right, a second bullet was sent on its way after a quick adjustment.

1,180 yards downrange, a target invisible to the naked eye was perforated by tiny speck of lead and copper, packing about the same energy as a 9mm ball round. Five shooters clustered around an iPhone that had been held up to the spotting scope and chuckled at the thought of Stoner's mouse-gun banging steel at almost three quarters of a mile. Everyone wanted to try their hand.

The 6mm AR-15 is a lot of work and not for everybody. If your day is constrained by work or family, or if you don't want to invest the time to learn to reload a wildcat cartridge, then we'd be the first to admit that your efforts and energies are much better spent elsewhere — a Grendel is a readymade long-range solution for the small-frame AR with plenty of factory ammunition options. But there's something about the non-existent recoil, super-efficient powder charge, and flatline trajectory of the 6mm that transcends the usual campfire caliber arguments. There are a lot of people who, for whatever reason, have elements of their identities tied up in their firearm purchases. Crunch the numbers in a ballistics calculator and make your own decisions.

We just want a better mousetrap. **R**

THE SCIENCE-Y BIT

We want our rifle bullets to be sleek and pointy, because everything we've learned by lusting after sports cars since third grade tells us that shape is the fastest and most efficient. Unfortunately, the science of fluid dynamics says that when passing through air, the center of pressure (CP) acting on a pointed projectile is in front of its center of gravity (CG). This produces instability and means the bullet wants to flip over — great if it happens inside a meat-filled target, but less than ideal if it occurs before it actually gets there.

We counter this by using rifling to spin the bullet fast enough to resist the tendency to tumble, and different twist rates are needed in order to spin different length bullets. As the bullet slows down and approaches the speed of sound, it enters the transonic zone — cue dramatic music — and here, things start to get a bit squirrelly. The nice, predictable supersonic shockwave around the bullet tip starts to decay, and the CP shifts even farther forward. Our bullet might have left the muzzle with enough spin to keep it flying straight, but now it needs more! Even if it has enough spin to resist tumbling end over end, the forces acting on it produce a phenomenon know as precession. Think of a child's toy spinning top; as it spins, it also wobbles gently around its axis. In the case of a bullet, precession throws another variable into the mix, further degrading accuracy while it's in the transonic zone. As the bullet becomes subsonic, precession is damped out, so long as it hasn't tumbled.

Different bullets handle the transonic zone in different ways. Some fly straight and barely notice its effects, while some become wildly inaccurate and start doing backflips. Because of the unpredictable nature of the beast, we can guarantee more consistent results if we limit ourselves to the supersonic zone which, just to make matters more difficult, varies according to air density. To be on the safe side then, we look for the maximum distance at which the bullet hits Mach 1.2, which should ensure it remains supersonic at all reasonable altitudes and weather conditions.

DROP IN INCHES

Range (yds)	.223 Rem	6.5mm Grendel	.240 Tomahawk
100	0.0	0.0	0.0
200	2.7	3.3	2.9
300	11.5	13.3	11.8
400	27.6	30.9	27.7
500	53.0	57.0	51.0
600	88.0	94.0	84.0
700	136.0	142.0	127.0
800	200.0	204.0	173.0
900	282.0	282.0	251.0
1,000	387.0	378.0	336.0
1,100	518.0	495.0	439.0
1,200	680.0	636.0	564.0

.223 Rem: 75-grain Hornady BTHP, 2,800-fps muzzle velocity; **6.5mm Grendel:** 123-grain Hornady A-Max, 2,600-fps m/v; **.240 Tomahawk:** 105-grain Hornady BTHP, 2,700-fps m/v.

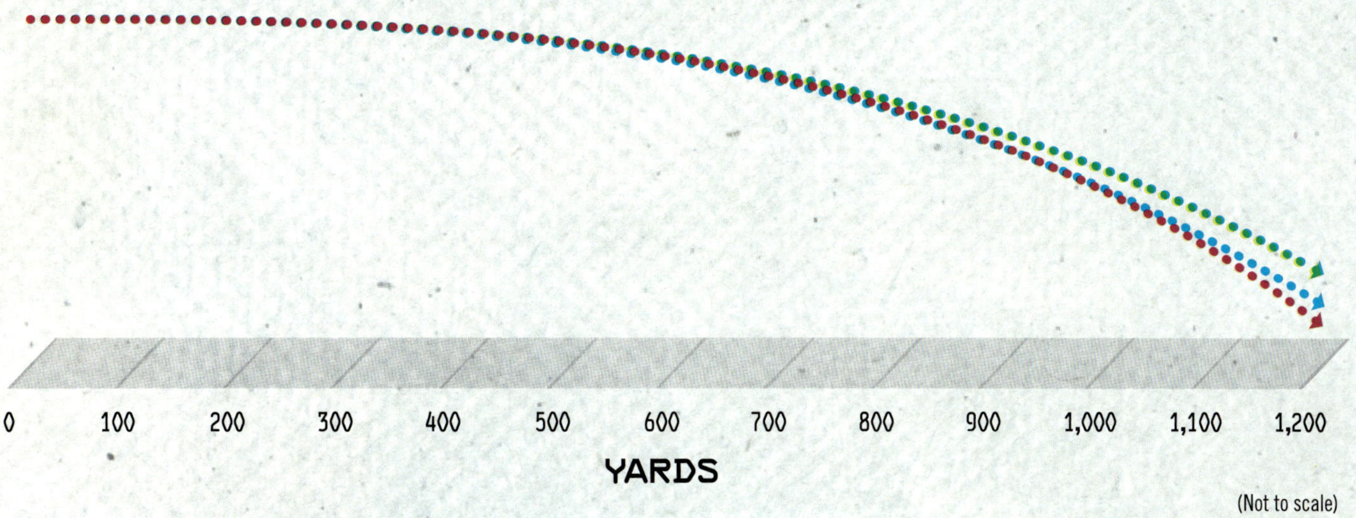

(Not to scale)

20% EFFORT, 100% AR

Complete Your Own 80-Percent AR-15 Lower Receiver

By RECOIL Staff

When we decided to embark on this project, deciding just how to approach removing metal from the 80-percent forging was a tough decision. Everyone wants to see impeccably machined billet and flawless surfaces, but clamping the piece in a CNC machine and pressing "play" would detract from the ethos of the build. In the end, we decided that despite having access to a Haas and a couple of Bridgeports, it was only right and proper that we use the most basic equipment possible. And that means a cheap drill press, handtools and, of course, the inevitable almighty Dremel.

Bear in mind that you're abusing the drill press more than just a little during this adventure. It's designed to be loaded only in the Z (vertical) axis, and even though you're only ever going to be plunging end mills into the work, you're still going to induce side loads. This means it's going to protest by vibrating like crazy and trying to spit the work piece into your soft parts, so wear safety glasses and take appropriate precautions.

WARNING!
When handling firearms, always observe safety rules and the precautions set forth in the firearm's owner's guide. Be certain that your firearm is unloaded and made safe before proceeding with this DIY.

SOME TIPS BEFORE YOU BEGIN

- Ensure that the work is clamped securely to the table
- Run the slowest speed possible and use a light hand
- Clear chips away often and use plenty of cutting fluid
- Take your time to set everything up right. It's all too easy to get drawn into making chips, but the key to a good result is in setup.

WHAT YOU NEED

Other than a drill press (the bigger and heavier the better), you'll need the following:

- C-clamps (again, bigger is better, though we ran out of room on our tiny drill press table to use more than one)
- 5/16-inch two-flute, center-cutting end mill
- 3/8-inch two-flute, center-cutting end mill
- 5/16-inch drill bit
- 3/8-inch drill bit
- 5/32-inch drill bit
- Cutting fluid (WD-40 works well on aluminum)

Not Pictured:

- 80-percent jig
- Dremel with sanding drum and 1/8-inch rotary rasp
- Hand files
- Hammer, trigger, and safety selector to test fit

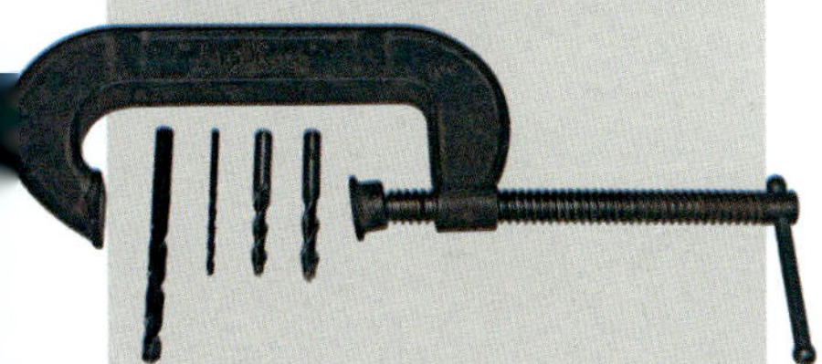

STEP 1

Set up your 80-percent lower in the jig and make sure it's well secured.

STEP 2

Insert a 5/16-inch drill bit into the chuck and tighten well. Clamp the jig and receiver to the drill press table and, with the drill not running, touch the tip of the drill to the top of the receiver. You'll now set the depth of cut, so that the tip ends up 1 1/8 inches from the top of the receiver. If your drill press has a scale, use it. If not, use a steel rule to set the distance from the stop nut to the surface it contacts on the drill press casting.

Look at a completed AR-15 receiver and you'll see the fire control pocket has two areas. We're going to first remove material from the deeper of the two. Line up the jig so that the side of the drill bit almost contacts the fixture. If you want to get super precise, you can hold it away from the wall .003 of an inch, simply by using a piece of printer paper as a feeler gauge.

STEP 3

It's now time to make the first cut. Take a deep breath and fire up the drill. Squirt a little cutting fluid and start drilling, raising it often to clear away chips. Drill all around the first area of the fire control pocket, removing as much aluminum as you can.

Now swap out the drill bit for the 3/8-inch end mill. Set the depth of cut 1/8-inch deeper, to 1 1/4 inches, in the same way as for the drill then go back around to finish hogging out the fire control pocket.

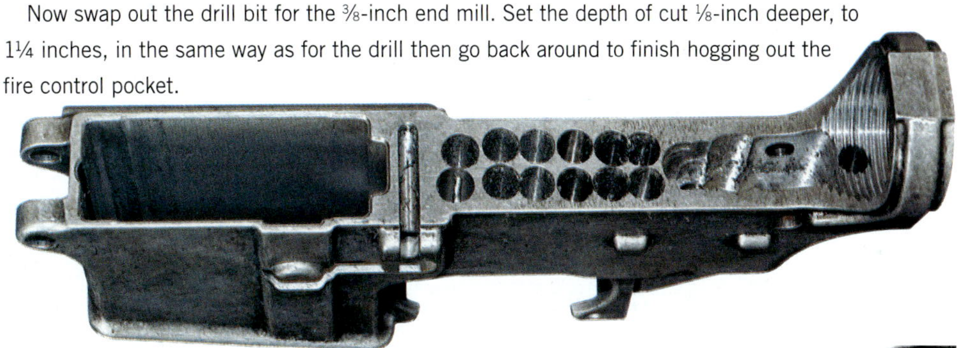

STEP 4

Now that you have the hang of removing material, reconfigure your jig as necessary and follow the same procedures used to set up to remove material from the shallower area of the fire control pocket. Once completed, the receiver should look like this.

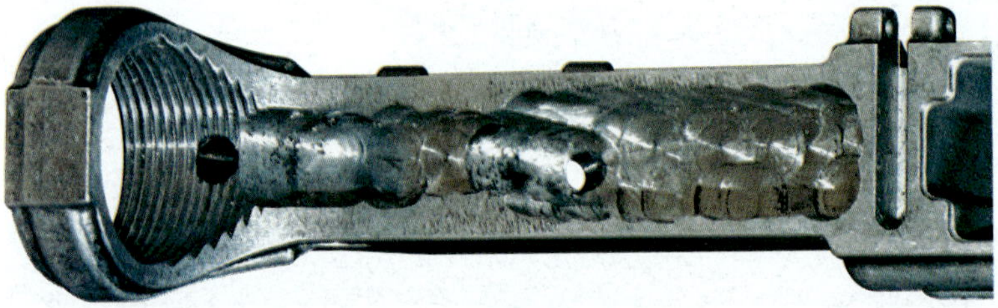

STEP 5

We're now going to mill out a slot for the trigger. Set up the jig so that the trigger slot is correctly positioned and chuck up the 5⁄16-inch end mill. Using the jig as a guide, run it completely through the receiver, breaking out into the trigger guard area. Reposition as necessary and repeat until you've removed as much material as possible.

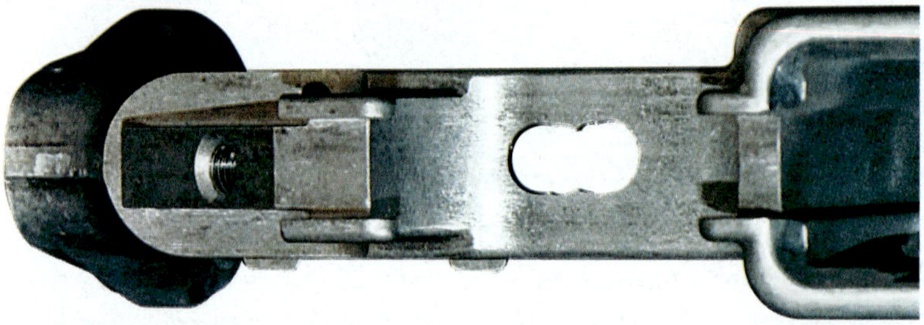

STEP 6

Flip the jig over onto its side and prepare to drill the hammer and trigger pin holes with the 5⁄32-inch drill bit. With the drill off, position the jig so that when you bring the drill bit down, it doesn't make contact with the holes in the jig, instead passing cleanly through them. Now hit the switch and start drilling, stopping when you break through into the fire control pocket. Flip the jig over and repeat the process from the opposite side.

Set up the jig as above and repeat the process with the 3⁄8-inch drill bit to drill the 3⁄8-inch hole for the selector.

STEP 7

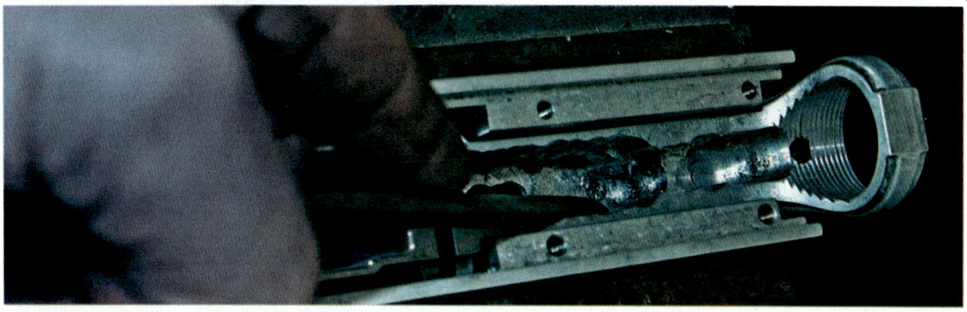

You can now break down the jig and remove the receiver in preparation for hand finishing. Using your Dremel with a sanding drum, flatten out the peaks and valleys caused by the end mills. Carefully use the rotary rasp to flatten out the very bottom of the pocket, then switch to hand files for a final finish.

STEP 8

Use the hammer, trigger, and selector from your lower parts kit to test fit as you go, removing only as much material as necessary.

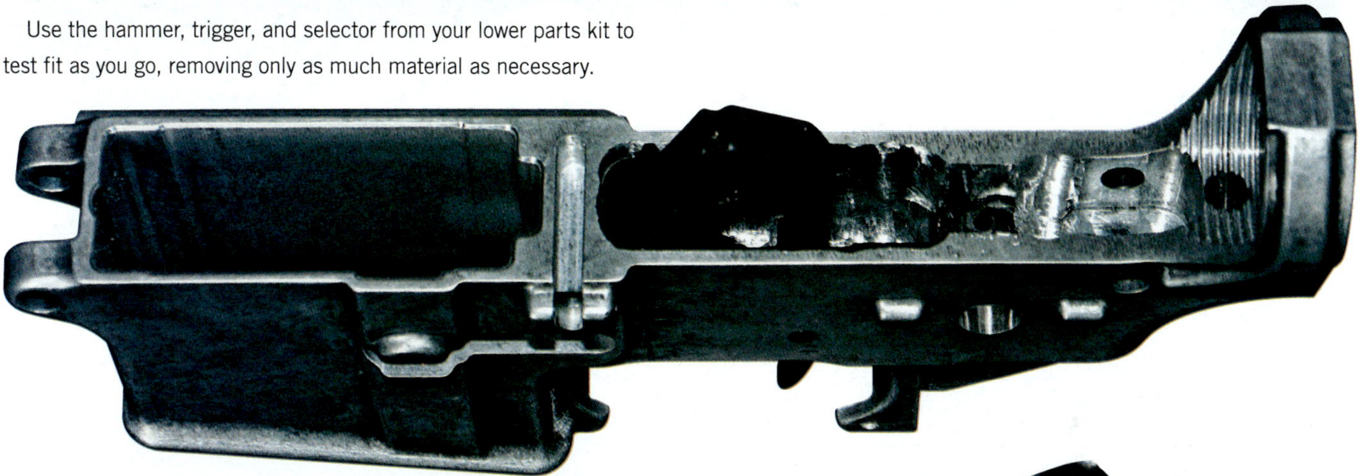

STEP 9

Once the three components will drop in, insert the pins and check for function. With the selector on Safe, press on the back of the hammer and pull the trigger. The hammer shouldn't drop — if it does, you have a problem caused by the relative placement of the fire control holes. If it passes muster, congratulations! You now have a choice to make as to what type of finish to apply to your fully functional Death Star AR-15 receiver. **R**

» SOURCES

Brownells | www.brownells.com
Grainger | www.grainger.com
McMaster-Carr | www.mcmaster.com
Billet Rifle Systems | www.billet80.com
American Spirit Arms | www.americanspiritarms.com

Carbon Fiber Equipment Case by ECS Case

FROM POLAND WITH LOVE

Building a Custom AKM From a Polish Parts Kit

Story and Photos by Rob Curtis

AKs are magical beasts. They come from foreign lands and are shrouded in mystery brought on by import laws and urban lore. One thing that might be unclear to the casual observer of the AK market, U.S. citizens can't buy authentic communist AKs made in authentic communist factories.

It's easy enough to buy an AK off the shelf, but it's likely a mash-up of imported components that were assembled stateside with some domestically manufactured parts to make it legal. More on those laws later. Some companies are building guns with all U.S.-made parts, too. Good guns, but no commie soul.

We wanted the best performing AK intertwining our American heart with a COMBLOC soul, so we decided to cut out the middleman and build our own AK from an imported parts kit.

We linked up with Gary Hughes and Mike Pappas from Dead Air Armament. The two are partners in a side project called MOD Outfitters, an outlet for their huge fascination with imported guns. And we mean huge; Pappas has his own BMP (see page 40). Hughes just likes the AK platform for its elegance.

"The thing I like about AKs is that they are brutally simple," says Hughes. "It's the assault rifle concept boiled down to its essential ingredients."

If the AR-15 is a Lego set for grownups, we guess that makes the AK something like an Erector set. Where an AR clicks, bolts, and pins together, building a reliable and accurate AK is more involved.

Gary Hughes helps out by test firing the newly built RECOIL Custom AKM.

AN AK IS NOT JUST AN AK

In fact, the Soviets learned this lesson the hard way. At the time Mikhail Kalashnikov was dreaming up the AK-47, he was likely looking at the German sheetmetal masterpieces, the MP40 submachine gun, the MG42, and the StG 44. The stamped-metal construction of these firearms was a departure from the milled manufacturing methods of the day.

In those stamped guns, the Soviets saw a way to equip their enormous army with easily manufactured, inexpensive, accurate, and reliable assault rifles. Turns out, the original AK-47 met only two of those four criteria and had issues with durability and accuracy.

According to Frank Iannamico in *AK-47: The Grim Reaper*, the Commies gave up on the newfangled stamped metal Type 1 AK receiver and replaced it with the more reliable and accurate milled-steel Type 2 and later, the Type 3 milled-steel receivers.

By the mid '50s, the Soviets finally figured out the sheetmetal thing, perhaps with help of the captured German weapon designer Hugo Schmeisser, the Kalashnikov of the StG 44. The original AK went the way of the Apple Newton and was replaced by the iAK, otherwise known as the AKM (AK Modernized). The AKM was everything the AK-47 should have been, incorporating the accumulated features of the AK-47 platform with lighter and less expensive riveted sheetmetal construction.

The AKM assumed the AK mantle and never looked back.

Soviet allies who adopted the AK retrofitted their operations and began pumping out AKMs by the hundreds-of-thousands through the '60s and '70s.

And that's the abbreviated story of how Kalashnikov's AK-47 resulted in the standard Polish AKM parts kit we have before us for our building pleasure.

Mike Pappas figured out a method to confirm and reset the headspace using a set of feeler gauges. Once headspace is set, gauge the distance between the front trunnion and the sight block. When reinstalling the barrel in the trunnion later in the build process, insert the gauges, and they'll act as a stop when the barrel reaches the previously recorded depth.

WHAT'S IN THE BOX?

We bought a Polish AKM parts kit from Arms of America. The kits include all the parts of a surplus Polish military AKM, minus the receiver, which is crushed or torch cut during the demilling process, and barrel. In place of the original barrel, AoA provides an American-made, button-rifled barrel they've populated with parts from the original kit. This saves us a huge amount of time and hassle.

We picked up an AKM receiver from Childers Guns. Childers developed a following (and a wait list) by providing excellent quality receivers that mate up with Polish parts as if it were the original murdered receiver's unbloodied identical twin found in a Polish orphanage and delivered to our doorstep. With only a couple of U.S. made parts in the mix.

SIDENOTE ON SCREWS VS. RIVETS

Screws suck. AKs can be built with screws in place of rivets, but a properly installed rivet is much stronger. Screws are faster and require fewer tools, but your rifle will look weird, Kalashnikov's ghost will weep, and your rifle will fall apart one day just like the Soviet Union. And, no amount of Loctite could've kept the USSR from falling apart.

WTF IS A TRUNNION?

For all you AR guys who are too ashamed to ask, a trunnion is a

BUILD PARTS LIST

ARMS OF AMERICA	Polish AKM Parts Kit with 4150V Nitrided Barrel	$539	www.armsofamerica.com
CHILDERS GUNS	AK47 Receiver with Straight Selector Notches	$90	www.childersguns.com
AK BUILDER	AKM Fixed Stock Rivet Kit	$9	www.ak-builder.com
ALG DEFENSE	AKT Trigger	$49	www.algdefense.com
TAPCO	Interfuse AK Slant Flash Hider	$10	www.tapco.com
AK BUILDER	Spare Rear Trunion	$30	www.ak-builder.com
CERAKOTE	Sniper Grey	$60/16 ounces	www.cerakoteguncoatings.com

Here's what we started with: a Polish AKM parts kit, a Childers Guns receiver, ALG Defense AKT trigger, and a Tapco muzzle device. The kit came with some auto parts that we tossed.

Tools of the trade. Clockwise from top left: AK Builder Riveting Jig and 7.62x39mm Headspace GO/NO GO Gauges (center), Robert Forbus Barrel Pin Tool, AK Builder AK Trunnion Rivet Hole Drilling Fixture, AK Builder Barrel Press Kit, Toth Tool and Engineering Rivet Jaws on Harbor Freight 24-inch Bolt Cutters, AK Builder AK47 Trigger Guard Rivet Jig.

TOOLS, JIGS & FIXTURES LIST

HARBOR FREIGHT	20 Ton Hydraulic Press	$200	www.harborfreight.com
CRAFTSMAN	12-inch Bench Top Drill Press	$150	www.craftsman.com
TOTH TOOL AND ENGINEERING	Rivet Jaws for Bolt Cutters	$75	www.tothtool.com
HARBOR FREIGHT	24-inch Bolt Cutters	$16	www.harborfreight.com
AK BUILDER	Riveting Jig	$250	www.ak-builder.com
AK BUILDER	AK47 Trigger Guard Rivet Jig	$65	www.ak-builder.com
AK BUILDER	AK Trunnion Rivet Hole Drilling Fixture	$73	www.ak-builder.com
ROBERT FORBUS	Barrel Pin Tool	$55	www.robertforbus.com
AK BUILDER	Barrel Press Kit	$95	www.ak-builder.com
AK BUILDER	7.62 x 39mm Headspace GO/NO GO Gauges	$65	www.ak-builder.com
AK BUILDER	5/32-inch (4mm) Drill Bit	$3	www.ak-builder.com
AK BUILDER	#16 (4.5mm) Drill Bit	$3	www.ak-builder.com
COMMON TOOLS	Automotive Feeler Gauges, Dremel Rotary Tool, Cordless Drill, Hand File, Rubber Mallet		

hefty union that supports the barrel (front trunnion) or the stock (rear trunnion) in the receiver. Rivets are used to connect the trunnion and receiver. The trunion also imparts structure and rigidity to the stamped steel receiver. The best trunnions are forged from steel; cast trunnions are best used in prop guns and as paperweights.

Our AKM kit came with forged trunnions, as do nearly all military surplus parts kits. The demilling process is a bit like a slaughter-house for guns, though. Popping thousands of rivets a day can get messy, and it's not uncommon for a kit to arrive with a cracked trunnion. Ours did, and we were glad to have a spare on hand to swap in.

TOOL TIME

Rivets are no joke. They're a commitment. You get a single shot at installing one, and the tools to work with them aren't common, either. Unless you're as lucky as we are, you're not going to be able to pillage your buddy's shop for riveting tools and alignment fixtures.

We used specialized AK gunsmithing tools from AK Builder, Toth Tools, and Robert Forbus Engineering Works in addition to a standing hydraulic press as well as some common tools, such as a drill press, rubber mallet, Dremel, and files.

BUILD

Our kit came with a populated barrel. We didn't have to worry about fitting the gas block and sight block, or drilling/reaming the barrel pinhole. Even so, if there's one thing, just one thing, that you take away from this article, it's the importance of checking the headspace.

There aren't many showstoppers on the AKM, but plus or minus a couple thousandths of headspace can result a rifle which fires out of battery or separates the heads from cases.

We checked our AK-Builder built barrel and used a set of automotive feeler gauges to record the gap between the front trunnion and the rear sight base. With the feeler gauges stacked in the gap during reassembly, they act as a stop, ensuring the barrel is seated at the same depth it was headspaced at.

The work begins with removing the partially installed barrel pin using the Forbus barrel pin tool guiding the hydraulic press' ram. With the pin out, the barrel is pushed out using the hydraulic ram with the trunnion supported by the AK Builder Barrel Press fixture.

Now that the trunnion is free of the barrel, it's fixed in the AK Builder drilling fixture. We line up the rivet hole in the trunnion with the bit and clamp the fixture in place, then we slide the receiver into the trunnion. The hole is drilled and so long as nothing moves, the rivet hole in the receiver and the trunnion are lined up for success. We repeated the same process for the remaining five front rivet holes.

We deburr and chase all the holes with a drill and a Dremel, and dimple the two bottom holes into a matching recess in the trunnion using the press and the riveting jig. This is an important step that locks the receiver and trunnion together when a swell-neck rivet is installed. Some builders mistakenly relieve material from the receiver here to get the swell-neck rivet to sit flush. Doing so leaves only the rivet core

Deburring the rivet holes ensures the rivets seat properly for the strongest connection while also looking good.

Our AKM parts kit came with a cracked rear trunnion (aka buttstock tang), right. Because it's not uncommon, we had a spare on hand and carried on without a hitch.

The AK Builder Riveting Jig properly supports and forms rivets for the front and rear trunnion and can dimple the AKM's lower front trunnion holes painlessly.

The rivets in the recessed area of the AKM's front trunnion are best reached with the Toth Tool and Engineering Rivet Jaws.

Filing the ejector in the Childress AKM receiver. It's part of the welded-in guide rail, and it's usually left long and needs to be filed down to allow the bolt to pass over it.

to hold the parts in shear, and the rivet will eventually snap.

The six short front rivets are crushed into place using the Toth Tool rivet jaws mounted on a 24-inch bolt cutter.

We move on to the trigger guard and spacer. The four front and one rear trigger guard rivets get the business. The fit of the trigger guard can affect the magazine release, so carefully test the parts with many different types of magazines. Some filing of the mag well is normal, but ours needed no attention.

The rear trunnion is treated similarly as the front trunnion, except we have two long rivets instead of shorties. Pappas suggests rounding over the end of the long rivets so they come out of mashing process with an aesthetically pleasing mushroom shape.

Drop the bolt in, and check the ejector clearance. On an AKM, the

We mount the front trunnion in this AK Builder AK Trunnion Rivet Hole Drilling Fixture, slide it around a drill press table to line up the holes with the drill bit, lock it down, then slide the receiver on to drill a rivet hole. Reposition the fixture for each rivet hole.

ejector is just a protruding tang on the left bolt guide rail. The bolt has a slot that runs its length for the rail/ejector. As the bolt retracts, the brass hits the tang and gets chucked. Childers leaves the rail tang long so it can be filed for a precise fit to the bolt. We were careful to take material off from back to front since we want as much ejector contact at the business end as we can get.

Time for the barrel. We grease up the barrel, put it in the AK Builder barrel press and line things up. A penny protects the crown from the ram as thousands of pounds of pressure force the barrel into the trunnion. As it gets close, we insert the feeler gauges we talked about earlier, taking the guesswork out of finding the stopping point. We pull the assembly, check the headspace, and drive the barrel pin in and, yep, check headspace again. All good. Phew.

The handguard and gas tube are installed and we head for the fire controls. We rough fit the ALG Defense AKT trigger kit then load it into the receiver. Check the feel and, whoa, that ALG trigger is sweet. A lot lighter than we expected.

We install the piston, bolt, and guide spring, snap on the cover, and run the bolt carrier, feeling for binding. There's a little hitch as the bolt slides from the guide rails to the front trunnion. We chase it with a Dremel and files down until the bolt travels smoothly ... or as smoothly as we think it should. It's an AK, after all.

After six hours in the shop, we take the AKM to the range and run a few boxes of steel cased, 7.62x39mm, 122-grain Tulammo through it. Boom. First five rounds and we're dancing inches from steel more than 200 meters out

It's a good idea to keep a lot of different mags at hand to test-fit the magwell and mag catch on a newly assembled receiver.

Proud papa in the shop with his assembled, but uncoated, Polish AKM.

922R
JUST WHAT THE HECK IS IT?

When people say "922R" they're referencing U.S. Code Title 18, Chapter 44, Section 922, Paragraph R with additional references to U.S. Code Title 27, Part 478.39. It applies to the assembly of rifles and shotguns from imported parts that would otherwise be banned from importation — you know, guns with evil features like detachable magazines and (gasp!) bayonet lugs.

These sections of the U.S. Code provide instruction pertaining to the number of foreign manufactured parts allowed during assembly. The short story is that you are not allowed to use more than 10 imported parts. The longer story is that not all "parts" are considered parts, at least legally speaking. The government provides a list of 20 items that can be counted as parts on any given rifle or shotgun. Seeing as how this is a general list and not specific to an AK, not all parts on the list apply. Confused yet?

You legally *can* assemble these guns, provided you have the requisite number of domestically manufactured parts. For this reason, the vernacular of 922R is "parts count."

Your standard AK with a threaded barrel is made up of 16 of the 20 items listed. In order to remain legally compliant, six of those 16 parts will have to be replaced with domestically produced items. For this build we used an ALG Trigger (three parts: trigger, hammer, disconnector), barrel (one part), receiver (one part), muzzle device (one part). Of course, you can always add more than six, as seen in some configurations with Magpul furniture.

Look on RecoilWeb for a longer piece on staying 922R compliant.

Rodney Backus, with the Park City Gun Club, Cerakotes the barrel of our AKM. PCG's fledgling Cerakote operation might be low volume, but it's high quality.

without ever touching the sights. Not bad. But, not done, either.

As mean and hungry as the gun looks in the raw, it deserves a finish. We head over to meet Rodney Backus at the Park City Gun Club. Backus Cerakotes gun parts in the back of the upscale, indoor range.

Bakus let us try our hand with Cerakote. Under his guidance we sandblast enough of the gun to realize it's worth paying someone else to do it. At press time, we've still got some of that abrasive medium stuck in our sinuses...

We head out for lunch as Backus stays to load the coated parts into his oven; but not before covering up all our sins with his far more skilled hands, we surmise. When we see the AK next, he's reassembled it for us, and it's glorious.

ADJUST

The AKM was mint. Spitting aces and looking sharp in its wood-toned commie uniform. Just the way Hughes likes it. We know he's a bit of a purist because he looked at us a little funny when we talked about swapping the wood for some Mag-

Our AKM, finished in Sniper Grey Cerakote and wearing its original Polish furniture.

» SOURCES

Park City Gun Club: www.theparkcitygunclub.com

MOD Outfitters: www.modoutfitters.com

The RECOIL Custom AKM ditches the original Polish AKM furniture and replaces it with modern parts from Magpul and Troy.

pul furniture and adding a railed gas tube.

His look communicated a clear message, "Don't put that shit on that gun here; I want to know that it was beautiful when it left."

We get it. Once we get the bone-stock Polish AKM home, our AR roots won't let us leave it be. Within days, it's wearing Magpul furniture and a Troy Industries front end. Serendipity struck when it came to the sling; Blue Force Gear had just made a run of its limited-edition AK slings in wolf gray that matches the gray Cerakote.

SHOOT

The only gripe we have with the whole build is with the Troy lower rail — installation was mildly challenging, and we found it loosened a bit even after applying blue Loctite to the jam bolts. Tightening those bolts through the M-Lok slots is a PIA, but once we retightened it and applied a more serious level of Loctite, we found it a solid and comfortable handguard.

With the new furniture in place, we took the gun out to see how it would shoot. We posted respectable 1.75-inch groups at 75 yards with the 122-grain Tulammo and an Aimpoint Micro T-1. We attribute the surprising (to us) accuracy to a combination of the button rifled barrel and a lot of care and attention during the build process.

On balance, our AKM shoots well, functions reliably, and feels great. Replacing the stock butt-stock with the MOE AK improved the shoulder pocket fit without affecting the length of pull. The MOE+ grip fills the hand far better than a skinny stock AK grip, and the Troy railed gas tube seems to be holding our low-mounted T-1 steadily in a lower fourth cowitness.

We're only 800-plus rounds into our relationship with this little Polish looker, but so far it's a love affair that's left our ARs wondering where we've been. **R**

RECOIL CUSTOM

MOD OUTFITTERS AKOLLABORATION AKM

CALIBER:
7.62x39mm

BARREL LENGTH:
16 inches

OVERALL LENGTH:
35.5 inches

MAGAZINE CAPACITY:
30 rounds

WEIGHT UNLOADED:
8.1 pounds

MSRP:
$1,150

URL:
www.aimsurplus.com

FEATURED ACCESSORIES:

Magpul MOE AK Stock	$60
Magpul MOE+ AK Grip	$25
Troy AK47 Rail, M-Lok, Bottom, Short	$125
Troy AK47 Rail, Top	$150
Blue Force Gear Limited Edition AK Sling Version 5	$65
Aimpoint Micro T-1 w/LRP mount	$800
SureFire M300 Mini Scout Light	$300
Price as featured:	**$1,150 + optic**

PIPE DREAMS

By Mike Searson

The Pop-A 410 DIY Survival Shotgun

CH. 2

Gun Kits

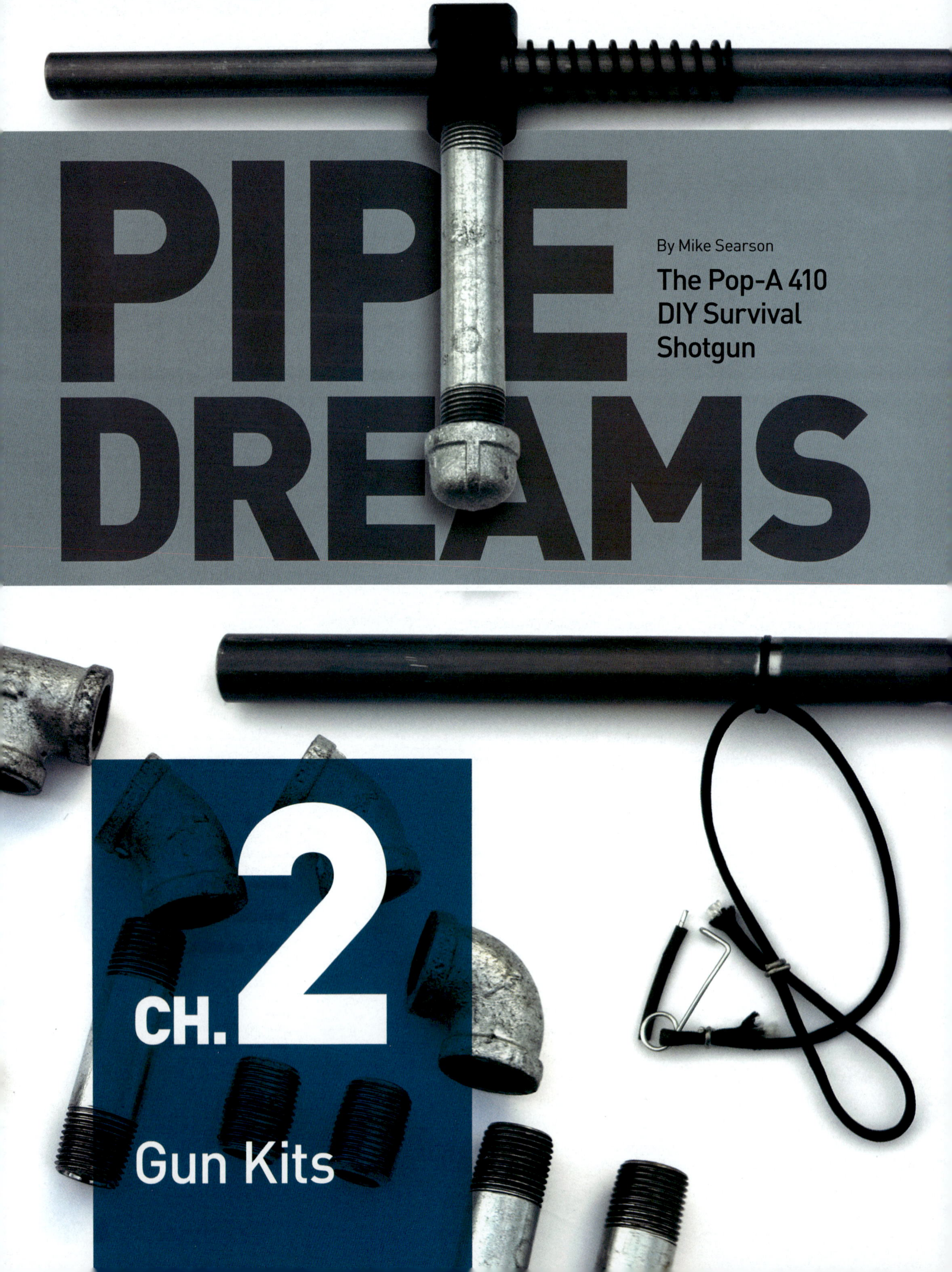

In the past few years, a "do-it-yourself shotgun" kit has become popular, billed as a "survival shotgun." It's made by Runway Sub-Cal, a manufacturer known for producing rifled flare gun inserts to allow shooters to fire pistol rounds through 26.5mm flare guns.

Essentially, they provide a barrel, firing pin, and an outer barrel sleeve; the rest of the shotgun is built from steel pipe fittings. I was a bit hesitant going into this assignment, but read on and see why it's important for a number of reasons.

BUILDING A SHOTGUN

It's legal for you to build your own firearms. Convicted felons or other prohibited persons, however, cannot. There may also be other applicable laws where you live, so double check the regulations in your jurisdiction. Several retailers carry this kit. We ordered ours from Bud-K for $130 with free shipping.

The package arrived via FedEx Smart Solutions, meaning that FedEx did all the heavy lifting getting it to within 20 miles of me and transferring final delivery to my local post office. Bud-K packaged it well enough, but after opening it up I realized there was no instruction sheet.

This wasn't a big deal, as the kit is straightforward with a video on the company's website. When compiling the items needed to finish it up, I went to double check on the website what a "jam nipple" was and realized the firing pin wasn't included in the shipment.

Bud-K's customer service desk began looking for extras in their warehouse. If they were unsuccessful, they'd have me return the kit in

exchange for a new one. A few days later, they advised that the manufacturer would ship out the missing components directly. After about a week, it arrived.

Completion of the kit requires the following ½-inch pipe parts:

- one ½-inch tee
- three ½-inch caps
- two 4-inch nipples
- one 3-inch nipple
- two 45-degree elbows
- one 90-degree elbow
- two jam nipples
- one 5-inch nipple

The only parts that were somewhat difficult to find were the so-called jam nipples. These are short nipples that are completely threaded. I found them under a variety of names such as zero nipples and flush nipples. Perhaps jam nipples are a colloquialism in certain parts of the country or an older plumbing term, but I couldn't find them under that name in stores near me.

After a quick trip to Home Depot and $32 later, I had everything I needed to put the Pop-A 410 together.

The Pop-A 410 kit consists of three basic components: a spring-loaded barrel, chamber/barrel sleeve, and a firing pin.

Assembly took about 15 minutes, if that long. One of the trickier things to get right was ensuring that the angled pipe connectors aligned properly. They might line up perfectly while only grasping two or three threads, but then would be totally out of alignment when tightened all the way down. Lucas Oil Gun Grease and a small pipe wrench in conjunction with a vise got everything properly into position. I threaded as far as I could, backed off, and repeated in order to grasp more threads to make it sturdy and straight.

Once you have an idea on how everything lines up, you can make it semi-permanent by using plumbers' tape, Loctite, or another adhesive. Some builders have used JB Weld for a more permanent fix, but that might be excessive.

A crucial piece of this assembly is the ½-inch tee piece. This acts as your receiver, for lack of a better term, and the firing pin should be able to pass completely through it.

When finished, you have basically a two-piece shotgun composed of a chambered, spring-loaded barrel, a homemade receiver containing a fixed firing pin, a homemade stock, and pistol grips. A "safety" is included in the form of a clip that inserts between the firing pin and primer, attached to a short piece of cord so you don't lose it. There's no sighting system, but the forward pistol grip has a Weaver-style base so you can mount a red dot or visible laser. A short sheetmetal screw with a rounded head could make for an improvised bead sight.

You can leave the Pop-A 410 unfinished, "in the white," or you can break out some Krylon, Duracoat, or even Cerakote if you feel fancy and coat it in a protective finish.

FIRING THE POP-A 410

To fire the Pop-A 410 you load the chamber, shoulder the shotgun, take hold of the forward grip, aim, and slam the barrel rearward. The impact of the firing pin against the primer of the shell causes ignition and fires the projectile.

The first round was a bit of a doozy. The thought that kept going

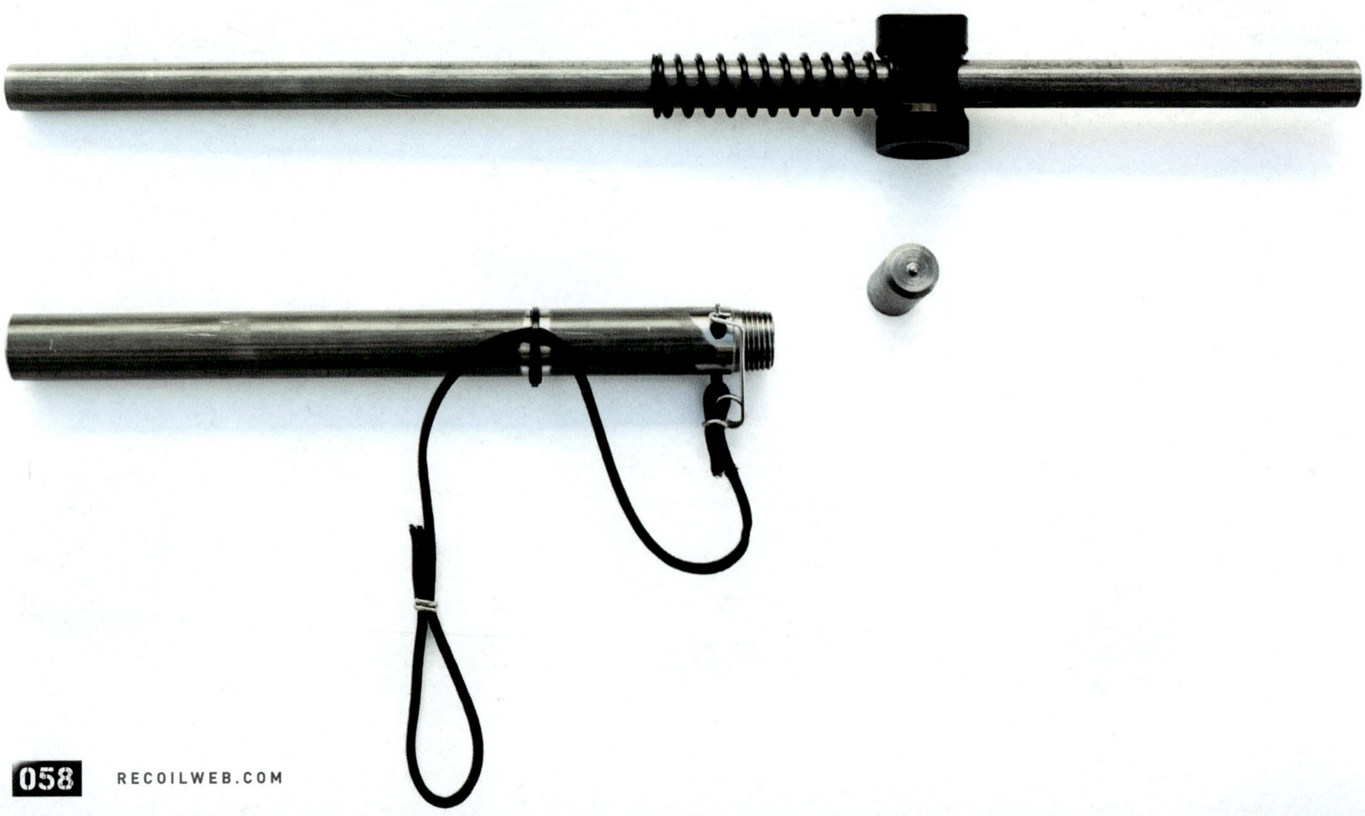

For the cost of a completed Pop-A 410, a shooter can pick up a purpose-built shotgun in the same caliber, many made before 1968 also lack a serial number such as this single-shot Winchester Model 37 and an over/under Stevens Model 420.

½-inch pipe parts:

(1) ½-inch tee: $2.95

(3) ½-inch caps: $1.95 each / $5.85

(2) 4-inch nipples: $3.35 each / $6.70

(1) 3-inch nipple: $1.95

(2) 45-degree elbows: $2.53 each / $5.06

(1) 90-degree elbow: $2.95

(2) jam nipples: $1.95 each / $3.90

(1) 5-inch nipple: $2.95

Total: $32.31

For less than $35 in pipe fittings, you can complete the Pop-A 410 kit and have a single-shot shotgun.

RUNWAY SUB-CAL POP-A 410

CALIBER: .410 gauge

CAPACITY: 1

BARREL LENGTH: 18 inches

OVERALL LENGTH: 28 inches

WEIGHT: 3.9 pounds

MSRP: $130

URL: www.budk.com

through my mind from the moment I loaded the chamber until I actually fired it was, *"Will that galvanized pipe fitting and that two-piece barrel assembly that I bought be enough to handle 12,500 to 15,000 psi?"*

It turned out that it was. The burst rating on this type of pipe is 21,000 to 26,000 psi, while pressurized. The short impulse of firing a single round of ammunition plays a role, too. Still, no one will get "millions of rounds out of this" weapon. I fired close to 150 rounds of assorted birdshot, buckshot, and buck and ball shot, without incident.

Beyond the fun factor, this is a firearm of last resort. For instance, if you had to build something out of scrap to forage for birds and squirrels in a survival situation. Or, in a darker scenario, if you needed to make something in order to take out the opposition and scavenge his weapon.

ALTERNATIVES

While this was a fun project, it's a bit expensive for what you get. It's basically a slam-fire .410. You can walk away with the satisfaction that you built a firearm on your own. Plus, putting it together may provide a better sense of how firearms work, along with the pride we all get whenever we build something with our own two hands.

If you're mainly intrigued by the fact that you can build a firearm without a serial number and wondering why you'd invest the time and money in one of these, there are other alternatives.

For example, there are untold numbers of firearms, mostly shotguns and rimfire rifles, built before 1968 that have no serial numbers and can be purchased for much less than what was invested in this project. These aren't just rusty single-shot cracked-stock scatterguns, either.

Over the years I've acquired single-shot shotguns made by Hopkins & Allen, New England Firearms, Harrington & Richardson, Iver Johnson, Savage/Stevens, and Winchester for less than $90 each — and in the case of the Hopkins & Allen and Stevens models, less than $40. An over/under Stevens 420 in .410 made 75 years ago set me back $110, and a bolt-action Sears 12-gauge cost me $60. None of these firearms have serial numbers; neither do a handful of 22 bolt-action and single-shot rifles I've acquired over the years.

Any prepared individual should have low-cost firearms on hand, because they're cheap, easy to learn how to use, and can be loaned, bartered, or otherwise utilized in the event of a large-scale long-term disaster.

The Pop-A 410 survival shotgun can fill this role, but it's more significant along the lines of a first-time gun-builder understanding how a firearm can be built, along with the inspiration for ideas on how to build your own, should you find yourself in a situation where making a firearm from scratch may be your only option.

Examine your ammunition types on hand and see how they fit in various pipes or tubes. Find pipes or tubes in a suitable grade of metal to contain the chamber pressure of the firing sequence and experiment. You can build something like the Pop-A 410 to gain an understanding of how the parts fit together properly and take those concepts to forge ahead on your own.

The key to preparing for a disaster and surviving is the knowledge we obtain and how we successfully apply it. A simple project like this can act as food for your mind. R

Opposite: Shooting the Pop-A 410 shotgun is a bit tricky due to the lack of a trigger, but not difficult. Shoulder it, take aim, and slam the shell rearward into the firing pin.

Below: It may not be a collector-grade, heirloom-quality firearm, but the Pop-A 410 survival shotgun represents a learning tool and a lightweight firearm that may be enough to keep you alive.

P80
P80

POLYMER 80 COMPACT LONG SLIDE

Roll Your Own Anti-G19X

By Rob Curtis and Steven Kuo

If you thought Glock got it wrong when they released the Glock 19X to the civilian market, you'll love what Polymer80 has in store for you. Intended as a military service weapon, the 19X has a 19-length slide and a 17-length grip. For the civilian market, with concealed carry as a primary use case, this is the opposite of what many might desire. The Glock 19 is already an immensely popular carry gun. And if you were to make a change to its overall packaging, you might rather eschew extending the grip — thus making it harder to conceal — and lengthen the slide instead. A shorter grip with longer slide is easier to conceal and provides a longer sight radius than the opposite configuration.

That's exactly what Polymer80 did with its new PF940CL frame. It's essentially a Glock 17-sized frame, accepting G17-compatible Gen 3 uppers and parts, with the grip chopped down to G19 length.

True to Polymer80's roots, the PF-940CL frame is first being offered as an 80-percent non-gun. In other words, the ATF doesn't consider it to be a firearm — it's simply a piece of polymer that needs additional machining before it becomes an actual gun. With Polymer80's clever V2 frame design, all it takes is to drill six holes and cut away extraneous pieces of polymer. With the included plastic jig and tooling (two drill bits and an end mill), it's a very straightforward process. Then, you simply install a front locking block that also incorporates front slide rails as well as rear slide rails that straddle the stock Glock trigger assembly. So, unlike Polymer80's original V1 frame, all four slide rails are metal. Beyond that, it's just like assembling a regular Glock. With the utter simplicity and low parts count of the underlying Glock architecture, the whole process is very straightforward.

And if you aren't seeking the satisfaction of building your own gun or don't care about a paper trail (subject to the laws in your jurisdiction), Polymer80 now also sells completed frames. In fact, because you won't need the jig or tooling, a 100-percent frame is actually cheaper than an 80-percent frame — $135 versus $150 at MSRP for their current offerings. They've started offering their current frames as completed firearms, with the 100-percent, serialized PF940CL expected to be available later in 2018. We're also seeing Polymer80 frames being used for semi-custom guns offered by other companies.

For now, we got our hands on a preproduction 80-percent PF940CL frame, built it up, and wrung it out.

MAKING FREEDOM IS SATISFYING

Place the 80-percent frame in the jig and secure it in a bench vise. We added clamps to ensure it was properly aligned — but don't crush it. With a hand drill and the M4 drill bit, drill the trigger pinhole. Drill the left and right side separately; don't drill all the way through from one side. Then, with the smaller M3 drill bit, drill holes for the trigger housing pin and the locking block pin. Again, drill the left and right sides separately. Clean up the flashing that remains around the new holes. Polymer80 advises using a hand drill for this operation, not a drill press, in order to align properly with the guide holes in the jig.

While a mill would be a more appropriate choice, the next operations are often performed with a drill press by the typical garage tinkerer. There are four tabs that extend above the top of the PF940CL frame; they need to be removed to clear the slide. The requirement is basically to machine them off, keeping it level. Drill presses are technically only designed to take load in the Z-axis — i.e. drilling holes. But with the help of an X-Y table, you can secure the workpiece and use the table's gearing to move it in the X and Y axes to machine away material. When working with softer material like polymer, this allows you to use your drill press like a mill. But you needn't even use a drill press; we've seen folks use everything from Dremels to whittling knives.

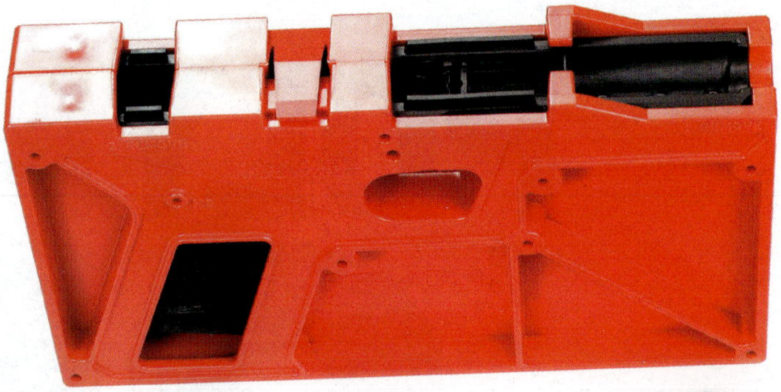

Polymer80 offers a ReadyMod version of the frame with a completely blank grip that you can stipple or customize however you'd like.

PF940CL compared to a stock Glock 19 from the early '90s.

In our case, we used a dirt-cheap \$135 bench-top drill press from Harbor Freight along with their \$75 cross slide vise. As it turns out, this bench-top drill press only has 10 inches of travel — in combination with the height of the vise, it was insufficient to accommodate the end mill and jig when placed on its end. Moreover, its table and base were small and only had two slots for attachments, lacking the typical X-shaped slots on larger machines that can secure items like a cross slide vise. It would have been easier to use a mill or a larger drill press, but we wanted to see how it could be done on a budget. So, in grand Wile E. Coyote school of gunsmithing fashion, we clamped the drill press and vise next to each other on a flat work surface, then rotated the drill press' head

Differences from a stock Glock include the front locking block, additional locking block pin, rear slide rails, and a coiled slide lock spring.

The end mill is used to remove tabs on the frame and cut a U-shaped area.

away from its base to instead hover over the vise. This gave us enough clearance and travel to perform the remaining operations.

Drill presses such as the one we were using don't have failsafe chucks. They're simply pressed into place; they're only intended to drill, not take loads laterally. So grab a mallet and beat the sh*t out of yours to secure it in place before inserting the end mill. And even then, keep an eye on it — more on this later.

Orient the jig horizontally with the tabs on the top and clamp it in the X-Y table. We again used additional clamps to keep all the corners secure, without overtightening. Set the appropriate depth for your end mill; we set it to bottom out when just kissing the jig. Familiarize yourself with how the gearing on the X-Y table moves the workpiece, so you know which direction to turn the cranks to make your cuts. Then machine off the tabs. Near the end of our last tab we noticed the drill press starting to complain. Just as we backed off, the chuck came loose. Fortunately, it didn't make much mess of either polymer or flesh, but it stands as a clear warning to exercise caution, especially with these small, light-duty drill presses.

The final step is to cut out a piece to clear the guide rod and barrel lug. Clamp the jig on its end and plunge the end mill into the dust cover of the frame to mill out a U-shaped opening, as indicated on the frame. Don't cut beyond this point and be careful not to hit the jig with the chuck. It's a little tricky to see what you're doing, so we cut short of the relief edges and finished with a round file.

Remember that you can take additional passes to remove material, but you can't add it back. So cut less rather than more — it's easy to file away any excess. That's exactly what we did, followed by a light sanding to smooth everything out.

Beyond this point, the only differences from assembling a regular Glock are the front locking block that doubles as front slide rails and rear slide rails that nest underneath the trigger assembly in the rear of the frame. Just insert them and secure with pins. We finished the build with a flat-faced curved trigger and PVD finished barrels and slides from Polymer80. We had two sets of the latter to try out. The match-grade barrels are available with and without fluting and made of 17-4 stainless steel. Barrels can be had in black DLC, titanium nitride, and stainless.

The Polymer80 frame won't fit in most Glock holsters, but a number of holster manufacturers make models specifically for it, including this one from Squared Away Customs.

Trigger guard is contoured wonderfully, gas pedals are cut into the frame, and a standard Pic rail adorns the dust cover.

Frank Proctor Shooting sights with a Y-notch on the left; NIght Fision sights with Accur8 single rear tritium dot setup on the right.

The Gen 3 slides are available in black, flat dark earth, and gray. Remaining parts were stock Glock, except for sights. On one slide, we installed Night Fision sights with tritium inserts — they were very bright; the company says they're 30-percent brighter than other makes. The other slide was treated to Y-notch sights from Frank Proctor Shooting, with a fiber optic front and a rear notch cut in a Y shape to combine the precision of a narrow notch with the speed of a wider notch.

BANG BANG BANG

The completed gun felt great in the hand, though the slide was quite tight on the frame, in particular on the front locking block. Still, it passed our safety and functionality checks, so we loaded up mags and got to work.

Polymer80 recommends several hundred rounds to break in a new build. We brought out a mix of ammo with brass, aluminum, and steel cases. We experienced some failures to eject early on that went away, followed by periodic failure to feeds where the slide didn't go completely into battery that persisted even as empty cases piled up, especially with steel cased Brown Bear ammo. The front locking block that came with our pre-production frame continued to bind on the slide and must have been out of spec. Fortunately, we had an extra locking block from an unfinished production Polymer80 PF940V2 standard frame. We disassembled the preproduction unit and compared the locking blocks. Our spare block slid easily and smoothly in the slide, while the original would bind. After swapping it into our build, the gun ran like a top with no further malfunctions. When you receive your kit, testfit the pieces; grab the loose locking block and your slide and make sure they slide smoothly.

Polymer80 incorporated great features on the frame, like those that custom builders toil for hours to add to stock Glock frames. The infamous Glock grip angle is mitigated by the flat backstrap, great for those who can't get used to regular Glocks. However, if you like Glock grips, the difference is subtle enough that it shouldn't be too hard to adjust. If you have large hands, the grip will feel almost as short as a G19 (helped by the undercut trigger guard), but you may appreciate the deletion of the finger grooves on the front strap. And if you have small hands, you'll be able to get better purchase on the gun. The texturing is grippy without being too aggressive, though like with any grip texture an undershirt helps reduce discomfort on the ol' beer belly. Contouring and undercuts on the trigger guard and the small beavertail allow you to get a comfortable,

The Polymer80 ran perfectly once we sorted out the front locking block. The frame's features are well-thought-out and contribute to a secure, comfortable, and high grip.

high grip on the gun. There's also a small gas pedal cut on the frame in front of the slide lock for those who like to get some extra leverage on the gun with their support thumb; it's also a nice spot to index your trigger finger. To make mag changes easier, the frame is scalloped out around the magazine release and the magwell is smoothly flared. All told, Polymer80 made excellent choices in their frame design that work together very well.

Most of the Glock holsters we had on hand wouldn't accept our Polymer80. However, an ever-growing list of holster manufacturers offer compatible holsters. We got a Squared Away Customs P80 holster in sexy MultiCam; it's stuffed down your pants, so no one will ever know, but like your favorite pair of underwear, it'll put an extra spring in your step. Strapping on the gun helps you appreciate the original concept behind the gun —vertically, you don't really notice the longer slide, and horizontally, the shorter grip doesn't print as much.

As a carry-focused build, we kept the trigger mostly stock Glock, except for the P80 trigger shoe. It broke at 4.75 pounds with a hint of grittiness. The trigger shoe felt good, and we shot five-shot groups at 25 yards of just under 3 inches. In general, it shot like a customized Glock — and that's a compliment.

Even if you forgo the premium parts on this build for more economical selections, when you add it up, you may lament that it costs more than an actual Glock. If a Glock fits your needs, we encourage you to get one — all of us here at RECOIL have many of them. But if you appreciate the extra features on the Polymer80 frames, price out the cost of getting custom framework done on your Glock before you make your decision. You should check out Polymer80's frames.

We really liked shooting and carrying the PF940CL; the anti-G19X format works well. If you want a Glock 17 with a 19-length grip — well, Glock doesn't make one of those. And if you want to build your own gun, Glock can't help you either. So, you know what you need to do. Build yourself some perfection.

POLYMER80 PF940CL 80 PERCENT PISTOL FRAME KIT

CALIBER: 9mm

CAPACITY: 15

BARREL LENGTH: 4.5 inches

OVERALL LENGTH: 7.6 inches

WEIGHT: 1.4 pounds

MSRP: $150 (frame kit only)

ACCESSORIES: Polymer80 PF940CL 80-percent Pistol Frame Kit $150
Polymer80 P80 G17 Gen 3 Standard Slide $325
Polymer80 P80 G17 Barrel (Standard) $165
(Fluted) $175
Polymer80 PF-Series Slide Parts Kit $100
Glock Recoil Spring Assembly $5
Glock Lower Parts Kit $50
Polymer80 P80 Curved Trigger $100
Sights
(Frank Proctor Shooting Y Notch Sights) $79
(Night Fision Perfect Dot Accur8 Night Sights) $117
Squared Away Customs P80 Holster (Multicam) $69

PRICE AS FEATURED: $1,043 - $1,091

URL: www.polymer80.com

CETME-L

Hill & Mac Gunworks Shows Off its DIY Vintage-Inspired Rifle

Story by Iain Harrison
Photos by Kenda Lenseigne

In the 1970s, most NATO countries saw the writing on the wall and started the process of converting their armed forces' main small arms over to the 5.56 cartridge. This must have been particularly galling for many, as they were originally forced to adopt the 7.62 NATO round by the U.S., despite several of them coming up with smaller, better, general-purpose designs. After being in service for less than a decade, America then did an about face and introduced the .22 caliber as its main infantry round, while keeping the 7.62 for crew-served weapons.

As part of the switch to the lighter caliber, Spain looked at its successful CETME 58 battle rifle and used it as a basis for a smaller version. If you think the CETME looks like an HK, well there's a very good reason for that. Ludwig Vorgrimmler designed the CETME in the early 1950s, in a career path that took him full circle from Mauser during WWII, to France in the immediate post-war years, to Spain, and finally back to Mauser in 1956. During his time in Santa Barbara, the West German government adopted and adapted the design of the CETME 58, producing it under license as the G3 in the factories of Rheinmettal and Heckler & Koch. HK then used it as the basis of an entire line of some of the most iconic 20th-century firearms.

The CETME-L then, while it isn't an HK design, certainly shares some DNA, and is an interesting side note in small-arms development. Due to asinine federal legislation, we mere serfs aren't able to own a real CETME-L, though Hill & Mac Gunworks has an interesting workaround that allows us to own a semi-auto version of the original. It's also a cool project to show off to your friends at the range, as you'll create a fully functioning rifle from a collection of scrap metal.

H&M CETME-L KIT

You'll need to get your hands dirty for this one, but isn't that part of the fun? The H&M non-rifle arrives as several bags of parts, some of which are recognizable as bits of a weapon. There's also a big, honkin' shiny steel assembly used to form the included stamped sheetmetal flat into a working receiver — but for now, resist the urge to just jump in and start screwing things together. Before you go any further, it's time to bust out the Dremel.

Chopped up bag o'bits. How the parts kit looks when it arrives — don't worry, we'll soon have this thing up and running.

Separating the good stuff from the crap. Note material removed from bolt carrier where it trips the auto sear.

STEP 1

In order to import the CETME-L kit, it first has to be chopped up so that it's no longer a firearm. Sad, but blame those paragons of virtue in Washington. The various destroyed chunks still have usable components welded to them, so you'll need to separate them from the remaining scrap. Using a rotary tool fitted with a fiber-reinforced cutting wheel, carefully cut the welds securing the rearsight assembly, cocking tube, and trunnion to the remains of the butchered receiver. Remember to cut on the waste side of the line, so that you don't remove steel from the components you want to save. Once the welds have been cut, you'll need to peel away the scrap receiver from them. We used a set of Vise Grips and a small prybar to do this, but there are many ways to skin this cat.

Once you've delivered up the usable components from the old gun's carcass, dress up their rough edges using a file or abrasive of your choice. We used a square to ensure the cocking tube would mate up with the forward edge of the new receiver.

SEMI-AUTO CONVERSION

As the original CETME-L was a select-fire weapon, you'll need to ensure that your newly built rifle won't land you a stay in Club Fed. This means you'll convert it to semi-auto only by removing a portion of the bolt carrier that trips the auto sear and grinding away a portion of the trigger group, so that it clears the denial block that you'll then weld into the new receiver. The latter ensures that you can't insert an original, unmodified trigger pack into the gun, keeping you nice and legal-like.

As a final measure, you'll also completely destroy the auto sear by cutting off its actuator arm, while retaining the sleeve that fits over the trigger pin. By completing these three steps, you remove any possibility of the rifle ever firing in rock-'n'-roll mode. Final legal niceties are addressed by H&M's including the requisite U.S.-made parts to ensure compliance with 922(r).

PRESS THE BARREL

Unlike in comparable HK designs, slapping the new, U.S.-made barrel into the trunnion is a fairly painless process. While a hydraulic press makes life easier, you can accomplish the same result with nothing more than a vise and a hammer, with the vise being optional. Simply coat the chamber area of the barrel with antiseize grease, insert it into the trunnion you peeled out of the old receiver and, using whatever method you favor, press it until the shoulder is fully seated against the trunnion's inner lip.

At this stage you can do a confidence check to ensure correct headspace. Using a 5.56 "Go" gauge, insert it into the chamber and close the bolt carrier group behind it. Check the gap between the bolt and carrier with feeler gauges — if you see a value of between 15 and 25 thousandths of an inch, congratulations, you're in spec. As your bolt and trunnion both came from the same functioning gun, the chances are that this step will be redundant, but a little caution never goes amiss when dealing with 65,000-psi pressure events, right in front of your face.

Once the barrel is seated to your satisfaction, it then needs to be secured in place with a 3/16 pin. A drill press is recommended for this procedure, but you may be able to get away with a hand drill if you're very patient and drill from both sides of the trunnion.

FORM THE RECEIVER

That honkin' chunk of steel you first found when unpacking your kit is the receiver forming die. Unlike most die sets, this one doesn't require the use of a hydraulic press, though if you have one handy, there's nothing to prevent you from using it. Instead, once you've welded in your denial bar, the receiver is folded by tightening down two bolts on either end of the die, also serving to locate the flat.

It's brilliantly simple and effective, and there's a locating pin to ensure you don't accidentally put the flat in the wrong way round. As with most important areas of human endeavor, a little lubrication on the important bits goes a long way to achieving a perfect result, so slap a little

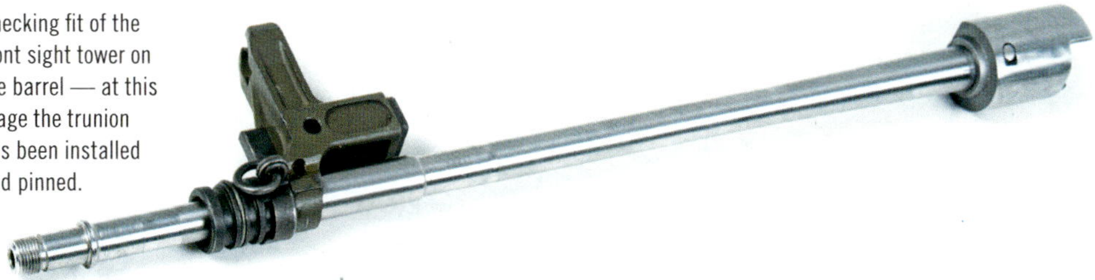

Checking fit of the front sight tower on the barrel — at this stage the trunion has been installed and pinned.

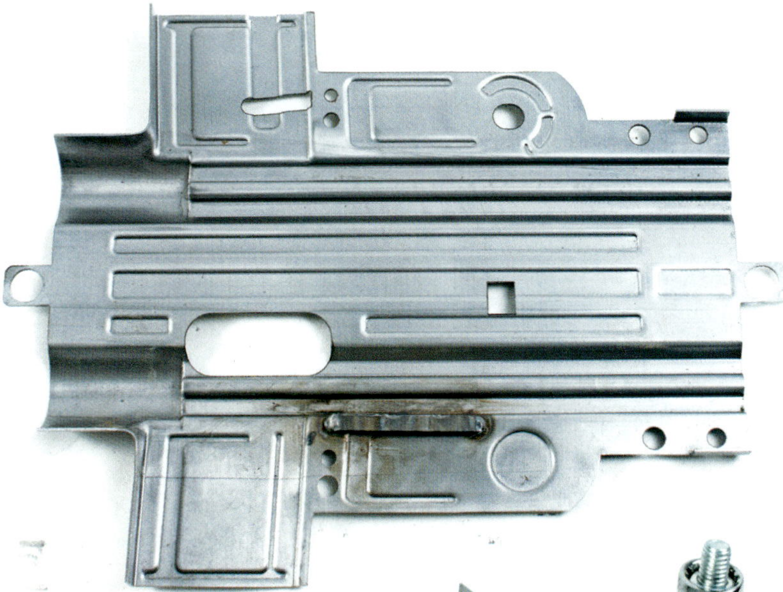

Denial bar welded into receiver prevents insertion of a full auto-trigger pack.

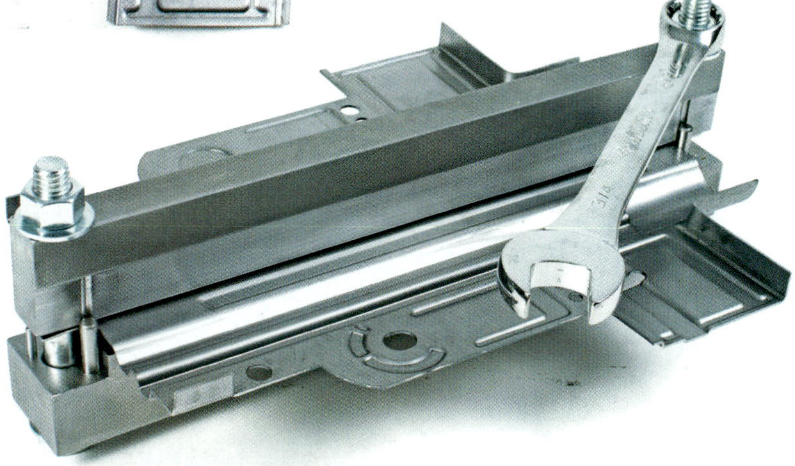

Receiver flat is pressed by means of a continuous rotary inclined plane. That's a screw, for the uninitiated.

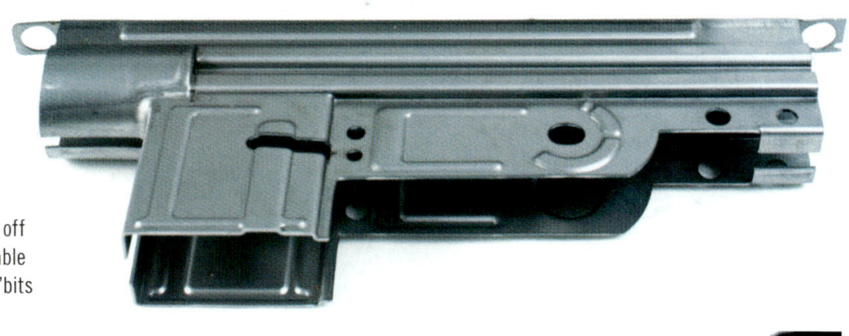

Once folded, it's off to the welding table to add the bag o'bits

high-pressure grease on the die and flat where they meet. Once you've formed the receiver, you can remove the locating tabs on either end.

BURN, BABY, BURN

Once you've formed the receiver, it's time to again fire up the welding gear. This is the most intimidating part of the build, especially if you've never welded before, but due to the design of the CETME-L, you can do a fairly crappy job and still end up with a working rifle. If, like us, your practice welds look like they were done by Michael J. Fox using a glue gun, then it might be a good idea to enlist the help of a competent friend, once you've tacked everything in place.

Top: We probably won't win any prizes for our welding abilities, unless there's a category for "welds that look like animal feces."

Use the front sight tower to locate and secure the cocking tube while it's tacked in place.

Note that due to the legalities of homebuilt firearms, you and only you must form and tack the receiver. Once it's capable of firing a round, you can then get help to dress the welds up and make them look pretty. If you're happy with your abilities to melt steel, then just remember to take it slowly, use short beads, and don't concentrate heat in small areas — otherwise you run the risk of warping your receiver. Welding in the magazine guide and rear trunnion are fairly straightforward, so don't sweat it.

Once the receiver is looking like, well, a receiver, you can then make it look like a rifle by seating the barrel, complete with the trunnion you assembled earlier. In order to ensure it stays put, plug weld it by drilling a hole or two and welding through them. We placed ours under the handguard to keep things neat and tidy, then dressed them up with a grinding wheel.

Installing the cocking tube is a little more involved, as it must be correctly aligned in order to function. As luck would have it, the rifle can serve as its own alignment jig, so long as you do your part by ensuring the cut end of the tube you separated from the original receiver is neat and square. Tap the front

sight tower into position on the barrel and use this to locate the forward end of the cocking tube while you use the cocking lever assembly to line up the rear.

SIGHTS

You'll want to add some means of sighting this baby, and, unfortunately, the CETME-L was devised in an era that predated widespread optics use — its as-issued iron sights leave much to be desired, being on par with those found on an M16A1. The bad news is that while it's hard to make them any better, if you don't pay attention to their installation, it's easy to make them worse. Starting with the front sight, tap it into place on the barrel and reach into the toolkit for a finely calibrated Mk1 eyeball. Once aligned, take a 3/16-inch drill bit and run it into the cross pinhole in the sight tower. A drill press is advised, but we wanted to make sure this build could be done with a minimum of tools, so we opted to grab the cordless DeWalt and carefully drill from both sides. Once you've created a slot in the barrel, drive in the supplied pin, and call it good.

The rear sight assembly also houses the CETME's bolt hold-open, and you'll use this to locate its position on the receiver. Once eyeballed in place and clamped, a couple of welds secure it in place.

As seen on objectives throughout Iraq, though ours is in better shape. Our local Cerakote guys did a great job with the finish.

TWEAKS

You're attempting to resurrect in your garage a piece of technology that took teams of professional engineers and craftsmen years to develop. There are going to be hiccups. We hit the following snags and worked through them to get the gun functioning, so don't freak out if you find the same problems — they're all fixable.

1. Denial block. This interfered with the trigger pack and hammer, requiring hand-fitting in order to

allow the pack to seat and hammer to strike with enough force to light off a round.

2. Bolt hold-open. Our cocking lever didn't have quite enough throw to allow the bolt carrier's slot to align with the BHO. A couple of file strokes were necessary to open the BCG's slot.

3. Stock pinholes didn't align with the receiver. We wound up clamping the stock in place and using the steel-reinforced stock pinholes to ream out the ones in the receiver. It's a kluge, but it works.

4. Our build's selector lever didn't quite line up with the corresponding hole in the receiver's flank. A little judicious grinding solved this.

FINISHING

We took the completed build, still in the white, out to Cowtown Range for its initial test firing. Once we'd proven that it cycled with SIG's 77-grain OTM loads, it was packed off to our local Cerakote guys for finishing in their newest coating, Cerakote Elite, covered in our last edition on page 92.

As the original kit was showing signs of hard use, rather than try to make the finished product look like it just rolled off the factory floor we opted for a battle-worn look. We Plead The 2nd worked its magic, and the results are on display here. So there you have it. An afternoon's work will give you a rifle sure to elicit inquiring looks from your shooting buddies. And if they echo our 44th president's words to the effect that, "You didn't do that," you can tell them that legally and practically, you did. So pound sand. R

Before investing in coating, we took the rifle in the white to test-fire it. It proved to be surprisingly soft-shooting, and, unlike an HK, brass didn't land in the next county.

YOU, DEAR READER

CETME-L SEMI AUTO

CALIBER: 5.56 NATO

MAGAZINE CAPACITY: 30 rounds

BARREL LENGTH: 16 inches

OVERALL LENGTH: 31 inches

WEIGHT: 8 pounds, 3 ounces

MSRP: $800

URL: www.hmgunworks.com/product/hmg-cetme-l-builders-kit/

Story by Iain Harrison
Photos by Straight 8

HILL & MAC GUNWORKS STG 44

A Modern Throwback to a WWII-Era Assault Rifle

In terms of its overall effect on small arms design, the StG 44 had more influence than any other single 20th century firearm. There, we said it. Despite an almost fanatical loyalty to the United States and a love of the guns it has produced, we're going to insist that this is the one model that changed everything. And it almost never happened.

As is usual when a revolutionary design comes along, its major components were already in use in other applications. The iPhone wasn't groundbreaking in the technologies it employed, but putting them all together in a way that solved the problems of many users ensured its place in technological history.

The StG 44 was the iPhone of the German Wermacht, and it solved problems with 125 grains of steel jacketed lead.

After encountering the Soviet mix of SVT-40s and Ppsh 41s on the Eastern Front during WWII, Germany's weapon designers dusted off plans previously shelved due to logistical concerns. It's not that they hadn't been working on an intermediate round and the rifle from which to fire it, it's just that bureaucracies as a rule take an inordinate amount of time to get shit done — in this case, about 20 years. There's nothing quite like a good ass kicking to concentrate the mind, however, and when the wheels came off of Operation Barbarossa in 1941, the German high command resurrected the program to use a chopped-down version of their 8mm Mauser service rifle cartridge in a select-fire, magazine-fed weapon. And lo, the assault rifle came to pass.

In order to circumvent Hitler's prohibition of new rifle programs, the assault rifle's development flew under the radar, camouflaged by the somewhat obvious means of renaming it as a submachine gun. The fruits of this scheme, designated the MP 43, was deployed in small numbers to infantry on the Eastern Front who loved it, as it provided greater range and power than a 9mm SMG, with way more firepower than a bolt-action rifle — all crammed into a handy package. Six months after its initial deployment, Hitler found out about the ruse, renamed the gun the "Storm Rifle" of 1944, and allowed people to congratulate him on his brilliance. Unfortunately for them, however, the weapon's arrival failed to make any difference to the war's outcome, a situation no doubt appreciated by GIs and Tommies throughout Europe.

Despite being too little, too late, the combination of features offered by the Sturm Gewehr has resonated with every significant military since 1945. Or at least, it eventually would, even with the mendacious seat warmers of the U.S. Ordnance Board, who inflicted the M14 on U.S. troops before finally tumbling to the fact that a full-auto 7.62x51 rifle is about as useful as a carpet fitter's ladder.

CONSTRUCTION

One of the design criteria of the StG 44 was that it was to consume less in the way of high-quality materials than the K98 service rifle. This was achieved through the use of a 1mm thick, stamped receiver, housing a short trunnion. The trunnion served as the locking mechanism to contain the pressures of firing, and, apart from the bolt, was the only piece of high-strength steel in the entire gun.

The StG 44 operates by means of a long-stroke gas piston attached to a carrier, which operates a tilting bolt. On firing, propellant gases enter a gas block located near the muzzle and push the piston rearward. After approximately 5mm of travel and before the bolt has unlocked, the piston exposes a series of vents that exhaust gas to the atmosphere. Being shoved to the rear, the carrier continues and a camming surface on its lower side engages a lug on the bolt, lifting it up 2 degrees and out of engagement with the trunnion's locking surfaces. Bolt and carrier continue rearward, extracting the spent case, compressing the recoil spring and eventually returning to chamber the next round, after which the carrier cams the bolt downward and into battery.

While it seems fairly standard when compared to modern designs, the StG's combination of engineering considerations was as revolutionary as the mix of intermediate cartridge, select fire, and detachable magazine. On examining an original StG, you get the impression that the engineers involved in the project kind of got the concept of making a high-quality rifle out of mediocre materials, but didn't fully trust the idea. As a result, they decided to do everything in their power to over-engineer the hell out of it. For example, the lower receiver consists of 203 individual parts. And yes, you read that correctly. You may think that a hammer or trigger pin would logically require the manufacturer to construct it from a single piece of steel rod, whereas in this case, five separate components are required.

Despite including many worthwhile modern improvements, the Hill and Mac gun looks almost identical to the original.

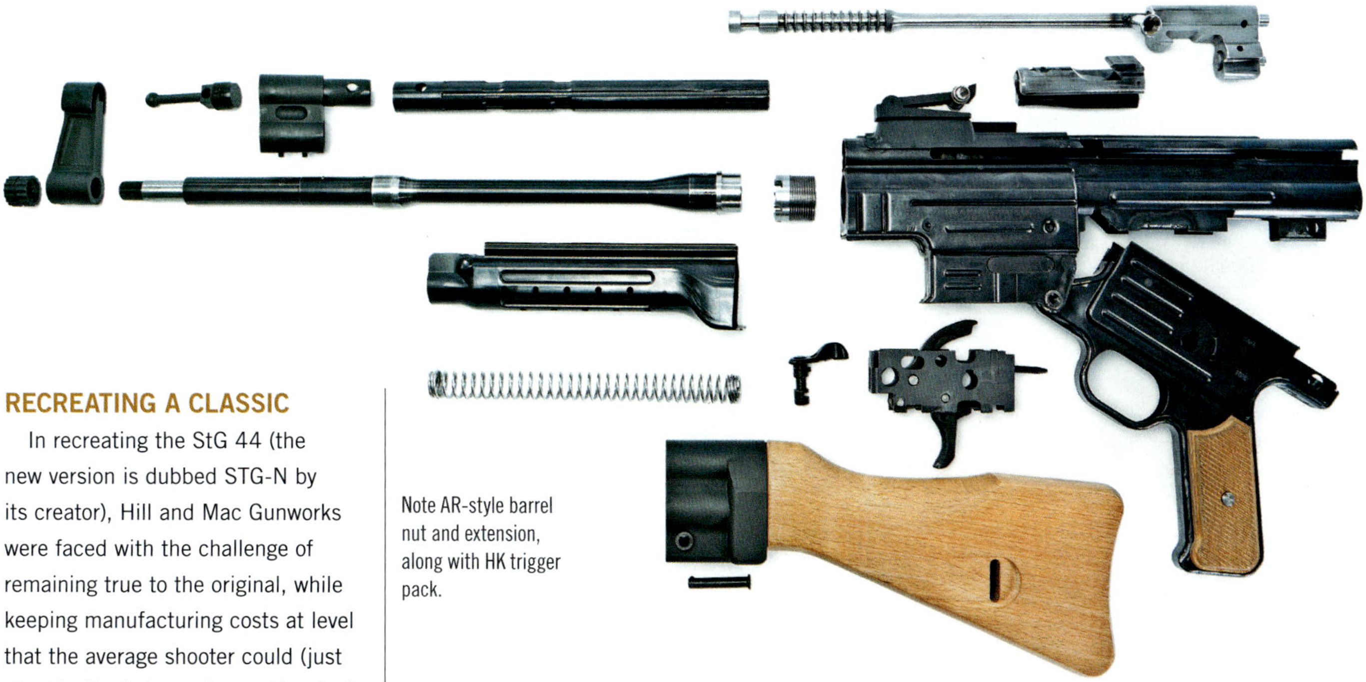

Note AR-style barrel nut and extension, along with HK trigger pack.

RECREATING A CLASSIC

In recreating the StG 44 (the new version is dubbed STG-N by its creator), Hill and Mac Gunworks were faced with the challenge of remaining true to the original, while keeping manufacturing costs at level that the average shooter could (just about) afford. According to Mac Steil, designer of the new version, copying just the German lower receiver would cost them $2,500 per unit, if full interchangeability was to be achieved. Instead, the decision was taken to maintain the external look and feel of the WWII model, while improving the design where feasible and keeping it within reach of as many people as possible. To this end, a standard HK trigger pack forms the heart of the fire control system, as it's proven, cheap, and reliable — and it fits inside a stamped lower receiver that looks almost identical to the original. Certain internal modifications were insisted on by the ATF, in order to ensure full auto parts couldn't drop in, but many others can be directly swapped with the German rifle.

Up top, the stamping that houses the new trunnion is both 50-percent thicker than the one that saw action against the Russians and made of stronger steel. As a result, the new version no longer suffers from the original's Achilles' heel, where lateral pressure in the center of the action — for example when the rifle was leaned up against a wall — would cause the receiver to buckle forward of the ejection port and effectively destroy it.

In order to economically feed the beast, changes were made to the magwell, so that instead of using StG magazines at over 200 bucks a pop, the same ones you use in your AR-15 will slide right in. The 7.92x33 Kurz ammo is likewise a little thin on the ground, so the U.S. recreation will also be available in 7.62x39 (which is apt, as the Russians were inspired to develop a short round that utilized the same bore diameter as their existing service cartridge by — you guessed it — the Germans). 5.56 NATO and .300 Blackout will also be available, and calibers can be changed by means of swapping out bolts, barrels, and gas pistons.

Due to increased precision afforded by CNC machining, the ability for the user to do their own barrel installation

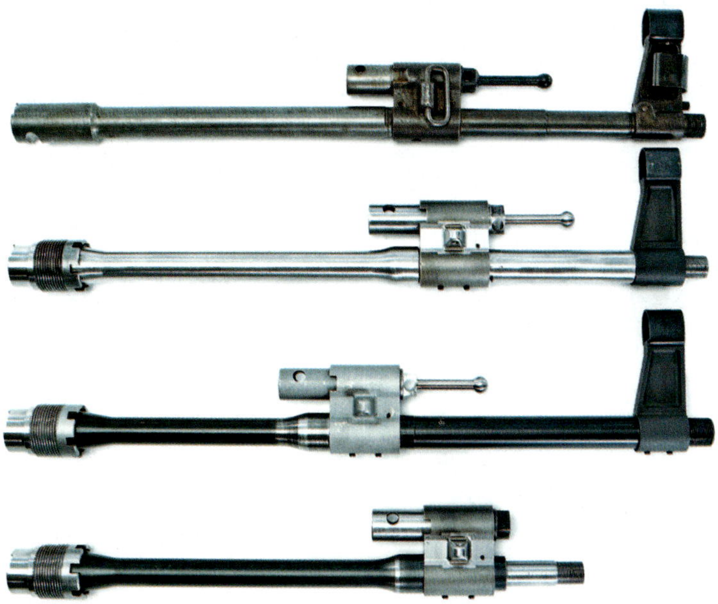

Gas port locations and piston lengths change according to caliber. Original World War II barrel is on top.

is now a reality. Under the stamped-steel handguard lies a castle nut, used to secure the barrel to the trunnion in much the same way an AR barrel is sandwiched to the upper receiver. In order to remove it, the user simply applies the supplied wrench and drops in a replacement. Factory barrels are turned from Lothar Walther blanks and match the WWII profiles on their exposed sections.

Despite the bolt locking up to a trunnion rather than a barrel extension, tolerances can be held to within a couple of thousandths of an inch, ensuring that headspace is consistent no matter which barrel is installed. Gas pistons are both replaceable and tuned for each caliber, which makes sense given the wide range of operating pressures encountered; 7.62x39 runs at around 45,000 psi, while 5.56 can go over 60,000, with 7.92 Kurz falling somewhere in the middle.

One significant departure from the original design can be found in the recoil spring assembly. Whereas the German guns used a recoil spring that extended into the buttstock, the new version keeps everything inside the upper receiver. Doing so allows the use of a folding or telescoping stock, and the receiver is profiled to match HK furniture. And speaking of HK, you'll no doubt recognize the way in which the StG field strips.

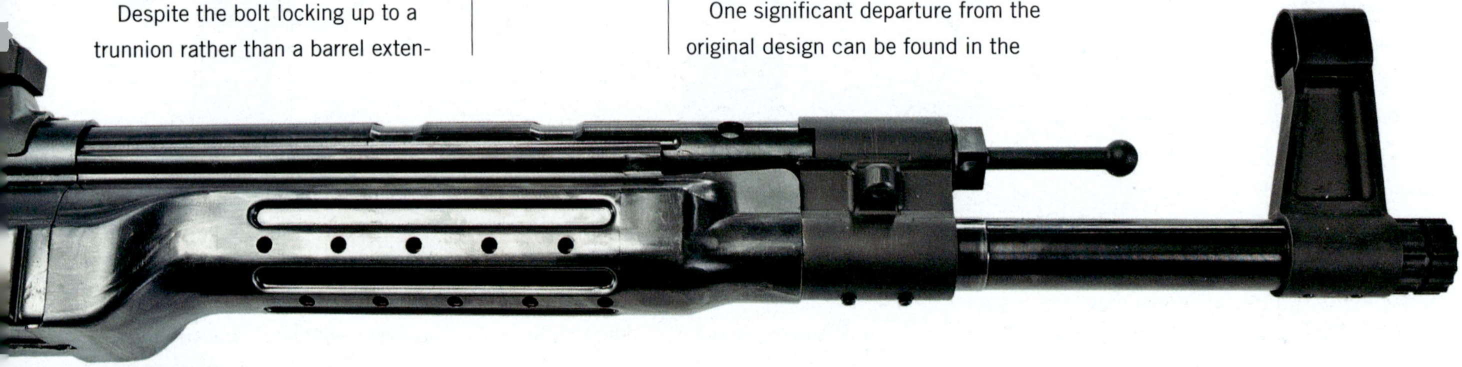

CONCLUSION

We'd love to be able to give you a glowing range report, comparing the American StG 44 with one made over 70 years ago, but at press time, the prototypes simply weren't ready for a full-on review. We did, however, manage to send three rounds downrange from a prototype weapon that embodies the most revolutionary/evolutionary advances in small arms design ever seen, but brought to a modern market by an American small business success story.

The structural and mechanical improvements wrought by Hill and Mac Gunworks are in keeping with the essence of guns made by Haenel, Sauer, and Steyr, but fix some of their problems while bringing down costs. If you've ever lusted after the original assault rifle, the one that started it all, then you can now own something that's almost indistinguishable from it. The best part of all is that it won't cost you 30 grand. **R**

Bored with ARs? Tired of AKs? Here's the OG that started it all.

HILL & MAC GUNWORKS STG-N

CALIBER: 7.92x33, 7.62x39, 5.56 NATO, .300 BLK

MAGAZINE CAPACITY: 30 rounds

BARREL LENGTH: 16 inches

OVERALL LENGTH: 37.5 inches

WEIGHT: 11 pounds 9 ounces

MSRP: $1,800

URL: www.hmgunworks.com

SELF-RELIANCE

Find an 80-Percent Lower You Fancy So You Can Build Your Very Own AR

By RECOIL Staff

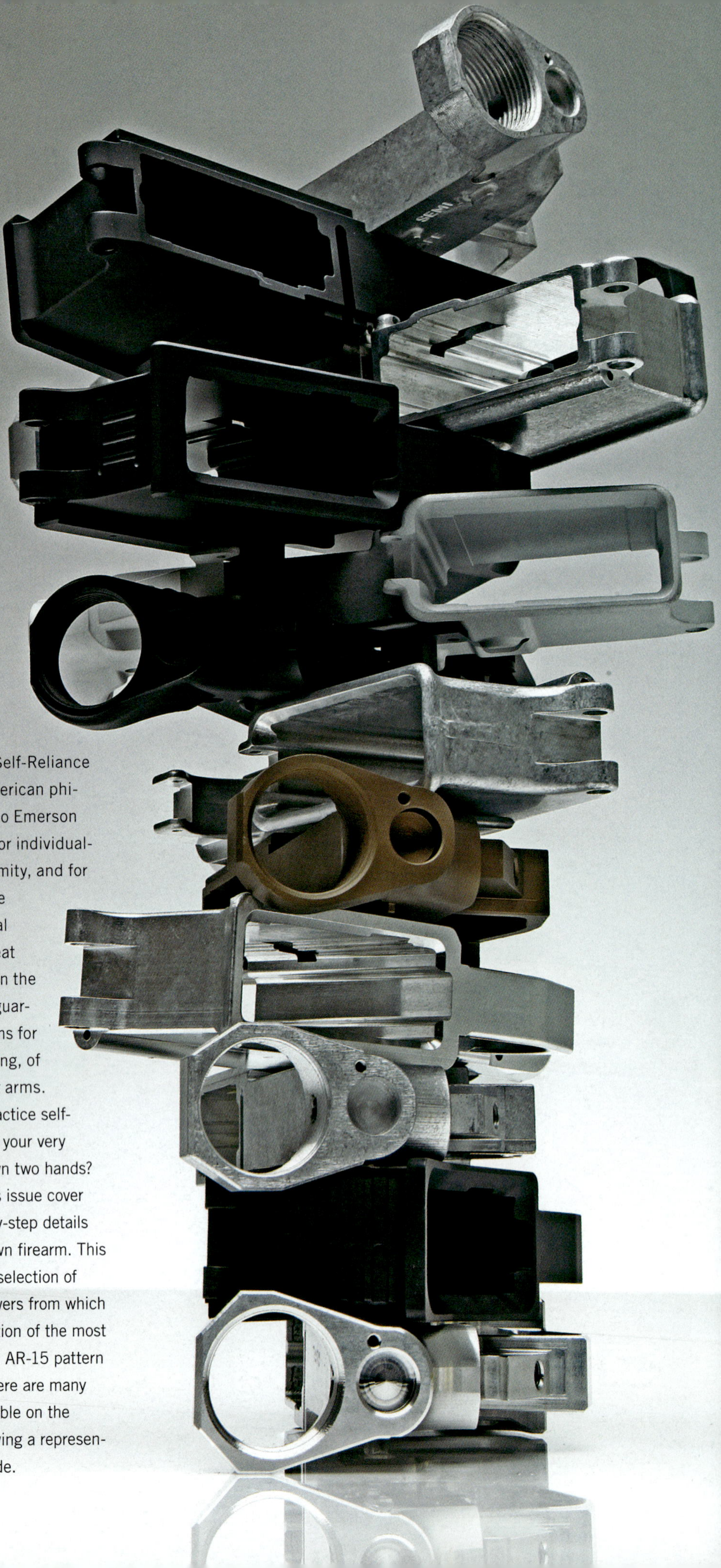

In his famous essay Self-Reliance from the 1800s, American philosopher Ralph Waldo Emerson argues strenuously for individualism, for avoiding conformity, and for all to "trust thyself." The strength of the individual is a core tenet of our great country, and no less than the Bill of Rights explicitly guarantees personal freedoms for all Americans — including, of course, the right to bear arms.

What better way to practice self-reliance than by building your very own firearm with your own two hands? Articles elsewhere in this issue cover the legalities and step-by-step details of manufacturing your own firearm. This buyer's guide outlines a selection of so-called 80-percent lowers from which you can build your rendition of the most American of firearms, an AR-15 pattern rifle. Please note that there are many 80-percent lowers available on the market; we are just showing a representative sample in this guide.

SOME THINGS TO KEEP IN MIND:

Billet versus forged: 80-percent lowers are available in both forged and billet (or more accurately, machined from aluminum billet) variants. Last issue, we covered billet upper/lower sets, and the same considerations apply here. Forged lowers will be stronger, ounce for ounce, than their billet counterparts — but both will function perfectly well. Forged lowers more or less all look the same, while billet lowers can be machined to incorporate additional features and aesthetic concerns. Billet lowers also command a price premium.

Material: Forged lowers are made from 7075 aluminum, but you may often have a choice between 6061 and 7075 for billet lowers. How to choose? 7075 is stronger, but more expensive. 6061 can be easier to work on for your average garage tinkerer and, when machined from billet, is more corrosion resistant than 7075 (forgings can be a different matter, hence issues that you may have read about during the early days of the AR-15 platform). Some may prefer to pay a little extra to get the strongest material, but don't feel bad about going with 6061. Both will get the job done just fine.

Features: Billet lowers may have various additional features, such as integral trigger guards, flared magazine wells, and pins with screws to replace those annoying roll pins. You can also shop around to find one that suits the look you want.

80-percent versus 95-percent: All of these partially formed pieces of aluminum have been determined by the ATF not to be firearms. Most lowers require you to hog out the entire fire-control pocket and drill holes for the trigger bow, safety selector, hammer pin, and trigger pin. Some (including the American Spirit Arms forged, which coined the phrase "95-percent lower" for its offering, Ares billet, and Colfax forged) have additionally started the fire-control pocket for you at the rear of the lower.

Finish: Those who like to "go big or go home" and do things by the book should probably choose a raw lower, complete the machining and any engravings, surface finish to preference (i.e. bead-blast), and anodize the finished product. While more expensive, this will ensure the entire lower (in particular, the pin holes) is hard-coat anodized. Alternately, many a home builder has started with an anodized 80-percent lower and either left the internal machined surfaces and pin holes in the raw, or hit them with something simpler and more economical like Cerakote. But if intended for hard use and high round counts, our preference would be to start with a raw receiver and then type III hard-coat anodize it afterwards.

Jigs: If possible, purchasing a jig from the same manufacturer as your lower ensures that they will be perfectly compatible. However, many jigs will work with a variety of lowers, as noted in the compatibility grid, and some are designed to be universal. Note that, generally speaking, forged lowers are compatible with all jigs, but billet lowers may or may not be, depending on their particular design characteristics. Exactly what makes billet lowers more unique and distinctive can make them incompatible with other jigs. For example, differences in the contour of the magazine release fence, flares in the magwell, or extra thickness in the lower's "tail" may interfere with the shape of certain jigs.

> "A FOOLISH CONSISTENCY IS THE HOBGOBLIN OF LITTLE MINDS, ADORED BY LITTLE STATESMEN AND PHILOSOPHERS AND DIVINES."

Some other considerations in choosing a jig include its rigidity and clamping surface, which help to stabilize the work piece and reduce chatter from vibration. This unfortunately can be difficult to assess, as it will depend not only on the lower you select, but also the mill vise or other equipment that you're using — we encourage you to think carefully about this; for instance, the integral trigger guards on billet lowers may protrude from the bottom of some jigs. Almost all of the jigs utilize the front and rear takedown holes for locator pins. Some jigs have replaceable bushings for their drill holes to enhance their ability to be used multiple times.

ARE THEY COMPATIBLE?

	80% ARMS JIG	ARES ARMOR JIG	BILLET RIFLE SYSTEMS JIG	CNC GUNSMITHING JIG	COLFAX TACTICAL JIG	JUGGERNAUT TACTICAL JIG	KE ARMS JIG	TACTICAL MACHINING JIG	TR ENABLING JIG
80% Arms Lower	✓	✓	✗	✓	✓	✓	✗	✗	✗
American Spirit Arms Lower	✗	✓	✓	✓	✓	✓	✓	✓	✓
Ares Armor Forged Lower	✗	✓	✓	✓	✓	✓	✓	✓	✓
Ares Armor Billet Lower	✗	✓	✗	✓	✓	✓	✓	✗	✗
Ares Armor Signature Lower	✗	✓	✗	✓	✗	✓	✗	✗	✗
Billet Rifle Systems Lower	✓	✓	✓	✓	✗	✓	✓	✗	✗
Colfax Tactical Lower	✗	✓	✓	✓	✓	✓	✓	✓	✓
Juggernaut Tactical Lower	✗	✗	✗	✓	✗	✓	✗	✗	✗
KE Arms Forged Lower	✗	✓	✓	✓	✓	✓	✓	✓	✓
KE Arms Billet Lower	✗	✓	✓	✓	✓	✓	✓	✗	✓
Tactical Machining Lower	✗	✓	✓	✓	✓	✓	✓	✓	✓
TR Enabling Lower	✓	✗	✓	✓	✗	✓	✓	✓	✓

DISCLAIMER: Fitment information based on our sample products. Manufacturers may update products from time to time, so please verify before purchasing.
NOTE: All lowers that are available with anodizing are type III hard anodized. Additionally, all magazine wells are broach cut unless otherwise noted.

THIS IS WHAT FREEDOM LOOKS LIKE

Building your own firearms is not limited to the province of AR-15s. There are other platforms that may suit your fancy, and 80-percent receivers and frames can be had in a variety of selections. Shown here are just a few examples.

The DPMS/SR-25-pattern AR-10 80-percent lower ($179 raw, $189 black anodized) and jig ($140) are from Juggernaut Tactical. Note that most of the companies in this guide sell AR-10 lowers. AK-Builder (www.ak-builder.com) offers a variety of AK receiver flats ($35 for the M92 under-folder shown). You can also build a pistol, with a Rudius 1911 80-percent frame ($250) from Ares Armor, and even an 80-percent frame ($219) from Limited 10 (www.limited-10.com/store) that you can use to build an STI 2011-pattern double-stack pistol. Also pictured are examples of AR-15 80-percent lowers with black and FDE anodizing. Polymer lowers are also available.

80% ARMS

80% Arms offers billet lowers in your choice of bead-blasted raw or black anodized, and in 6061 or 7075 aluminum. Shown is the black anodized receiver. The lowers have engraved selector markings ("SAFE" and "SEMI") on both sides, integral winter trigger guards, and flared magwells.

MODEL:
80% Lower Receiver Billet AR-15 Lower

MATERIAL:
6061 or 7075

FINISH:
Raw (bead-blasted) or black anodized

MSRP:
$105 (raw 6061)
$110 (anodized 6061)
$125 (raw 7075)
$130 (anodized 7075)

80% Arms' Easy Jig is designed specifically for its own billet lowers, allowing you to complete your build more quickly and using only commonly available hand tools. Milling is done with a router; you need not have a mill or drill press. The jig features hefty side plates with various adapters and templates to complete each operation. It completely encloses the integral trigger guard on the billet lower. Note the Easy Jig is not compatible even with standard forged lowers — 80% Arms does offer a universal jig for $90 (raw) or $98 (anodized), which we did not have to test.

MODEL:
80% AR15 Easy Jig

FINISH:
Anodized

MSRP:
$150

URL:
www.80percentarms.com

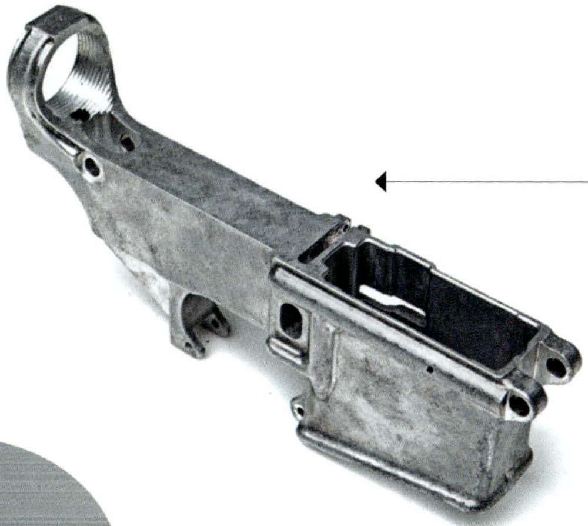

AMERICAN SPIRIT ARMS

American Spirit Arms offers a traditional forged receiver with a nice tumbled finish. They were one of the first to push to machine as much as they could while still receiving a favorable determination from the ATF. ASA calls it a 95-percent lower — compared to many other 80-percent lowers, the fire-control pocket has been started with the upper receiver rear lug pocket already machined. The magazine well is EDM wire cut and moderately beveled.

MODEL:
ASA M4 AR-15 95-Percent Lower Receiver

MATERIAL:
7075 (forged)

FINISH:
Raw (tumbled)

MSRP:
$150 ($120 on sale)

URL:
www.americanspiritarms.com

ARES ARMOR

As one of the more high-profile purveyors of 80-percent receivers, Ares Armor offers a full line of products, including forged, billet, and premium Signature billet lowers. They can be ordered raw (usually tumbled, though the forged lower shown is bead-blasted) or black anodized for a $10 surcharge. The forged lower's magwell has additional beveling. The billet receiver has subtle styling with a mag fence similar to the forged; it also features an integral trigger guard and, like the ASA, the rear of the fire-control pocket has already been machined. The Signature billet lower (shown anodized black) has more squared-off lines and an integral trigger guard — note that it is only compatible with a few jigs in this guide.

MODEL:
80-Percent AR-15 Lower Receiver

MATERIAL:
7075 (forged)

FINISH:
Raw (tumbled) or black anodized

MSRP:
$80 (raw)
$90 (anodized)

MODEL:
80-Percent Billet AR-15 Lower Receiver

MATERIAL:
7075

FINISH:
Raw (tumbled) or black anodized

MSRP:
$160 ($90 on sale, raw)
$100 (anodized)

MODEL:
Ares Armor Signature Billet

MATERIAL:
7075

FINISH:
Raw (tumbled) or black anodized

MSRP:
$160 ($125 on sale, raw)
$135 (anodized)

The Ares Armor jig has thick anodized aluminum side plates with drill bushings for the trigger pin, hammer pin, and safety selector hole. The top plates are made of steel and provide templates to hog out the fire-control group. The jig completely encloses the integral trigger guard on the billet lowers and was compatible with all but two of the billet lowers in this guide.

MODEL:
80-Percent AR-15 Jig

FINISH: Anodized

MSRP: $135

URL: www.aresarmor.com

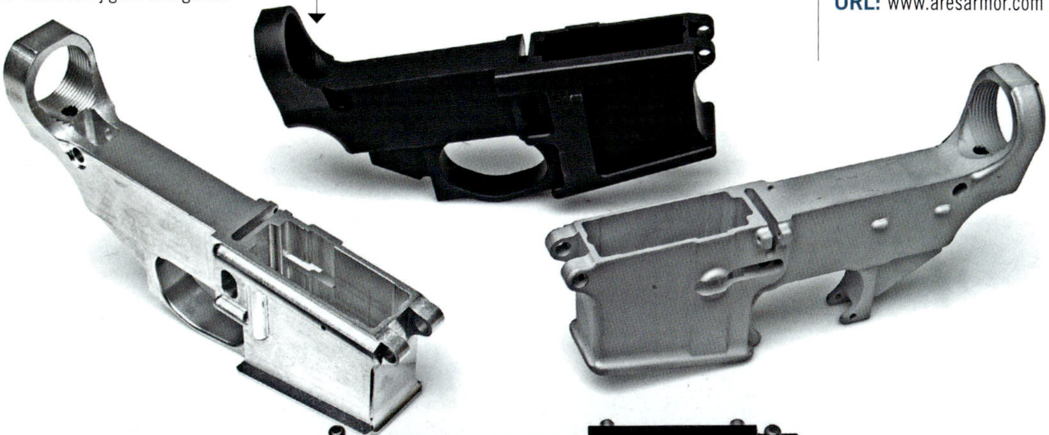

BILLET RIFLE SYSTEMS

Billet Rifle Systems makes 80-percent lowers in 6061 and 7075, available raw or black anodized. The lower has an angular look and features an integral trigger guard, flared magwell, threaded rear takedown pin detent hole, and a pin and set screw for the bolt catch. The magwell is machined. The lowers also have engraved pictograms for the selector markings on both sides.

MODEL:
Independence

MATERIAL:
6061 or 7075

FINISH:
Raw (steel ball burnished) or black anodized

MSRP:
$129 (raw 6061)
$159 (raw 7075)
$174 (anodized 7075)

The Jig Set from Billet Rifle Systems is very affordable and comes with two side plates and a single top plate that screws in and flips around to accomplish each operation. The jig fully encloses the integral trigger guard. The side plates are not as thick as some of the other jigs, but they get the job done (see it used in the DIY on page 40). Tooling is optional and includes a 3/8-inch end mill, 5/16-inch end mill, 5/32-inch drill bit, and 3/8-inch drill bit.

MODEL:
80-Percent Jig Set

FINISH:
Raw

MSRP:
$72 (jig)
$64 (tooling)

URL:
www.billet80.com

CNC GUNSMITHING

The other jigs in this guide use the takedown pin holes to align the jig and work piece, and are designed to be used with typical 80-percent lowers sold today which require the fire-control pocket to be milled out. The CNC Gunsmithing jig doesn't utilize the takedown pin holes, requiring more care during setup, but also allowing more flexibility, as it is designed to handle other operations on 80-percent receivers of differing styles, including those that originally had the fire-control pocket already milled out. You can infer from this just how long CNC Gunsmithing's track record has been in the 80-percent business; they focus primarily on jigs and do not sell AR-15 lowers (though they do offer an interesting 80% AR45 lower that accepts grease gun magazines). The jig also does not fully enclose the lower, which you should keep in mind as you consider how you will secure it. The CNC jig has two side plates and several adapters, sold separately; it was compatible with all of the lowers in this guide. In sum, though, it is the most expensive jig set.

MODEL:
AR-15 drilling jig, milling adapter kit, trigger slot adapter kit

FINISH: Anodized

MSRP:
$100 (jig)
$125 (milling adapter)
$50 (trigger slot)

URL:
www.cncguns.com

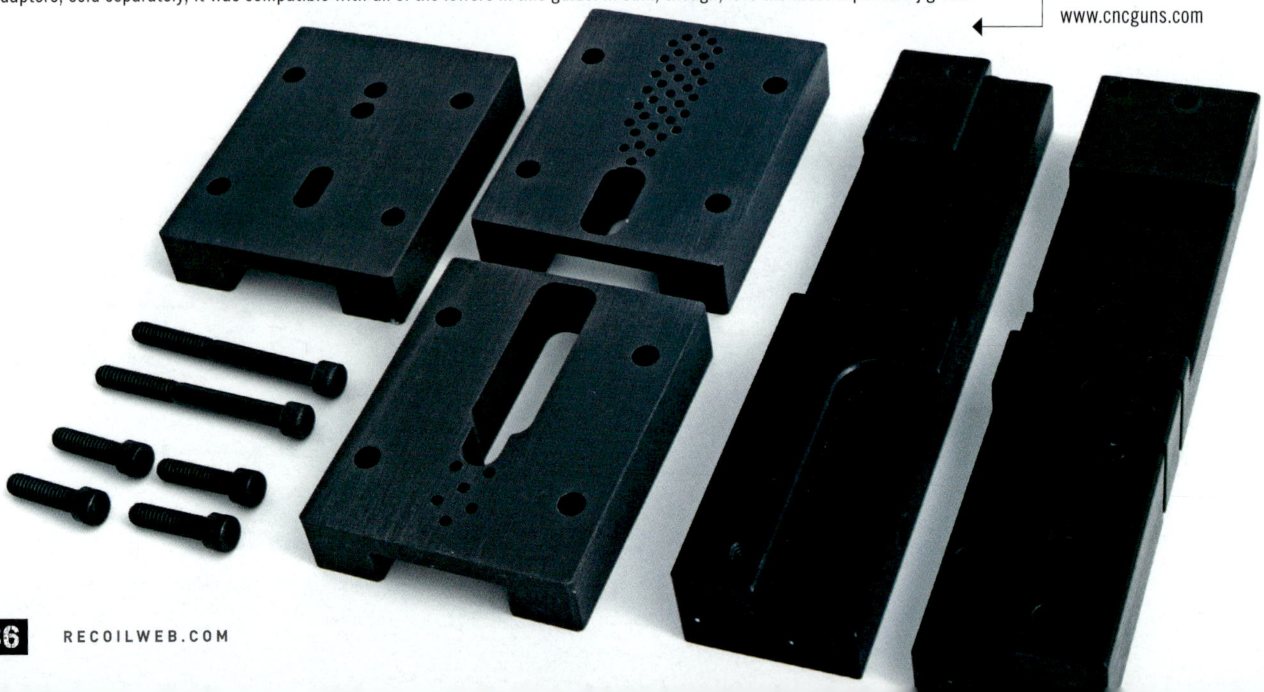

COLFAX TACTICAL

Colfax has a very nice forged receiver in your choice of raw tumbled (shown here), raw bead-blasted or black anodized. The magwell has extra beveling. Similar to the ASA, the rear of the fire-control pocket has already been started.

MODEL:
AR-15 80-Percent Lower

MATERIAL:
7075 (forged)

FINISH:
Raw (tumbled or bead blasted) or black anodized

MSRP:
$79 (raw)
$90 (anodized)

Colfax's jig includes side plates and a single top plate. Note this jig does not include a template to drill the trigger hole — because Colfax is concerned about potential drift when drilling from the top, given the distance between the template and the bottom of the receiver. Thus, they advise customers to use coordinates instead. The jig does completely enclose integral trigger guards on the other billet lowers.

MODEL: AR Jig

FINISH: Raw

MSRP: $99

URL: www.colfaxtactical.com

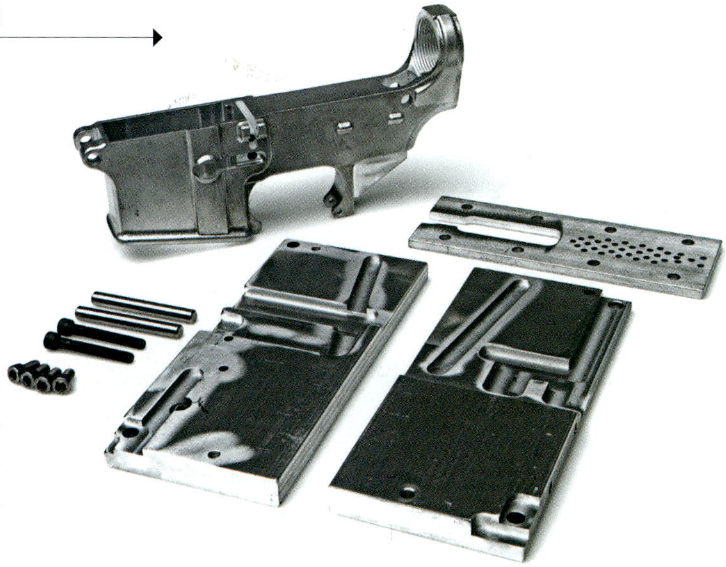

JUGGERNAUT TACTICAL

Juggernaut Tactical's billet lower has a unique look with an integral winter trigger guard, flared magwell, and grooves on the front of the magwell. The distinctive fence around the mag release and bolt catch also makes it incompatible with most of the jigs in this guide. Selector markings ("SAFE" and "SEMI") are engraved on the left side, and anodized models wear Juggernaut's logo. Available in 6061 or 7075, and either raw or anodized in several colors. Note that Juggernaut does not charge extra for 7075.

MODEL:
AR-15 80-Percent Lower Receiver

MATERIAL:
6061 or 7075

FINISH:
Raw (tumbled) or anodized (black, FDE or camo)

MSRP:
$155 (raw)
$165 (anodized black, FDE, or pink)
$175 (anodized camo)

Also unique is Juggernaut Tactical's Universal Jig, which sports an indexing system that utilizes the threading for the receiver extension to accept an adapter, ensuring proper centering of the receiver in the jig. There are three top adapter plates for the various operations. Laser-etched markings on the side plates indicate which drill bits to use. The jig was compatible with all of the lowers in this guide; however, it may not completely enclose all billet lowers — for instance, the TR Enabling's rather large trigger guard protrudes beneath the jig. The optional tooling package includes four Cobalt drill bits (0.484, 0.375, 0.156, 0.125-inch) and two carbide end mills (3/8, 1/4-inch).

MODEL:
Universal Jig

FINISH:
Anodized

MSRP:
$140 (jig)
$85 (tooling)

URL:
www.jtactical.com

KE ARMS

KE Arms makes forged and billet 80-percent lowers. The forged receiver is exactly what you'd expect, and the billet lower looks pretty similar. As a result, of the billet lowers in this guide, KE Arms' offering was compatible with most other jigs, working with all but Tactical Machining's jig and 80% Arms' Easy Jig (which is not intended to be universal).

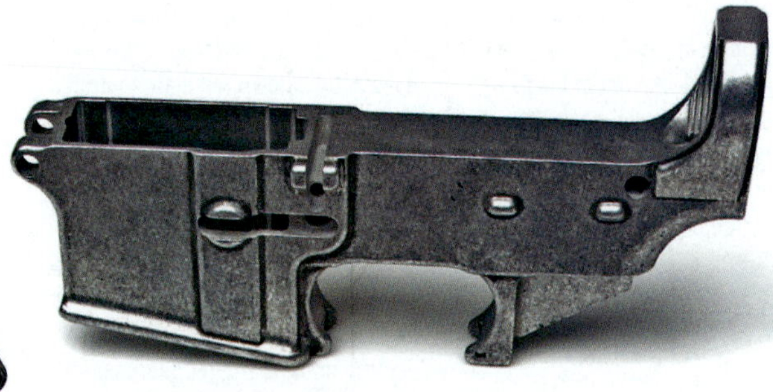

MODEL:
Forged 80-Percent KE-15 Lower

MATERIAL:
7075 (forged)

FINISH:
Raw (tumbled) or black anodized

MSRP:
$100 ($80 on sale, raw)
$120 ($100 on sale, anodized)

MODEL:
Billet 80-Percent KE-15 Lower

MATERIAL:
7075

FINISH:
Raw (tumbled) or black anodized

MSRP:
$130 ($96 on sale, raw)
$150 ($120 on sale, anodized)

The jig from KE Arms has steel bushings to drill the pins and selector hole. It also has convenient etchings to remind you of which tooling to use. There are two top plates with templates to mill out the fire-control pocket and trigger hole. The jig does completely enclose integral trigger guards on the other billet lowers and is compatible with all but three of them.

MODEL:
KE-15 80-Percent Finishing Jig

MATERIAL: 7075

FINISH: Anodized

MSRP: $150 ($130 on sale)

URL: www.kearms.com

TACTICAL MACHINING

Tactical Machining sticks to the basics with a traditional forged 80-percent lower — there's nothing wrong with that, and it's also the cheapest in this guide to boot. You can select between raw and black anodized. The magwell features additional beveling.

MODEL:
80-Percent AR-15 Lower Receiver

MATERIAL:
7075 (forged)

FINISH:
Raw (deburred and tumbled) or black anodized

MSRP:
$78 (raw)
$88 (anodized)

Tactical Machining's jig features side plates with steel drill bushings for the hammer pin, trigger pin, and safety selector hole. It has three top plates for the additional operations. The side plates are a little too short to enclose the integral trigger guards on billet lowers.

MODEL:
80-Percent AR-15 Jig Gen 2

FINISH:
Anodized

MSRP:
$90

URL:
www.tacticalmachining.com

TR ENABLING

TR Enabling's billet 80-percent receiver has some chunky styling cues with a large integral trigger guard, flared magwell, pin and set screw for the bolt catch, and a threaded rear takedown pin detent hole. It is only available in raw tumbled form and is the cheapest billet lower in this guide.

MODEL:
80-Percent AR-15 Lower Receiver

MATERIAL:
6061

FINISH:
Raw (tumbled)

MSRP:
$95 ($79 on sale)

TR Enabling's jig is also the most economical of the bunch. It has two side plates and a single top plate with templates for the milling operations. The top plate secures with screws and flips around to line up for the appropriate cuts. The jig fully encloses the integral trigger guard.

MODEL:
AR-15 Milling Jig

FINISH:
Raw

MSRP:
$69

URL:
www.tr-enabling.com

CH. 3

Gun Modifications

LEGAL WARNING:
The following article is for educational and informational purposes only. Building a machinegun in the United States is illegal unless the builder happens to be a Special Occupation Taxpayer and manufacturer.

GOING FAST

Turning your AR-15 into an M-16

By Mike Searson | Photos by Dave Merrill

In late 2017, in Issue 34, I wrote an article about the various "full-auto simulators" that were available at the time. In a tragic twist of irony, I was putting the finishing touches on the story when a mass shooting occurred that made some of the products (bump stocks) I wrote about illegal under federal law. In the time since then, I've seen thousands of examples of folks on both sides of the debate consistently misuse terms and get basic facts about the operation of a machinegun, or even a semi-automatic rifle completely wrong.

This is somewhat understandable as manufacture of new machineguns for civilian use has been illegal since 1986 when the NFA (National Firearms Act) registry for new machine-guns for the civilian market was closed. As a result, most firearm enthusiasts and even experts who came of age in the past 35 years have little firsthand knowledge of inner workings of machineguns. There are exceptions of course when it comes to military armorers, federally licensed Special Occupational Taxpayers (SOT) who manufacture new machineguns for government use and those of us who read old books and manuals to grasp a better understanding of how things work. To educate the public, I thought I'd take a look at the various ways to make an M-16 from an AR-15.

Again, unless you're an SOT building a post-1986 machinegun, using the information presented to make one of these items is illegal. Please do not break the law. I'd hate to see good gun people go to prison and, even worse, they don't let you read gun magazines in the federal prison system, or even browse our website!

Apart from enlisting in the military, becoming a SWAT officer, or signing up as a private security contractor where you may be issued an M-16 for an extended loan period, your sole recourse for obtain-

ing an M-16 is buying one. With the cost of a transferable M-16 being more than $20,000 these days, this article is intended as a guide to what you're buying if you're considering a registered conversion part over a dedicated machinegun.

THE RIGHT WAY

When most SOTs want to build a real M-16, they obtain the appropriate parts, source a proper receiver, file a Form 2 with the ATF, and when it's approved, they put it all together.

The most common way is to buy the actual M-16 fire control parts and install them in a lower receiver. Most lower receivers need to be modified by drilling a third hole to fit the auto sear.

Before you start thinking about becoming an SOT "just to build machineguns," bear in mind you have to be an actual firearms manufacturer, obtain a Federal Firearms License, Taxpayer ID Number, apply for an SOT status, pay all the appropriate fees, including a couple grand to the State Department for ITAR compliance, and be in the business full time. When you give up your license, however, you need to sell, destroy, or turn in your post sample machineguns.

DROP-IN AUTO SEAR

One of the first methods of converting an AR to fire full auto is to install a part known as a Drop-In Auto Sear, or DIAS. The DIAS requires an M-16 hammer, selector, and bolt installed in the rifle. The "Drop In" appellation refers to the end-user not having to machine the lower to accommodate an auto sear, as a secondary auto sear is part of the DIAS.

When a DIAS is installed in an AR lower, the hammer is cocked by the bolt carrier as it moves rearward. The carrier pushes the hammer down and the sear of the DIAS grabs the hammer. As it moves forward again, the carrier catches the top of the sear in the DIAS assembly and releases the hammer.

Transferable DIAS are referred to as Registered Drop-In Auto Sears or RDIAS.

What separates a registered receiver from the left is that third pin hole to house the auto sear.

THE "PRE-81" DIAS

Back in the late '70s and very early '80s (in retrospect this may have been the real "glory days" of firearms ownership in the United States), a DIAS was commonly advertised in Shotgun News or Gun List for about $150.

They more or less flew under the radar as ARs were still not a very common rifle; there was no internet, and there weren't legions of firearms owners writing letters to ATF's Tech Branch for clarification on what was and wasn't legal.

In 1981, BATF issued a formal ruling that these sears ("AR-15 Auto Sear," "Drop-In Auto Sear," and "Auto Sear II,") were considered machinegun conversion parts and any sears made after November 1, 1981, had to be registered as a machinegun. Unlike the recent bump stock fiasco, BATF grandfathered unregistered sears made prior to November 1, 1981.

However, before you start searching for one on the various auction

M-16 fire control parts are legal to purchase and own, though we nary see a point unless you have an M-16.

sites or local gun shows, there's a catch. You cannot legally possess one if you own an AR-15.

Essentially, this rule makes possessing a Pre-81 DIAS useless. It's kind of like owning a set of keys to a Lamborghini without being able to drive the actual car. A further legal problem is compounded because none of these sears were engraved with a serial number to substantiate a date of manufacture. In court, the burden of proof is placed on the defendant and, thus, the prosecution presume the sear was made well after the grandfathered date. This makes the possession of this little part akin to owning an unregistered machinegun. Bottom line: Don't do it!

LIGHTNING LINK

Another fantastic invention that came about in the 1980s was the "Lightning Link." Properly known as the S.W.D., Inc. Auto Connector, fewer than 900 of these were manufactured and in the NFA registry prior to May 1986. This is a drop-in part to convert a commercial AR-15 into a fully automatic firearm. There are a few caveats regarding the use of this device, however.

In theory, it's a two-way conversion that's installed and removed easily. It doesn't give the shooter select-fire capability. Once it's installed, the rifle is full-auto only. Not that there's anything wrong with that! There are several manufacturers making kits to allow select-fire with the Lightning Link by modifying real M-16 fire control parts to function with the Lightning Link in an AR-15.

The other downside is that it's a product of its time, meaning when it was designed, there were less than a handful of companies making AR-15s. The market share was largely owned by Colt and their SP-1 version of the AR.

That's right, the Lightning Link only drops into a Colt SP-1 and a handful of other, early civilian-made AR-15s such as Bushmaster, Sendra, Essential Arms, and most PWA pre-ban lower receivers. After 1989, Colt added what's known as a "High Shelf" in their lower receivers to prevent installation of a Lightning Link. These lowers can work if the shelf is milled away.

Many modern lowers have a "Low Shelf," but they may not be rounded in the rear properly to accommodate a Lightning Link. Again, milling will be required to make it fit.

The other challenging part is that the Lightning Link only works with Colt SP-1 bolt carriers. If you have a commercial bolt carrier, or even an M-16 carrier, these will need to be milled to work with the Lightning Link.

Lastly, the Lightning Link is a fragile piece of hardware. It's not uncommon for these to break

A DIAS doesn't require a third pin hole, though some receivers will require some milling.

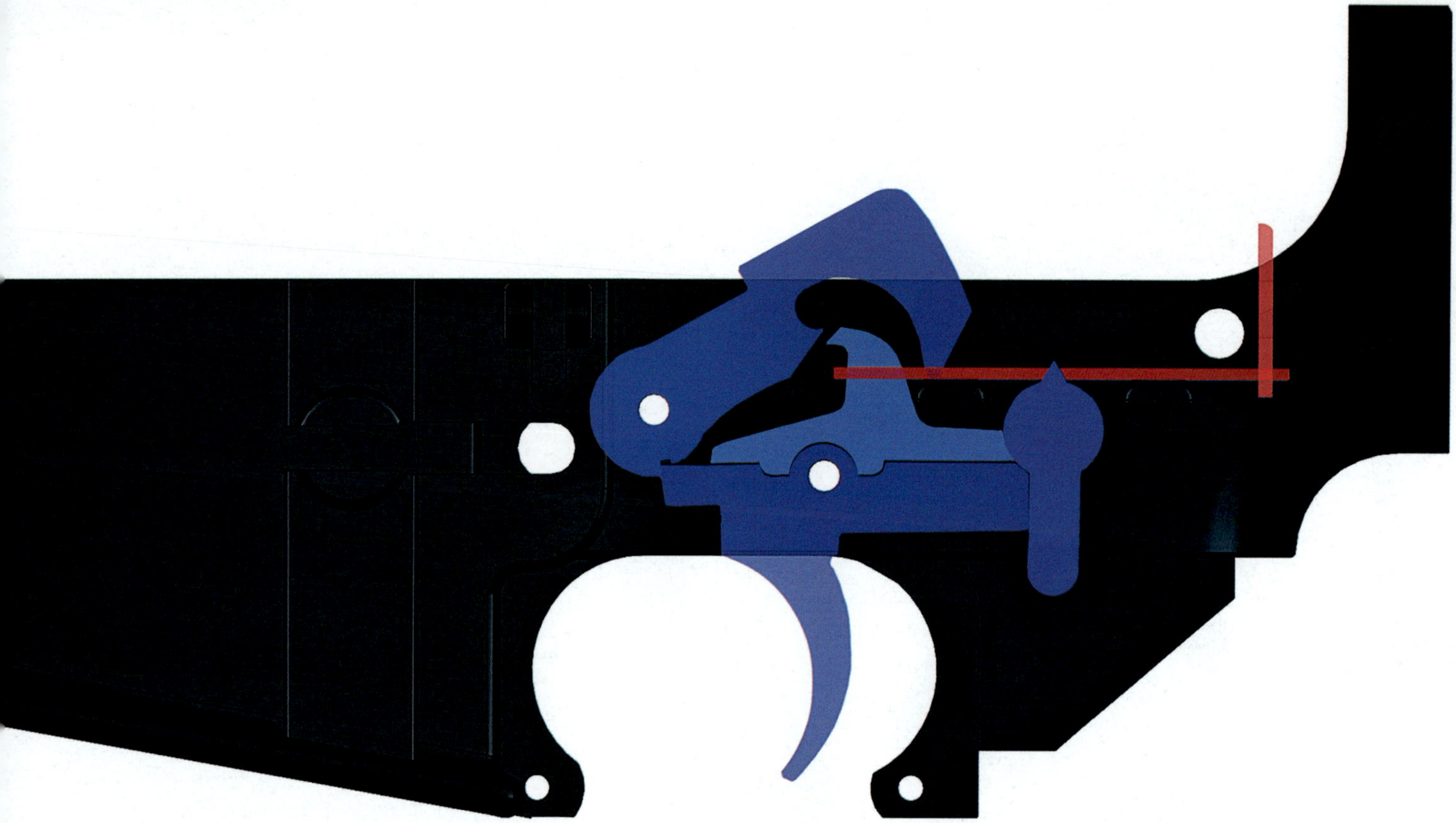

on extended use, and once they break, you've just destroyed an over $15,000 machinegun.

HOW IT WORKS

In semi-auto mode, the hammer is cocked as the bolt carrier moves rearward. When the carrier returns to the forward position, the hammer is held in this position by the sear located on the forward part of the trigger catching in the sear notch, on the hammer.

If you hold the trigger after a shot is fired, the sear will not catch in the hammer's sear notch when the hammer cocks because the sear is depressed below the arc of the hammer notch.

Because the trigger is being held back, the disconnector hook is pushed forward and in position to catch the hammer, preventing it from following the carrier forward. When the trigger is released, the hammer moves from under the disconnector hook and is caught by the trigger sear in the hammer sear notch, making it necessary to pull the trigger for each shot.

A Lightning Link looks like something someone came up with in their garage, which isn't terribly far from the truth. Using one in a modern rifle requires some tuning and other work.

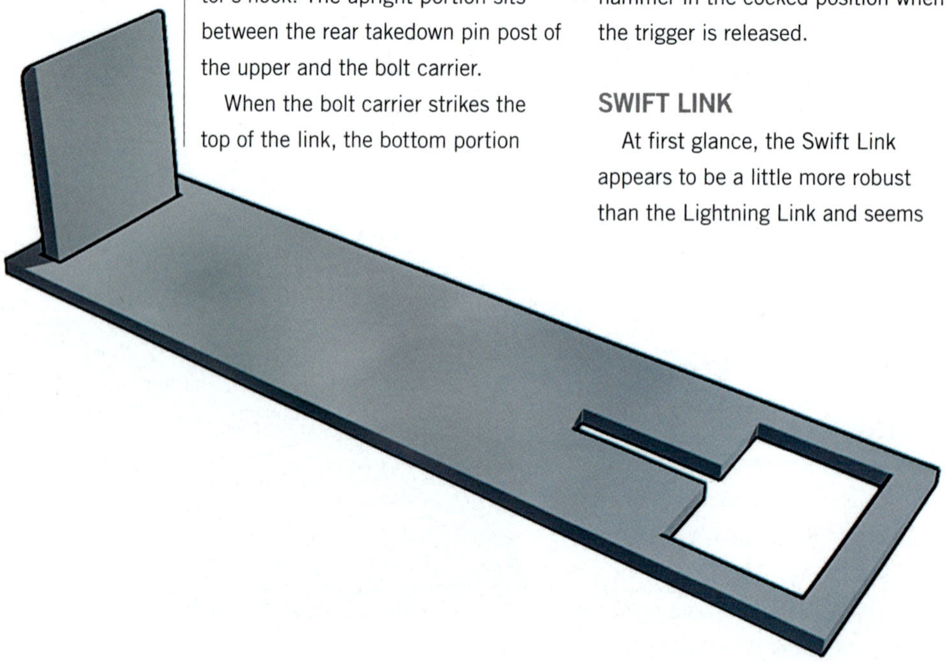

When the trigger is held rearward, the sear is below the arc of the hammer notch. The only thing holding the hammer in the cocked position is the disconnector. The Lightning Link works by pulling the disconnector to the rear and releasing the hammer. The Lightning Link lays flat in the lower receiver, with the forward opening fitting over the disconnector's hook. The upright portion sits between the rear takedown pin post of the upper and the bolt carrier.

When the bolt carrier strikes the top of the link, the bottom portion swivels to the rear, moving the body of the link rearward and releases the hammer from the disconnector hook. As the trigger is held rearward, the movement of the bolt carrier continuously cocks the hammer and forward movement of the carrier strikes the upright of the link as the bolt locks in battery, releasing the hammer, and firing the rifle. The sear stops the hammer in the cocked position when the trigger is released.

SWIFT LINK

At first glance, the Swift Link appears to be a little more robust than the Lightning Link and seems to be more user-friendly regarding compatibility with different lower receivers. The only other requirement seems to be the use of an M-16 bolt carrier. It looks like it would work much like the Lightning Link by depressing the disconnector when the bolt closes and releasing the hammer to enable full-auto fire.

I have no hands-on experience with this device and don't even know the name of the original manufacturer. It showed up on a few internet forums and appears to have an origin as the result of a Freedom of Information Act request submitted to the ATF with regard to the October 2017 shooting at Mandalay Bay in Las Vegas.

Questions remain as to whether the shooter actually made the device and used it, or if it was the image of a prototype found on his computer.

Yes, that is a bent-up coat hanger, properly registered as a post-sample machinegun.

THE COAT HANGER MACHINEGUN

An interesting variation of the Swift Link is one that was made by Ben Winslett (a firearms manufacturer and SOT) out of a coat hanger. Inspired by images of the Swift Link, Winslett was searching for suitable materials in order to make a version of his own. He twisted a coat hanger into shape and registered it on a Form 2 as a "Coat Hanger Machinegun."

This type of device has been talked about in NFA circles for years, with some referring to it as the "Afghan Coat Hanger Conversion," presumably because some Afghani fighter got tired of firing AKs and happened upon an AR that he converted using this method; or perhaps it was a PMC that was issued an AR. Regardless, Winslett's conversion is the first case we've seen of one being made legally and legitimately.

HOW IT WORKS

The Swift Link and the Coat Hanger Machinegun work on a principle similar to the Lightning Link, but in reverse. Rather than playing the game of "capturing the hammer," these conversions simply push against the disconnector.

As you can see, there are a variety of ways to convert an AR-15 to full auto. Many of these methods were conceived pre-1986, because contrary to popular belief, M-16s weren't exactly the staple of every gun shop as some revisionists might have you believe. In the past 33 years very little has been done as far as experimentation, and what little that's been accomplished is almost solely in the realm of SOTs.

While some of these registered conversions may sound a little wonky when it comes to timing them, fitting them to a receiver, and in some cases, their fragility, finding one may be the only route to legally obtaining a full-auto AR for under $20,000. These prices are escalating in step with everything else full auto as there are only a finite amount in the NFA registry. At one time, a registered Lightning Link could be had for as little as $200. Thirty years later, they're going past $15K. R

We really, really wish 100,000 of these were made and registered before May of '86.

HAPPY FINISH

Cerakote Your Guns — Like Building a Model Airplane, But it Lasts a Lot Longer

By Peter Palma
Photos by Tom Evans

WARNING!
When handling firearms, always observe safety rules and the precautions set forth in the firearm's owner's guide. Be certain that your firearm is unloaded and made safe before proceeding with this DIY.

The firearms industry currently produces an overwhelming amount of products to customize your firearms, and while some companies offer finishes in limited colors, most follow Henry Ford's rule of "You can have any color you want so long as it's black." Historically, there has been no simple and cheap way for small shops or individual shooters (who are usually limited in resources) to apply a finish to their firearms and accessories that rivals the colors and quality available from the manufacturers.

Cerakote is a ceramic-based coating that is available in hundreds of colors, is moderately flexible, offers wear protection and chemical resistance, and has a corrosion resistance similar to those clay pots you see dug up on The History Channel from ancient civilizations. Cerakote can be used to provide a professional coat to a modified part, as a basecoat for a rattle-can camouflage job that can be removed at will, solely as a protector from corrosion, or to simply change the color of your firearm's parts. Check RECOIL Issue 13 for an in-depth comparison on how it stacks up against other finishes. The makers claim that it's easy to apply, but we'll let you decide that after you read this article. That being said, anyone who has the patience and the mechanical ability to assemble a model airplane should be able to apply a professional-level finish of Cerakote.

PREPARATION

Before starting, you'll need some basic materials that are available at your friendly local hardware store and a discount retailer like Harbor Freight. These include masking tape, metal coat hangers, floral wire, acetone, a blasting cabinet, aluminum oxide, an air compressor, paint gun, oven, and of course Cerakote.

If this seems like a lot of equipment, consider it an investment, because once all of your friends find out you're Cerakoting you'll have plenty to do. We also solicited some help from our buddy Tom Evans, who has been Cerakoting for some time and works in a paint yard on an industrial plant. He's required to follow much more rigorous coating application procedures than we show here. He lent us his expertise, as well as his photography skills.

STEP 1: DISASSEMBLY

The first thing you need to do is disassemble your firearm completely, because you'll ultimately be heating it up in an oven. Any polymer parts or springs can be damaged, so they should be removed. You then have to decide how many of the small parts you're going to coat. It can be far more time consuming to prep and coat a ton of small parts, but having corrosion resistance and matching aesthetics may be worth it to the individual — it's your call. Since Cerakote is a coating and adds thickness, it should not be applied to parts with very tight tolerances, such as the inside of AR takedown pins or the slides of match pistols. If it's tight to start with, you should avoid coating it.

STEP 2: DEGREASING

Now that you've selected the parts you wish to coat, it's time to start prepping them. Any lubricant or fouling should be removed with acetone, though brake cleaner is a quick, cheap, and easy substitute. Any remaining lubrication will negatively affect the coating's ability to adhere to your metal part, so be thorough.

STEP 3: BLASTING

Now that we have the parts cleaned, they need to be blasted so that the part has what is called a "blasting profile." A light blasting profile will leave the factory finish mostly intact, so that when the Cerakote wears there will still be some level of protection. However, the deeper the blasting profile, the stronger the bond

between the part and the Cerakote. The amount of blasting needed will depend on the part's function and your personal preference. Two items that aren't strictly necessary, but are *strongly* recommended are an in-line water/oil separator (to reduce moisture going on the parts) and a large air compressor, because the blasting

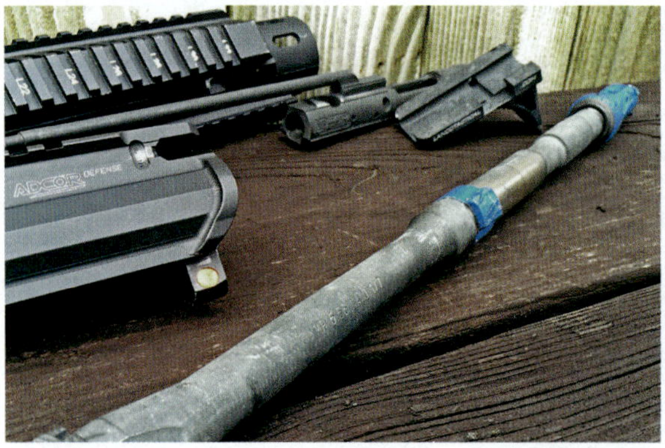

process uses copious amounts of air. For those looking to save some money, the aluminum oxide in your blasting cabinet can be substituted with play sand, but it doesn't last as long and is a bit messier. Always, but always, wear a respirator when blasting, as media can lead to long-term havoc with your lungs.

Steel parts that are not treated with a ferritic nitrocarburizing process (e.g. Melonite or Tenifer) should be taken down to bare metal, and if money is no object, you can send them out to be surface treated so that when the Cerakote wears off the parts will still be protected. This author's favorite nitriding company is H&M Metal Processing, but there are other options, as well. After the blasting process, clean the parts again with acetone or brake cleaner to remove any remaining blasting media, oils, or debris.

STEP 4: HANG YOUR PARTS

Suspend the large parts with coat hangers, or copper wire if you're concerned about damaging the part, so they can later hang in the oven. Set up your hanging wire so the part can be coated completely. Floral wire is ideal, because it can easily wrap around small parts or be creatively bent to form hooks or loops. We suggest attaching multiple small parts to each strand of wire for ease of hanging and painting. Note that you'll be spraying the parts with compressed air while coating, so if you make a wide loop the part will move around during this process. This problem can usually be resolved by tightening the wire so that the part is secure — however this will leave behind a "tan line." With some forethought and attention to detail, this can be minimized by trying to secure parts with two points of contact.

STEP 5: PRE-WARM

Hang your components in the oven you intend to use at 300 degrees. This serves two purposes. First of all, we need to pre-heat the metal parts to burn up and purge any residual oils and to give the coating a warm surface to initially bond to. The second reason is to require you to carefully plan out how you're going to place and remove parts from the oven once they're coated.

STEP 6: COATING

We're now ready to paint. But before we start going full steam ahead, let's go over some simple tricks to make our life easier. We suggest purchasing the Cerakote refill kits and a bunch of disposable plastic 50-mililiter centrifuge tubes for mixing paint and hardener. These are available at low prices online.

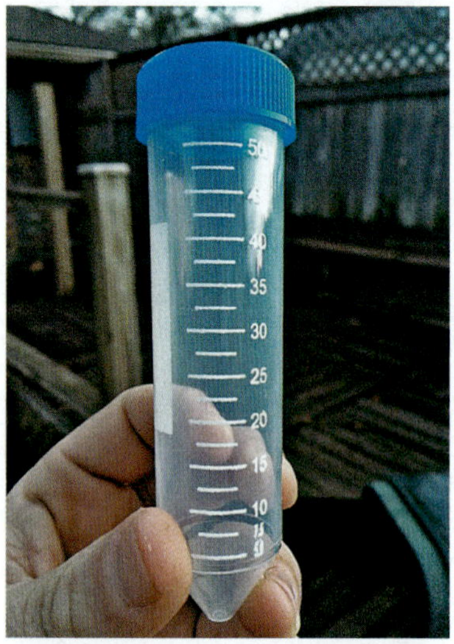

Refer to the manual for the ratio of hardener to achieve your desired finish — e.g. matte, satin, or semi-gloss. Note that the less hardener used, the less resistant to wear the coated part will be. Thoroughly mix the two components. Like a good martini, they should be shaken, not stirred. Set your paint gun and air compressor to the desired settings. Pour your mixed coating into the sprayer, and test the spray pattern. Once the hardener is mixed in, you only have two hours of working time, so make sure you set yourself up for success and prep as much as possible. The limitation will be the size of your oven.

Spray small, hard-to-reach areas of the parts first, and work your way to the easier areas. As you spray the multitudinous crevices, you'll inevitably get some overspray on the easy-to-reach areas. Start with a very thin coat and slowly work your way thicker. You can always make another pass, but removing excess Cerakote is a royal pain in the ass — you'll need to strip the part down and start over. Begin with a light coat and work your way up to a monolithic, even coat. We said that twice for a reason.

It's always a good idea to have some extra parts you'd like to coat on hand at this time. If you mix too much product, you'll end up throwing it out because once mixed it's only usable for a finite amount of time — and this stuff isn't cheap. We usually end up with leftover product because it's sometimes difficult to judge how much coverage a part will need. It varies with size, shape, and intended use. Moreover, you don't want to run out of mixed Cerakote in your paint sprayer, because not only is it a pain to mix more, mid-spray (ask us how we know this), but you may not exactly match the hardener ratio, resulting in an uneven finish.

STEP 7: CURING

It's now time to place the parts in the oven. Refer to the manual for the required temperature, depending on material and desired cure time.

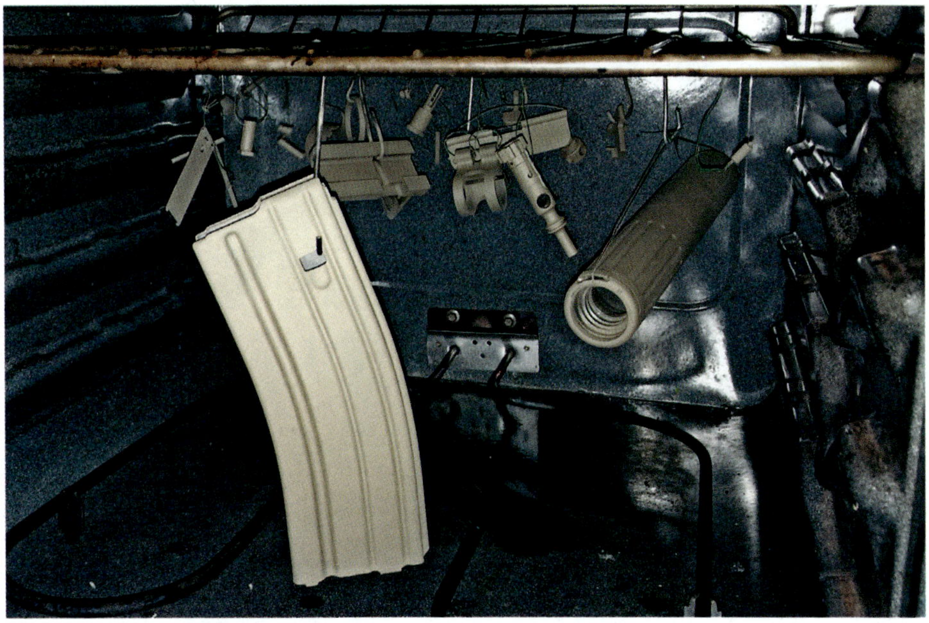

It's safer to have the oven turned off when placing your parts, so the worst thing that can happen is you smudge the finish and turn the air blue. If you're brave, you can pre-heat parts at the same time you're spraying; thus shortening your critical path and making the most efficient use of your two-hour working time. Just remember the inside of your oven and every part of your gun will be 300 degrees, so it'll be like playing a high-stakes, life-size game of Operation.

Allow your parts to cool before handling. You'll only make this mistake once. Hopefully you put the small parts you didn't paint in carefully labeled bags so that you know where all those springs and roll pins go when you reassemble your firearm. Promptly clean out your equipment after painting, so that the Cerakote doesn't start to gum up in your paint sprayer, thus ruining your next Cerakote day.

Now that you know what you're getting into, you can decide if this is something you want to tackle on your weekend or just pay someone to do for you. Don't start out on your favorite and most expensive gun. You may want to start out on some of your older guns with no intrinsic value, or just use some old car parts to refine your skills.

Like anything else you'll improve over time. For example, the parts of our firearm that our pro sprayed looked a lot better than some of the parts we sprayed. It was a bit of a learning curve, but we've definitely tackled some more difficult things — such as putting together a model P-47 Thunderbolt. **R**

GLOCK TRIGGER DISCIPLINE

Oh, The Things You Can Do With A Dremel and A Punch...

By RECOIL Staff

For all the jokes we could make about Gaston Glock's idea of "perfection," the Glock design has withstood the test of time as a truly remarkable achievement. With only 34 component parts, the design is simple, robust, and — importantly — very easy to work on. In this edition, we'll explain the Glock firing cycle and show you how to do a basic trigger job, as well as install an over-travel adjustment.

TOOLS

- 3/32-inch punch
- Rotary tool
- Felt polishing wheel
- Metal polishing compound
- #43 drill bit
- 4-40 tap
- 4-40 1/4-inch set screw
- 0.050 Allen wrench
- Drill press and drill press vise (recommended)

INITIAL PREP

- Remove the magazine from your weapon. Clear the weapon and ensure it is unloaded. Visually and physically inspect the chamber to ensure it is empty.
- Remove all ammunition from your work area.
- Field strip your Glock, removing the slide from the frame. Remove the recoil spring assembly and drop the barrel out of the slide. Place those items aside.

We have not illustrated these steps, because if you are not already familiar with them, you probably shouldn't be attempting the following DIY!

WARNING!

When handling firearms, always observe safety rules and the precautions set forth in the firearm's owner's guide. Be certain that your firearm is unloaded and made safe before proceeding with this DIY.

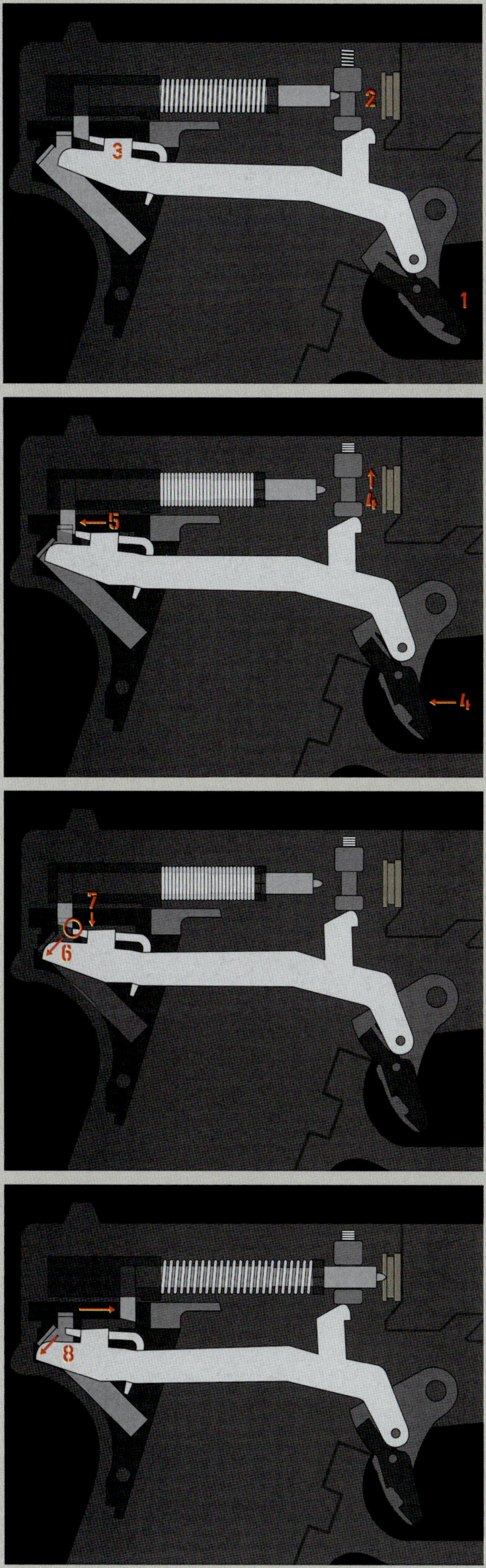

THE GLOCK FIRING CYCLE

1. The trigger safety protrudes from the trigger shoe, preventing the trigger from moving rearward until the trigger safety is depressed.

2. The firing pin safety blocks the firing pin from moving forward until the safety has been pushed upward.

3. The rear portion of the trigger bar (commonly referred to as the "cruciform" because of its shape) sits on top of the trigger housing. The part of the trigger bar that angles upward butts against the lug on the firing pin. This blocks the firing pin from moving forward, thus serving the function of a drop safety.

4. As the trigger is pressed to the rear, the trigger safety moves out of the way and frees the trigger to move. The protrusion on top of the trigger bar pushes the firing pin safety upward, clearing the way for the firing pin to move forward later in the process.

5. The rear of the trigger bar also retracts the firing pin, under spring tension.

6. The rounded tail of the trigger bar slides against the connector as it approaches its rearmost travel, causing it to displace downward.

7. The side tabs on the cruciform sliding on top of the trigger housing prevent the trigger bar from moving downward until this point.

8. The rear of the cruciform then acts as a sear as it moves downward and releases the firing pin to travel forward — through the firing pin safety and striking the primer.

In particular, take note of where components are interacting with each other. Glock parts are made of stamped steel, so they are hardly finely crafted and fitted works of art. If you have a number of rounds on your stock Glock, inspecting these areas provides the most tangible education on what you need to do to improve your trigger pull — you will see where parts are wearing on each other.

FRAME DISASSEMBLY

1. With a $\frac{3}{32}$-inch punch, push out the locking block pin.

2. Push the trigger pin out from left to right (this is really the only pin where it matters what direction you push it, but it's a good idea to just get in the habit of removing all of the pins from left to right and reinstalling them from right to left). It will help if you jiggle the slide stop lever while pressing on the pin. The slide stop lever will now be freed, so remove it and place it aside.

3. Remove the locking block. You may need to pry it out of the frame.

4. Push the plastic trigger housing pin out of the frame.

1

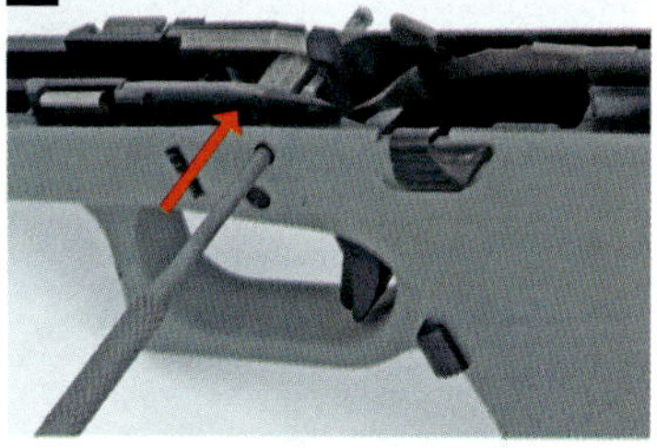

2

3

4

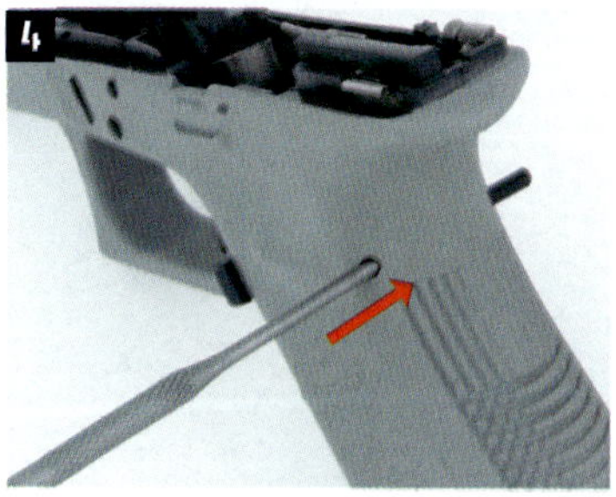

5. Remove the trigger group by prying it up and pulling the trigger housing up out of the frame (you can use your punch against the frame and pry up on the ejector).

6. Remove the trigger bar from the trigger housing by pulling it forward (against spring tension) and twisting the cruciform out of the housing.

7. Carefully examine the orientation of the trigger spring (the hooked ends will form an "S" shape) so that you can reassemble it in the same manner.

8. Unhook the trigger spring from the trigger bar and then the trigger housing.

9. Remove the connector by using your punch to push it out via the hole on the other side of the trigger housing. Be careful not to bend the connector.

Reassembly is the opposite of the steps above, except that the trigger pin should be the last one that you install. You will once again need to jiggle the slide stop as you push the trigger pin back in place. Remember to go from right to left.

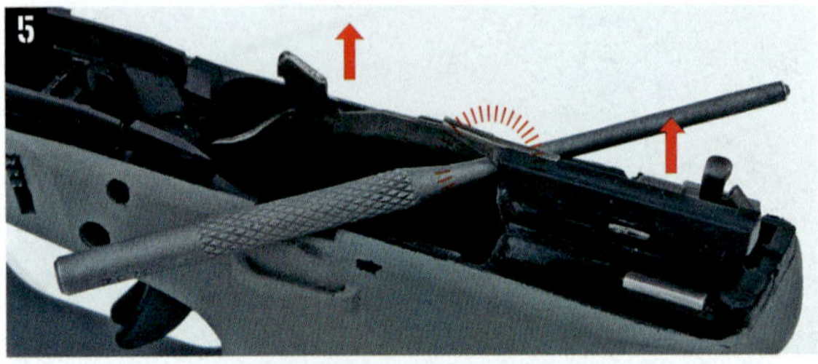

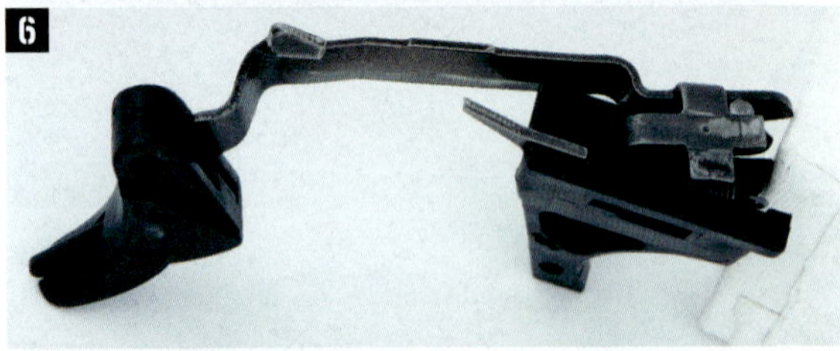

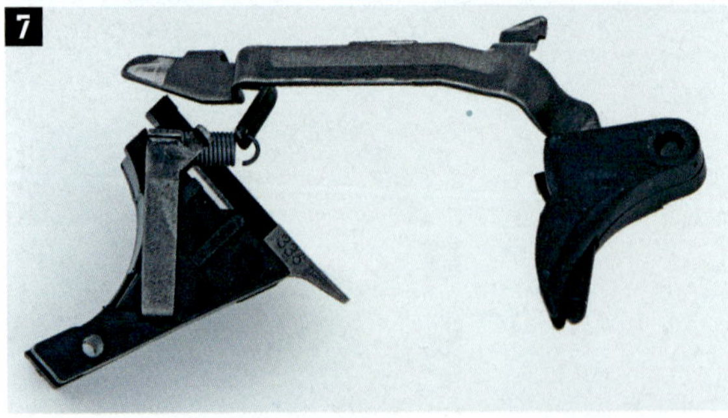

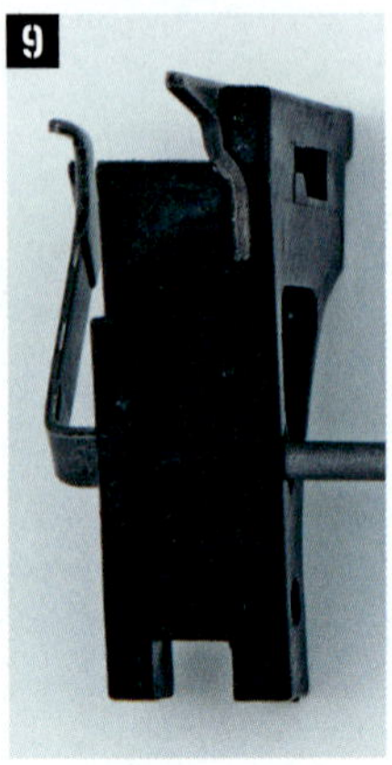

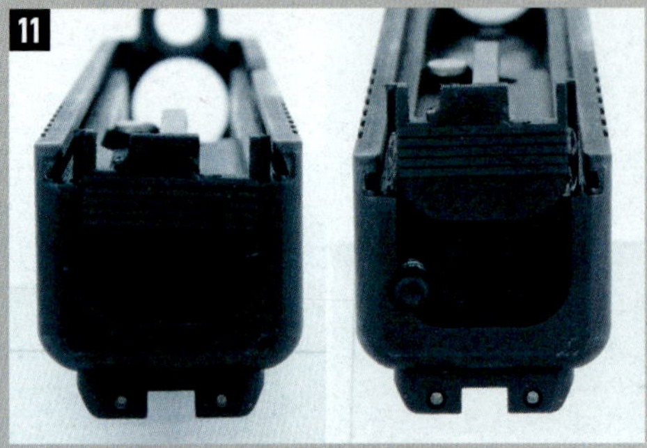

SLIDE DISASSEMBLY

10. Press down and hold the plastic spacer sleeve with your punch.

11. Then push the slide cover plate down while continuing to hold the spacer sleeve down. Be careful as the extractor depressor plunger assembly and firing pin assemblies are under spring tension below the cover plate. So as you slide the cover plate down, use your thumb to capture them.

12. Remove the firing pin assembly.

13. Remove the extractor depressor plunger assembly.

14. Depress the firing pin safety and remove the extractor. Once the extractor has been removed, the firing pin safety and spring will fall out of the slide (though they might need a little help).

15. Stand up the empty slide and insert the firing pin assembly backward into its channel. This makes it easier to disassemble it.

16. Compress the firing pin spring and remove the spring cups (two halves). Be careful, it's easy to slip up and send parts shooting around the room.

17. The spring and spacer sleeve slide right off of the firing pin.

Reverse these steps to reassemble your slide; there's nothing tricky here.

TRIGGER JOB

The parts that we will be working on are the trigger bar, connector, firing pin safety and firing pin. Carefully examine the wear that already exists on these parts, and you'll see where they are contacting each other. On the trigger bar, note the tab for the firing pin safety, the cruciform, the sear surface and the tail which rides on the connector. In turn, you'll also see the contact points on the connector. On the firing pin safety pictured here, you can actually see where the tab on the trigger bar has traversed back and forth hundreds of times across the firing pin safety, leaving linear marks on the surface. The lug on the firing pin similarly has marks from sliding back and forth.

Tab for the firing pin safety

The cruciform and tail

Connector

Firing pin safety

Firing pin lug

We used our trusty Dremel fitted with a felt polishing wheel and Flitz polishing compound. On the trigger bar, polish its tail (on the outside that rides against the frame, as well as the inside and curved surface that contact the connector), the cruciform, and the protrusion for the firing pin safety. Polish the connector where it contacts the trigger bar and the firing pin safety itself. Also polish the lug on the firing pin. Once done, be sure to thoroughly clean the parts to remove the polishing compound.

Advanced users: If you are more adventurous, there's more that you can do. Before you polish the parts, note the contact areas and how the stamped steel parts often do not have smooth and straight surfaces. There are typically lumpy or rough spots, such as on the tail of the trigger bar. Pay particular attention to the existing wear, and carefully stone those areas. We urge extreme caution around the sear surface and tip of the firing pin lug. We advise against removing material from these spots — you do not want to break through the surface hardening, as bad things can happen. The good news is that these parts are rather cheap to replace if you butcher them...

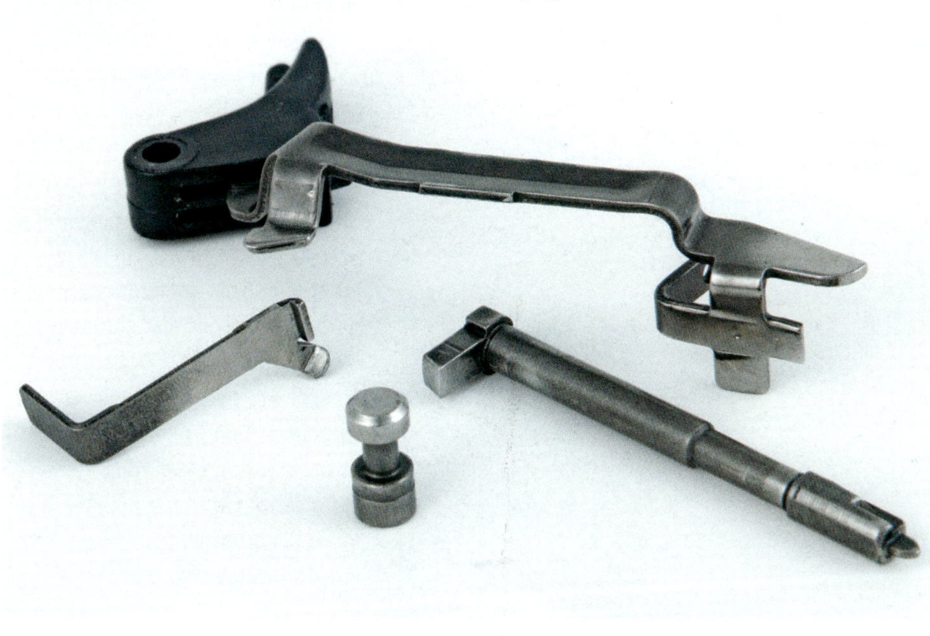

OVER-TRAVEL ADJUSTMENT

Examine the side tabs of the cruciform, and you'll note that they slide forward and back in the trigger group housing (in part to provide a drop safety). We can take advantage of this to add an over-travel adjustment, by adding a set screw to the trigger housing.

A drill press is recommended to properly align your work. In the spirit of good old fashioned Dremel engineering, we used a rotary tool drill press workstation and drill press vise. Very carefully align your drill and the workpiece in the vise. Chuck up a #43 drill bit and bore a hole on the ejector side of the housing, from the rear of the trigger group to the channel where the side tab rides. Then tap the hole with a 4/40 tap (tap handle omitted in photo for clarity).

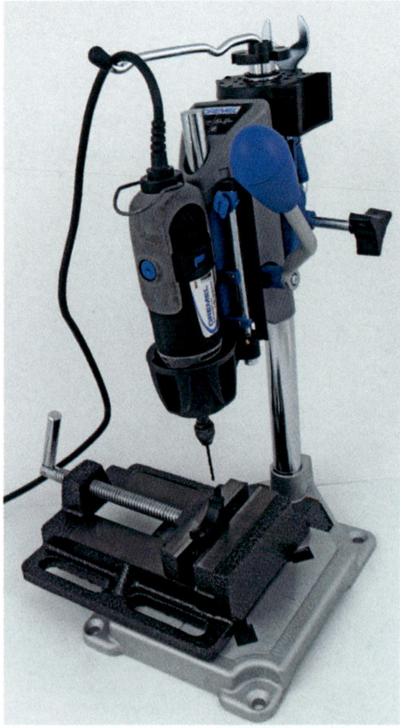

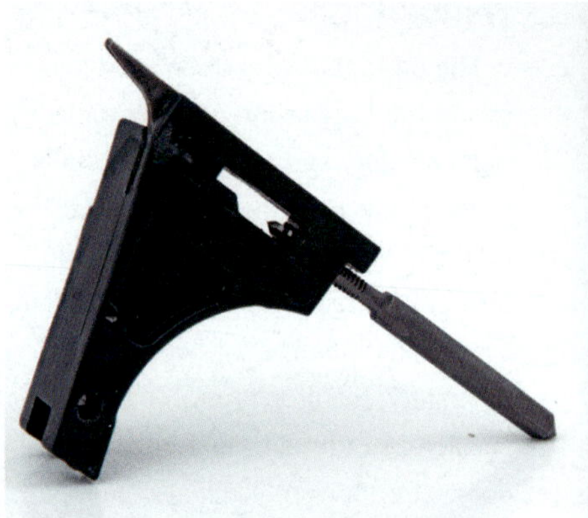

Install a 4-40 ¼-inch set screw with a 0.050 Allen wrench. Now comes the trial-and-error part, where you continually tweak the set screw, reassemble, and disassemble the gun to find a setting that you like. Don't go overboard; getting too aggressive with the over-travel set screw may interfere with proper function.

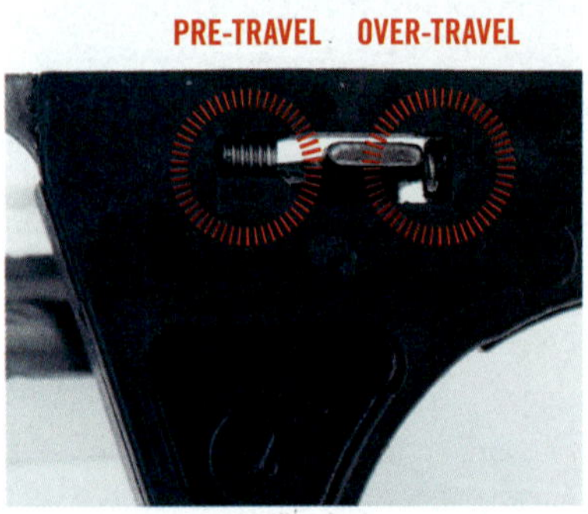

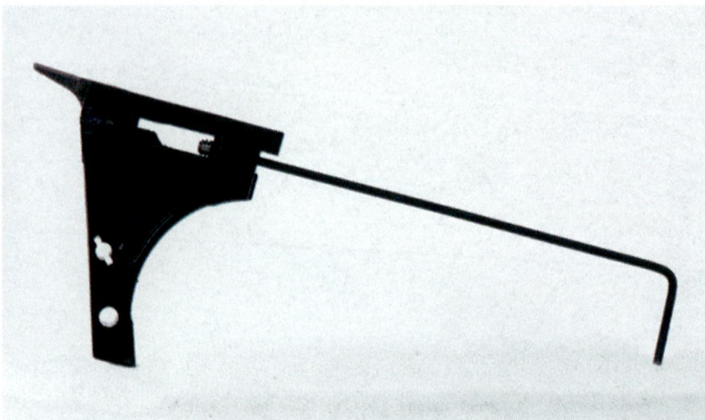

Advanced users: You can similarly add a set screw (2/56) on the front side of the trigger housing (as pictured to the left) to reduce pre-travel. However, proceed with caution, as you can easily disable the trigger safety and cause other problems if you go too far.

Congratulations! You've done your own trigger job with just a little bit of elbow grease and very little expense. Now that your trigger is much smoother, you can experiment with different connectors and springs to find a pull weight that suits your preference. Before the work, our test gun measured in with a trigger pull of 4.7 pounds and, given its round count, had a reasonable trigger with a bit of roughness. Afterward, it was 4.4 pounds, but more importantly, it's smooth with reduced over-travel and better trigger reset. Not too shabby indeed. **R**

DIY GUN PAINTING

Rattle-Can Your Way to the Perfect Grunge Camo Job

By RECOIL Staff

Take a look at your AR. Now look at the world outside. Who knows, maybe you live in factory that produces black, stick-like objects, but we're guessing there probably isn't anything out there that matches your rifle's shape or color. If you want your gat to blend in, you could send it to one of the companies listed in our buyer's guide (located elsewhere in this issue), or you could become the next Banksy of the gun world and pick up a rattle can. Who knows, *The New York Times* might write a fawning review of your efforts. But probably not.

We used Brownells Aluma-Hyde II for this job, but you can choose to go up or down the spectrum of price and durability. If cost were no object, we'd send the gun off to Andy Munson at Guncoat Northwest and have him work his Duracoat or Cerakote artistry. At the other end of the scale, Krylon makes a range of flat camo paints that will run you around $5 a can, but be aware that these aren't particularly resistant to common gun-cleaning chemicals.

WARNING!
When handling firearms, always observe safety rules and the precautions set forth in the firearm's owner's guide. Be certain that your firearm is unloaded and made safe before proceeding with this DIY.

STENCIL MATERIALS

To create camo patterns, you could randomly spray paint blotches on the gun's surface. Although this would fulfill the requirement of breaking up its outline, the result would not win you any gold star stickers or admiring glances from your shooting buddies. If practicality is your only metric (and oftentimes it's certainly ours), then skip this step and go right on to degreasing.

The number of potential stencil materials is only limited by your imagination, but here are three that we've found work well.

- **Netting:** Laundry bags, packaging materials, sports netting, and even expanded metal can all be used to create a snakeskin effect when they're employed as stencils. For use on a rifle, try and find material with a 1/16-inch-or-thicker web.

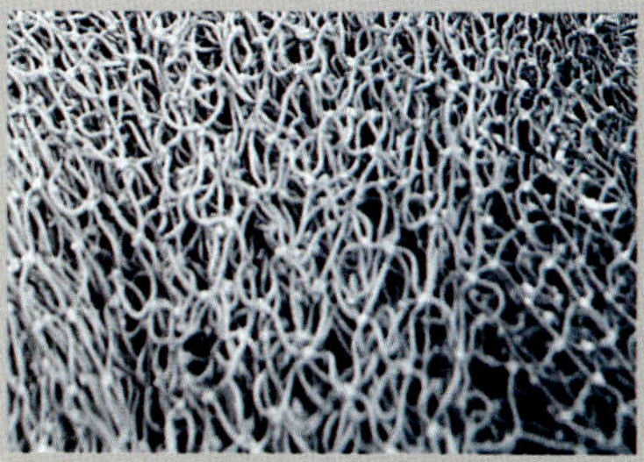

- **Foliage:** Leaves and grass, when laid on the surface of your firearm, make for an effective and natural-looking camo pattern.

- **Fabric:** Take an old T-shirt and cut it up into large rags. Using a pair of scissors, cut out holes or slits and lay the rag over the firearm. Then spray through the holes.

STEP 1

Assemble your supplies. No matter how much you plead with it, paint has a finite drying time, so you'll want to have everything in place before pressing that little white button on the top of the can. Have all your gear readily at hand, including whatever stencils or masking materials you've chosen.

STEP 2

After following normal safety procedures, thoroughly degrease your firearm. Because while napalm might stick to kids, paint definitely won't adhere to CLP. You can use denatured alcohol for this, but most of the time we reach for a can of brake cleaner as it's both cheap and effective.

STEP 3

Mask off any area you don't want painted using, appropriately enough, painter's tape. We covered the optic lenses, brightness knob and trigger, as well as the rubber butt-pad on our VItor stock. As paint doesn't work well on rubber components, we removed the Ergo grip — you could go to the trouble of stripping your gun down further, but most of the time this won't be necessary.

STEP 4

Apply your basecoat. Usually, this is the predominant color you'd encounter in your area. As we shoot a lot in the baked and sun-blasted Southwest, we opted for a light sandy base. To make it simple to achieve complete coverage, insert a length of wire through the muzzle device or rail and hang it up. Rotate it as you lay down the first coat, being careful not to create any runs. If you do, simply refer to them as "viscosity indicators" and you'll feel a lot better.

STEP 5

Apply your stencils (if used) and shoot the second and third colors. Once your basecoat has dried to the touch, grab whatever stencil materials you've chosen and spray your next color. If you're using netting, it helps to pull it tight against your firearm to avoid overspray getting underneath the netting. Remember, you can always add paint to areas that look a little light, so be sparing at first.

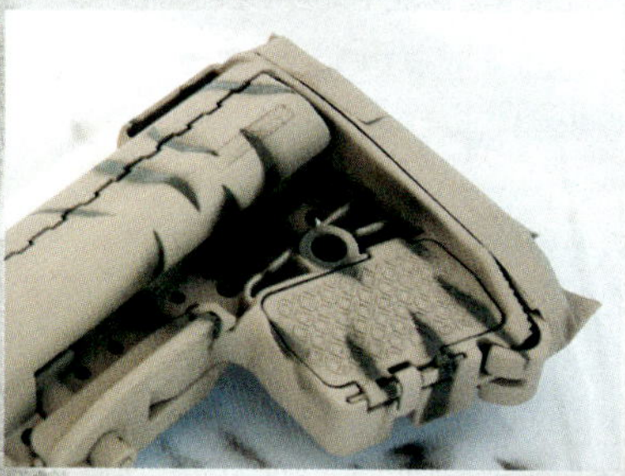

STEP 6

Blend it. If your stencils have left sharp edges to the pattern, you may want to soften them — we skipped this step as our pattern's edges were pretty soft, though if you want to create a more subtle outline, be our guest. Using the same color, hold your can about 12 to 16 inches way from the gun and spray a very light coat to break up their outline. If using Krylon, hit the entire gun with a flat clearcoat to give additional durability. Once you're happy with the result, allow the paint to dry according to the instructions (if impatient, you can speed up the process with a hair dryer). Then remove all masking tape and reinstall any parts you removed. Now go and enjoy your work.

RIFLE BUILD

The AR we chose as our test mule for this project was a work gun — a quality build, but fairly basic nonetheless, with everything you need in a practical carbine and zero fat. It comprised the following:

- **Receivers:** Spikes Tactical
- **Barrel:** 14.5-inch Daniel Defense CHF lightweight with permanently attached Troy Claymore brake
- **Bolt Carrier Group:** Rubber City Armory
- **Trigger:** Geissele Super Dynamic Combat
- **Stock:** Vltor EMOD
- **Pistol Grip:** Ergo
- **Rail System:** 15-inch Troy Alpha
- **Optic:** Aimpoint Micro
- **Mount:** GDI Precision
- **Charging Handle:** Rainier Raptor
- **Lower Parts Kit:** Strike Industries
- **Weapon Light:** Rosch SL1 with integral front sight
- **Rear Sight:** Magpul MBUS

FINISH STRONG

A Guide to Coatings, Finishes, and Treatments

By Steven Kuo | Photography by RECOIL Staff

For all the advances in polymers and Hollywood nonsense about plastic guns, metal remains the fundamental building block of firearms. And where there is metal, there are many options for how to treat, finish, or coat it. Long past the days of just choosing between bluing or raw stainless steel, we now have a veritable bounty of options. Each has a certain set of characteristics that may make one or another more suitable for your particular needs. Read on to learn more.

Key characteristics to keep in mind include the following:

› **Hardness, durability, and wear resistance:** For hard-use weapons, whether for duty or competition, this is often a key consideration. Thousands of presentations from a Kydex holster, thousands of rounds through a barrel — this will take a toll on metal. For our purposes, standardized hardness testing is measured on the Rockwell C scale (HRC) and Vickers (HV) scale, by putting an indenter under load against the surface.

› **Corrosion resistance:** Carry guns pressed up against your sweaty body or used in maritime or humid environments are often a recipe for rust and corrosion. It's not a license to abuse your gear, but finishes intended for this purpose can help a lot — of course, remember that nothing is truly corrosion proof. While it cannot exactly represent real-world performance, a standardized salt spray test (ASTM B-117) is typically used to assess the corrosion resistance of materials and finishes. In such a test, the parts are placed in a testing chamber and exposed to a dense fog of 5-percent salt solution until they begin to show the effects of corrosion.

› **Lubricity:** Many finishes have very low coefficients of friction — a measure of how easily one surface slides against another. For example, the coefficient of friction for steel on steel is over 0.7 dry and around 0.2 with lubrication, which is more than halved by many of the coatings. Again, we always advocate properly lubricating your weapons, but these finishes enhance performance and provide you with more leeway in your maintenance schedule.

These bolt carriers highlight various different finishes and treatments (left to right): Parkerizing – BCM Mil-spec; NP3 – Robar Mil-spec; BlackNitride – Rubber City Armory's Mil-spec and lightened variants; Titanium Nitride – Salient Arms International; Hard chrome – Young Manufacturing's super lightweight carrier. And a cool stapler in Duracoat from GCNW.

› **Type of material/substrate:** Some finishes or treatments can only be applied to steel, not to aluminum or polymers. Note also that many companies and end users mix and match finishes on different components.

› **Method of application:** Some require high heat, electrical processes, or specific characteristics, which may completely preclude use on, or else not be desirable for certain components (e.g. an optic, non-ferrous metals, non-conductive materials, etc).

› **Tolerances:** All except one of these processes add material to the substrate, so this must be considered for specific applications (e.g. slide rails, sears, etc). In some cases, high-tolerance areas might be masked off, require some fitting afterwards, or even be considered in advance during gunsmithing or manufacturing, given an intended coating (e.g. with custom 1911 builds or chrome-lined barrels).

› **Aesthetics:** Desired color, surface treatment, and ability to mask flaws in the base metal all come into play here.

› **Cost:** Specialized equipment and solutions, as well as how time and labor intensive the process is, factor into the cost. Note also that many processes benefit from economies of scale, so one-offs can be more expensive or your job might need to be batched with others to bring down the cost, potentially increasing wait times.

It may go without saying, but a reliable, experienced shop with talented craftsmen is important to achieving satisfactory results. As with so many things, prep work is absolutely critical, equipment can set limitations on what a shop is capable of, and attention to detail (whether in properly maintaining the chemistry of solutions or achieving proper tolerances) will ensure a good end product. Do your research and set realistic expectations.

Be honest with yourself about what you want from a finish — aesthetically and functionally — and how picky and anal retentive you will be. Bob Cogan of APW Cogan has been plating, coating, and 'smithing guns for nearly 40 years. He advises that every process has its strengths and the key challenge is getting a blend of desired qualities at an affordable price. So set your budget and choose accordingly — it may even make sense for you to pick and choose certain key components to be treated. For instance, for an AR-15 build you can literally piece together a build with your desired finishes utilizing only off-the-shelf components: e.g. hard anodized receivers and handguard in the color of your choice, hard chromed bolt carrier group, nitride-treated barrel, and so forth. The possibilities are endless. We are truly fortunate to have so many fantastic options these days.

BASIC FINISHES

Let's start with an understanding of the baseline finishes that you'll typically come across.

› **Bluing (black oxide)**: Bluing is a time-worn chemical process that forms black oxide on the surface of steel, typically by dipping the part into various tanks. Essentially controlled rust, the result is aesthetically quite pleasing and more corrosion resistant than bare metal — but that doesn't always get the job done. If you're not careful, blued components can still rust pretty easily depending on the conditions.

Top: Juggernaut Tactical lowers racked up and ready for anodizing at Dunham Metal Processing.

Above: Anodized lowers coming out of the electrolytic solution.

Photos courtesy of Juggernaut Tactical and Dunham Metal Processing.

› **Parkerizing (phosphating):** Widely used on military issue weapons, this process transforms the surface of the steel to a more corrosion-resistant manganese or zinc phosphate. The coating also traps oils and lubricants.

› **Anodizing:** Aluminum in the raw is relatively soft when compared to steel. Hailing from the 1920s, anodizing is a process that converts its surface to aluminum oxide, essentially resulting in a ceramic coating — in the case of Type III anodizing (the hardest and typically found on firearms and their accessories), roughly 0.002 inches, with half penetrated into the material and half built up above the surface. An electric current is passed through an electrolytic solution, in which the work pieces are placed. Aluminum oxide then forms on the surface. The result is quite hard (approximately 60-70 RC) and corrosion resistant (1,000+ hours in the salt spray test).

Above: Richard Utas of 2A Concepts preps and blasts a Glock slide prior to Cerakoting it.

CERAKOTE

Cerakote is a very popular ceramic-based finish that can be applied to metals, polymers, and wood. It is available in over 100 different colors and provides exceptional corrosion and chemical resistance, as well as improved scratch and wear resistance. It is widely utilized by firearms and accessories manufacturers as well as end users.

Parts are prepped by cleaning and blasting them. Then they are racked, pre-heated, and sprayed with the coating solution. Cerakote's C-series coating can then be air-cured, while its much stronger H-series coating is then heat-cured. Cerakote H salt spray test results are stratospheric.

Our sample rifle was Cerakoted by 2A Concepts in OD green, coating the Spike's Tactical upper and lower receivers and Geissele Super Modular Rail MK I handguard. The smooth and hardy finish provided the owner with his exact desired color. The AR has a 14.5-inch Daniel Defense barrel with a BCMGUNFIGHTER Compensator Mod 1 permanently attached to keep it legal. An Aimpoint PRO and Spike's Tactical micro folding backup iron sights keep it on target, and the build is finished off with a B5 Systems buttstock, BCM GunFighter Grip Mod 1, Geissele DMR trigger, and Inforce WML light. A no-nonsense working gun, ready to rock.

DURACOAT

Duracoat is a two-part chemical epoxy coating that is sprayed on like paint and then air cured. No heat is required, so it can applied to anything, including optics or other sensitive gear. Besides this, Duracoat's greatest advantage is aesthetics, as it is available in a large variety of colors that can be mixed to generate infinite variations and applied precisely with an airbrush. With proper prep and application, Duracoat is quite durable, though not as much as the other processes in this guide aside from hydrographics. It also provides strong corrosion resistance. Duracoat withstands temperatures up to 500 to 600 degrees F; for applications where you might exceed this, an oven-cured variant called DuraBake is available in a more limited selection of colors.

A turnkey finishing kit with everything you need to coat a few firearms

Photos courtesy of Gun Coat Northwest

costs just $60, for those willing to roll up their sleeves for some DIY fun. Alternately, Duracoat now also comes in an aerosol can that automatically mixes in the hardener with the press of a button. No sprayer required, and no measuring and mixing, for $35 per can (with a selection of a couple hundred colors).

This proud Akdal semi-auto shotgun belongs to our very own Iain Harrison. It is truly a work of art, having been the recipient of carefully crafted stencils and strokes of the airbrush, courtesy of Andy Munson with Gun Coat Northwest. The shotgun is an Akdal MKA 1919 that was thoroughly rebuilt by Firebird Precision, modified with a side-charging railed handguard, buttstock assembly, compensator, new trigger group, 10-round magazine, and other internal refinements. The colorful design is a nod to Iain's homeland, with album artwork borrowed from the Sex Pistols' *God Save the Queen* single, which was banned by the BBC in 1977 for being "anti-establishment." This 12-gauge smoothbore brings the house down in every possible way.

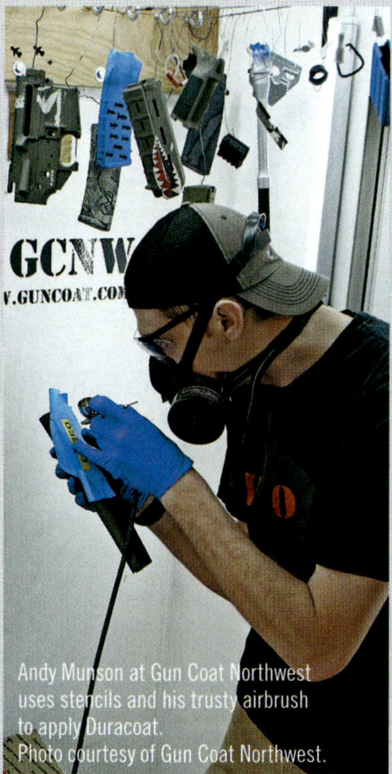

Andy Munson at Gun Coat Northwest uses stencils and his trusty airbrush to apply Duracoat.
Photo courtesy of Gun Coat Northwest.

ELECTROLESS NICKEL — NP3

Electroless nickel is a process that can actually be done in your own home, utilizing an auto-catalytic reaction to deposit a nickel phosphorus coating onto a substrate without the need for electric current. Put on your Dr. Frankenstein lab coat, prep your parts, mix the solution, heat it to the specified temperature and submerge the parts as long as necessary to achieve the desired plating thickness. The key characteristics of this process are strong corrosion resistance and consistent coating thickness, regardless of the shape of the workpiece. Wear resistance is enhanced, but the hardness rating is not as high as other alternatives. A home kit can be had for $150 and up.

The Robar Companies took this process to the next level by adapting processes from the oil and gas industry, introducing PTFE (polytetrafluoroethylene) particles, more commonly known as Teflon, into the solution. This results in a coating that is even more corrosion resistant, self-lubricating, and wear resistant. With over 30 years under its belt, Robar is well-known for its NP3 coating (not to mention their custom gunsmithing). Notably, NP3 has an extremely low coefficient of friction, whether wet or dry. In fact, Robar's Freddie Blish explains that with Teflon embedded in the coating, NP3 actually gets smoother and slicker with use. Their latest formulation, NP3 Plus, boasts even more outstanding corrosion resistance, with a rating of over 1,000 hours in salt spray testing, matching the polymer coatings.

The sample gun shown here is a Kimber Classic 1911 with frame work by Chuck Rogers of Rogers Precision. Rogers machined the front strap and mainspring housing with his scallop design and high cut the trigger guard; the rest of the gun is stock. Robar coated the gun in NP3, resulting in a beautiful, rich silver tone that is literally slicker than snot. We even sneezed on the gun to scientifically test this assertion. OK, not really, but trust us, NP3 is really slick — you could scramble eggs on this thing. The Kimber is topped off with Crimson Trace LG-401 laser grips — a great choice on the 1911 platform, with a conveniently located pressure switch and the electronics and batteries totally contained within the grip panels, thus avoiding any tumor-like protuberances on control surfaces as on some other applications. For disciples of the almighty .45 ACP, this is a perfect carry gun.

Meticulously maintained tanks at Robar; the chemistry of the solutions is critical.
Photos courtesy of The Robar Companies.

FERRITIC NITROCARBURIZING

H&M Metal Processing does a high volume of work for firearms manufacturers such as Smith & Wesson, Beretta, Adams Arms and LWRC.

Known by many names, such as Tenifer, Melonite, and quench-polish-quench (QPQ), ferritic nitrocarburizing differs from other processes listed here since it is a thermochemical treatment that affects the surface of the metal, rather than a coating of additional material (however thin) layered on top of the substrate. The part is immersed in a molten salt bath at around 1,000 degrees F, diffusing nitrogen and carbon into the surface and forming an outer iron nitride layer with a nitrogen diffusion layer below it.

It results in increased wear resistance, corrosion resistance, and lubricity — with no dimensional change. Since every other process adds some thickness, this is nitriding's trump card, making it worth considering for any applications where critical tolerances are an issue. Hardness is roughly on par with hard chrome, but salt spray test results are much higher. And the coefficient of friction while lubricated is very low. The tradeoffs of this process are that it only works on ferrous metals like steel, and you better love black — because like the Ford Model T, that's the only color you'll get.

The nitride treatment is gaining popularity, on barrels in particular — chrome-plated bores are widely employed in hard-use weapons for their many benefits in longevity and ease of maintenance, but it has always been a tricky procedure to get just right given the potential impact on accuracy. The properties of nitriding combined with the lack of dimensional change make it a great choice for this application.

Les Baer is known for building extremely tight and accurate 1911s with precise tolerances. The Thunder Ranch special here is no exception; it came out of the box with such a tight lockup that simply racking the slide was a challenge. It loosened up a bit with use, but was still extremely tight, presenting a great example of the benefits of a process that maintains all dimensional properties of the parts. H&M Metal Processing expertly finished the weapon with their BlackNitride treatment, which came out scoring very high on the badass scale with a deep black color. In business since the 1940s, H&M works with many manufacturers and has even treated parts that are out of this world, literally — BlackNitrided gears are on the NASA Mars Rover. The increased lubricity of the treated parts on our Les Baer was immediately felt when working the slide; the gun is still just as tight, but is smoother than before. We also fitted new FRAG G10 stocks from VZ Grips with the Super Scoop cut out to make the mag release more accessible. The end result is an intimidating 1911 that Darth Vader himself would be proud of.

HARD CHROME

Electroplating dates back to the mid-18th century, and commercial chrome plating came into its own in the US starting in the 1920s. Hard chrome plating has long been a staple in the firearms industry, and you will find it in any issue M4, lining the barrel and bolt carrier group. Hard chroming is an electrolytic process whereby the work piece is submerged in a chromic acid-based electrolyte solution, electric current is introduced, and chromium metal builds on the surface.

At APW Cogan, parts are prepped and wired. Beneath the foam, in the murky burgundy solution, is where the plating magic is happening. Photos courtesy of APW Cogan.

The resulting coating is very hard, corrosion resistant, and features a low coefficient of friction. Hard chrome is available in various different colors, including silver, silver, and silver — and, for all its other advantages, this has historically been its Achilles heel over the years, as end users do not always want their guns to be silver. Bob Cogan advises that light-colored plating such as hard chrome make any flaws more apparent in the final finish, so careful prep work is critical. Furthermore, Cogan emphasizes that plating can be as much art as science — different types of parts may be plated with different current levels or lengths of time to achieve optimal results.

Featured here is a race gun built for USPSA Open division by Jojo Vidanes of JV Industries. Built on an STI frame, it features a Caspian slide, T2 Trubor barrel and compensator, C-More optic, Cheely Custom Guns scope mount, and various components from Aftec, Matt McLearn, Ed Brown, EGW and JV Industries. Competition guns are run hard with extremely high round counts, week after week. So durable hard chrome finishes like the one found on this gun are very common on the competition circuit. Open guns don't get much better than this one.

HYDROGRAPHICS

Also known as hydrodipping, this is a process to apply a film with printed designs onto a three-dimensional object. This can be done on metal, plastic or wood (among other things). The part is prepped and basecoated, then the film is suspended in water and sprayed with chemicals to dissolve it, leaving the ink behind. When the part is lowered into the water, the ink wraps around and adheres to the part. It's then sealed with a top coat. Hydrographics provides beautiful results with designs only limited by your imagination, but they are not as durable with hard use as other finishes.

Our example is a KAC SR-15 hydrodipped in Crye Precision's MultiCam pattern by WW Guns. It is outfitted with an Aimpoint Micro T-1 on a KAC mount, Surefire M620V Scout Light, and a Laser Devices DBAL-A2 (AN/PEQ-15A). Complete and total badassery.

POLYMER-BASED — BLACK T, ROGUARD, POLY-T2

There are several polymer-based coatings that can be applied to steel, aluminum and polymer (so long as it can handle the heat cure requirements). Some well-known options are Black-T from W.E. Birdsong and Roguard and Poly-T2 from Robar. Generally speaking, parts are cleaned, degreased, and sandblasted, then pre-heated. They may also be other surface preparation before the coating is sprayed by hand, then heat cured.

Black-T was developed in the early 1980s by Walter E. Birdsong Sr., and has been used extensively over the years by custom gunsmiths as well as manufacturers such as Springfield Armory. The polymer formulation contains Teflon, resulting in a very slick coating that boasts high corrosion resistance and low friction. Notably, Black-T is also flexible and can even be applied on springs, unlike other finishes. Birdsong offers several base colors, but is capable of creating any color. Similarly, Robar's Poly-T2 is available in several colors and is composed of a Teflon based epoxy. Their Roguard coating is a little harder and only available in black. Note that Robar only offers these coatings on steel and aluminum.

Rena at W.E. Birdsong applies the Black-T Teflon polymer coating. Photo courtesy of W.E. Birdsong.

The Browning Hi-Power (circa 1970s) shown here was customized by Don Williams of The Action Works. He dehorned the pistol, stippled the front and rear strap, and beveled the magwell. He fitted a Bar-Sto match barrel, tightened and lapped in the frame to slide fit, did a trigger job, and worked it over for reliability. Then he had it finished in Black-T. The work was done almost 10 years ago — the finish is as slick as ever and looks nearly brand new except for wear on the barrel hood.

PHYSICAL VAPOR DEPOSITION (PVD) — TITANIUM NITRIDE AND VARIANTS

Physical Vapor Deposition (PVD) processes utilize a vacuum chamber to deposit thin films of various substances onto the surface of a work piece. Specific methods and equipment for doing so can vary. Coating Services Group, for instance, uses cathodic arc deposition, which was developed around 1960 to 1970. Jeff Grant, its president, describes the process: "In the middle of the vacuum chamber is a rotating planetary geardrive carousel on which the parts are wired and racked. There is an electrical bias on the parts so that the coating will be attracted to them. In the corners of the chamber are four large titanium targets that are hit with an arc, vaporizing the titanium and creating plasma. Reactive nitrogen gas is introduced and a titanium nitride (TiN) coating builds up on the parts. Our chamber and carousel design ensures the coating is applied evenly to all surfaces. But there is a limit to it, so we wouldn't coat the bore of a rifle barrel, for instance."

Above: Desert Eagle frames, fresh out of the vacuum chamber. Photo courtesy of Coating Services Group.

The result is a dense, extremely thin coating which is extraordinarily hard and wear resistant. TiN has a unique gold color that is familiar to machinists, since it is widely used on cutting tools. That should give some indication of how hard it is — in fact, it is so hard that it exceeds the Rockwell hardness scale typically cited for metals (and is thus measured using the Vickers scale). Slightly different properties and colors can be achieved by adjusting the reactive gas and target elements, including coatings such as TiAlN (dark gray) and TiCN (gray to purple). Exceptional wear resistance, hardness, and tight tolerances are the key advantages of this process; the coefficient of friction is decreased (though not as much as some other finishes), but corrosion resistance is not significantly enhanced, per Grant.

If you thought chrome was eye-catching, the gold TiN color is downright flashy. This example from the GlockStore has a Lone Wolf slide as well as other components coated with TiN. This look might not be for everyone, so remember other colors are also available.

PHYSICAL VAPOR DEPOSITION (PVD) — DLC

Another coating that utilizes PVD is known as DLC, or diamond-like carbon. A similar process as described before deposits layers of amorphous carbon, providing lower coefficients of friction, tight tolerances, and off-the-scale hardness and wear resistance — as the diamond reference implies, this is DLC's key characteristic. Corrosion resistance is somewhat improved, but again does not approach that of other finishes.

The Glock featured here was built by ZEV Technologies and is a good example of incorporating several different types of finishes that ZEV offers for their builds and components. Their Dragonfly stainless steel longslide features a black DLC coating (and a slide racker). The fluted barrel has an iridescent — ZEV calls it "chameleon" — TiN finish. And the Trijicon RMR optic is hydrodipped with a MultiCam pattern. They stippled the grip and installed a ZEV Tech Fulcrum trigger and ZEV magwell. The result is one awesomely tricked-out race gun. This is one of the great delights of Glock pistols — you could put together your own version of this gun, by selecting the exact components that you want with the finish you prefer, then assembling it yourself. R

Photo courtesy of Richter Precision

TALE OF THE TAPE

	Hardness Rating (HV)	Corrosion Rating (ASTM B-117 Salt Spray Test)	Coefficient of Friction (lubricated & unlubricated)	Coating Thickness	Sample Pricing (Complete 1911 pistol)
Cerakote	n/a	550 hr (C) 3,500 hrs (H)	n/a	0.001 in	$200
Duracoat	n/a	300 hrs	n/a	0.00025-0.0015 in	$165
Electroless Nickel – NP3	480-530	250 hrs (NP3) 1,000+ hrs (NP3 Plus)	0.07-0.1 (wet) 0.17-0.21 (dry)	0.001 in (NP3) 0.0003 in (NP3 Plus)	$315 (NP3) $395 (NP3 Plus)
Ferritic Nitrocarburizing	1,075+	336 hrs	0.06 (wet) 0.4 (dry)	None (not a coating)	$200
Hard Chrome	830-1,075	88-100 hrs	0.1-0.2 (dry, may vary by level of polish)	0.0005-0.001 in	$200
Hydrographics	n/a	n/a	n/a	n/a	$165
Polymer – Black-T	n/a	1,500+ hrs	0.08 (dry)	0.0003-0.0006 in	$165
Polymer – Roguard / Poly-T2	n/a	1,000+ hrs	n/a	0.0005 in	$315
PVD – TiN / TiCN / TiAlN	2,200-3,100	n/a	0.4-0.6 (dry)	0.00004-0.0002 in	$500
PVD – DLC	1,500-3,000	50 hrs	0.02 (wet) 0.1 (dry)	0.00004-0.0002 in	$350

» SOURCES

2A Concepts
www.2aconcepts.com

APW Cogan
www.apwcogan.com

Bravo Company Manufacturing
www.bravocompanymfg.com

Coating Services Group
www.titaniumgun.com

Crimson Trace
www.crimsontrace.com

Crye Precision
www.multicampattern.com

Dunham Metal Processing
(800) 386-4262 ext. 107

Firebird Precision
www.firebirdprecision.com

Guncoat Northwest
www.guncoat.com

GlockStore
www.glockstore.com

H&M Metal Processing
www.blacknitride.com

JV Industries
www.jvindustries.net

OC Armory
www.ocarmory.com

Richter Precision
www.richterprecision.com

Rubber City Armory
www.rubbercityarmory.com

The Action Works
www.theactionworks.com

The Robar Companies
www.robarguns.com

Rogers Precision
www.rogersprecision.com

Salient Arms International
www.salientarmsinternational.com

VZ Grips
www.vzgrips.com

W.E. Birdsong & Associates
www.black-t.com

WW Guns
www.wwgunsinc.com

Young Manufacturing
www.youngmanufacturing.net

ZEV Technologies
www.glockworx.com

CHEAP TRICKS

Three Quick and Easy Ways to Improve Your AR

By Recoil Staff
Photography By Henry Z. De Kuyper

Almost any stock AR can be tweaked to improve its performance without breaking the bank. Here are a few modifications that are so simple, they should be part of every rifleman's toolbox. Start off by following normal safety procedures and separating the upper and lower receivers.

TOOLS & PARTS:
⅛-inch punch
Hammer
Dremel with cutoff wheel
Bearing grease
High-temp barbecue paint
JP Enterprises Reduced Power Spring Kit

1. ELEVEN-DOLLAR TRIGGER JOB

A Mil-spec AR trigger is rated to come in at a pull weight of anywhere between 5.5 and 9.5 pounds. An improved trigger is one of the best modifications you can perform on an AR — in Issue 9 we presented a buyer's guide to AR triggers that featured aftermarket triggers priced from $60 to $300. But you needn't break the bank to improve your trigger.

Get thee to JP Rifles, pilgrim, and purchase a reduced-power fire control spring kit for the princely sum of 11 bucks. Dally not at the website, lest thy bank account be smitten a mighty blow, the likes of which would rattle the gonads on a bull elephant.

1 Use a punch to drive out the hammer pin, then the trigger pin. Remove the hammer (along with the hammer spring).

2 Press firmly on the right side of your safety while moving the lever through its arc. The safety will pop out, allowing you to remove the trigger. You can also loosen the screw inside the pistol grip to relieve pressure on the safety detent, if you wish.

3 Lay out your fire control components and note the orientation of the trigger and hammer springs.

4-5 Remove your stock trigger return spring and replace with the JP yellow version. Be careful not to dislodge the disconnector spring in the tail of the trigger.

Disassemble your fire-control components and install your new yellow hammer and trigger springs, following the steps on these pages. Congratulations, you now have a much-improved trigger pull that will allow for greater practical accuracy. Best of all, it didn't cost an arm and a leg.

For those home tinkerers who are tempted to grind and stone sear surfaces to further improve their triggers, remember that standard trigger groups are typically only surface hardened. So by removing material, you may be shortening its useful life.

6. Do the same for the hammer. Make sure you orientate it properly, or you'll get light strikes.
7. Place the trigger back in the lower. Thread your punch through the lower, trigger and disconnector (which will be under tension from the disconnector spring). Use the punch as a slave pin to hold the components in place while you replace the trigger pin from the opposite side.
8. Now use your punch to press down on the safety detent while you insert the safety. If you loosened the pistol grip earlier, tighten it back down now.
9. Replace the hammer, ensuring that the tails on the hammer spring sit on top of the trigger pin
10. While holding the hammer under spring tension, align the holes and press in your hammer pin. Function check thoroughly — cock the hammer and make sure it only releases when you pull the trigger, then test the disconnector by cocking the hammer while holding the trigger back. The disconnector should first catch the hammer and then release it to the sear when you let go of the trigger. Also test that the safety functions properly.

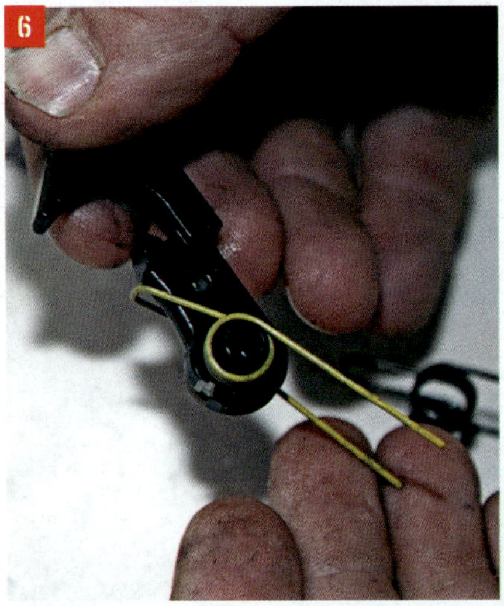

2. LOSE THE "SPROING"

One of the AR's less endearing features is the annoying twang the recoil spring makes inside the buffer tube (receiver extension for the purists) as the bolt carrier group cycles back and forth, right next to your ear. This is perhaps the easiest of faults to remedy. Using a punch or bullet tip, depress the buffer detent and pull out your buffer and recoil spring. Apply a dab of heavy bearing grease on the spring coils. Reassemble and voila! No more annoying "sproing."

For those who actually like the sproing noise, hearty congratulations are in order. Revel at being soulmates with your AR.

3. LOW-PROFILE GAS BLOCK

Let's say you're planning on installing an extended free-float tube or rail system, or you're just tired of the shadow in your optics. That front sight tower has got to go, but your piggy bank is empty and you can't buy a new low-profile gas block. Time to break out the Dremel.

Start off by removing your flash hider or muzzle brake by unscrewing it from the muzzle. If you have an upper receiver block and a vise handy, it will make this step easier. Now drive out the taper pins that secure the front sight base (FSB). Look carefully — one end (usually the left side) is a smaller diameter than the other, and this is the one you'll want to tap with a punch and hammer to remove the pin.

Slide the FSB off the muzzle end of the barrel. If you've got more than a few hundred rounds down the tube, you may need to use a little persuasion from a soft head hammer to get it to move.

Place the FSB in a vise and get to work with a cutoff wheel, following the dotted lines shown. Once you've removed all the extra steel, switch to a coarse sanding drum and contour the rough edges. A quick shot of black barbecue paint, and you're ready to reinstall. R

TAKE HOLD

Shape Your Glock's Grip to Fit Your Hand

By RECOIL Staff | Photography by Iain Harrison

WARNING!

When handling firearms, always observe safety rules and the precautions set forth in the firearm's owner's guide. Be certain that your firearm is unloaded and made safe before proceeding with this DIY.

One of the biggest criticisms of Glock's handgun lineup is its backstrap profile — despite the Gen 4 models featuring replaceable inserts, the infamous "Alien Bulge" remains. However, the backstrap can be reshaped, and there's a small cottage industry offering grip-reduction services, which shouldn't be beyond the abilities of an average DIY enthusiast.

While hacking on a perfectly good Glock might seem a little scary, simply follow our step-by-step guide and you'll be able to custom fit it to your preferences by adjusting the profile of the backstrap.

TOOLS AND MATERIALS

You will need the following:

1. Glock armorer's tool or ⅛-inch punch
2. 80-grit sandpaper
3. Plastic epoxy kit (with syringe)
4. Masking tape
5. Mixing container
6. Modeling clay
7. Degreaser (not pictured)
8. A 25-cent coin (not pictured)
9. Black nail polish (not pictured…it's OK, you're allowed to buy this as it's for a manly purpose)

STEP 1

Unload pistol and detail strip.

Remove the slide, press out the three frame pins, and pull out the locking block, slide release, and trigger group. Set them aside.

STEP 2

Flush out the trigger group recess at the rear of the frame with a degreaser, such as brake cleaner. Look for something labeled as "not for sale in California" to make sure you're getting the good stuff. Turn the frame upside down and flush out the grip cavity, then use some 80-grit sandpaper to rough up the cavity's interior, paying particular attention to the plastic that separates it from the magwell.

STEP 3

Carefully and completely fill the sear block recess with modeling clay, to make sure that no epoxy can enter the space. If it does, you'll have the next-to-impossible task of getting it out. While you're doing this, familiarize yourself with just where the sear block fits inside the frame and how far down it goes. When you grind later, you'll need to stay away from this area, as well as the trigger group pin hole. Use the same clay to fill in the lanyard hole.

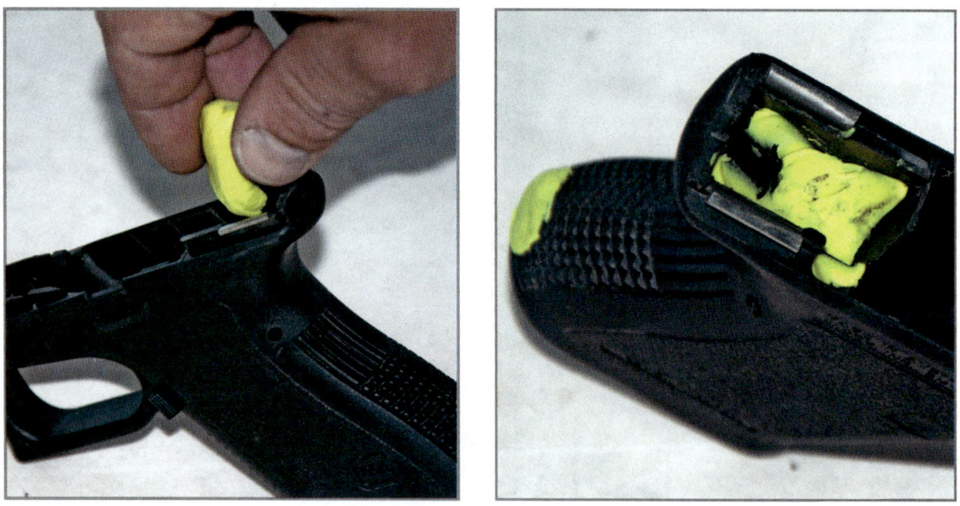

STEP 4

After you turn the frame over, you may notice a little half-moon cutout at the rear of the magwell. This will prevent epoxy from completely filling the grip cavity, so you'll have to seal it off. Stick your hand in your pocket and pull out a quarter — it's exactly the right size to jam into the mag well and block off the cutout. Wipe a tiny bit of gun oil on the side that will face into the grip cavity and then tape it in place securely, filling in any tiny gaps with modeling clay. Seeing as you already have the masking tape in your mitt, wrap it around the bottom of the grip frame to form a wall where the frame angles upward at its base.

STEP 5

Cover the bottom of your mixing container with nail polish and quickly mix up a batch of epoxy, following the manufacturer's directions. Use your pocket knife to split apart the two empty epoxy syringes — fill one with your epoxy mixture and use it to inject resin into the grip cavity. The fewer bubbles in the mixture, the better the end result will be.

STEP 6

Test the epoxy remaining in the mixing cup to determine its hardness. While still just a little soft, peel off the masking tape and remove your quarter. Allow the epoxy to cure to full hardness.

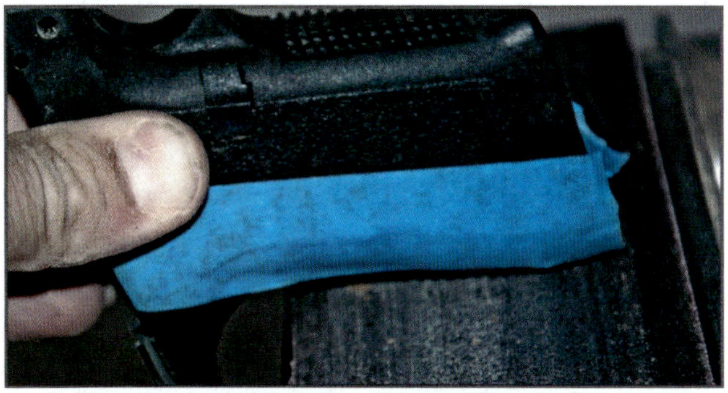

STEP 7

Use masking tape to create a line, marking out the "no grind" area of the magwell, sear block cavity, and trigger group pin. You can now start removing material from the grip to fit your hand. A belt sander is ideal for this job, but you can use a Dremel or coarse file, as well. Be patient — remove a little and test for fit each time until you achieve the desired result. Bevel the magwell area to allow for a smoother reload.

STEP 8

Use the stippling techniques detailed in the "DIY" from RECOIL Issue #1 to blend in the backstrap and create a better gripping surface. Reassemble the pistol and function check it.

Congratulations, you've now got yourself a custom Glock, fitted exactly to your hand. **R**

PATTERNED POLYMER DYEING

Give Your Component a New Look That Lasts

By Staff

With so many firearm parts and accessories available in plastic and polymer nowadays, it's a great time to be a do-it-yourselfer. Not everyone may have the skills or tools to hand-checker a 1911 metal frame, but most anyone willing to try can shave, stipple, or, in this case, dye polymer.

Dyeing polymer furniture and other firearms-related gear is a fun and inexpensive way to give your equipment a durable and unique, custom look. It's not all for show, either. Dye can be applied as a semipermanent, identification marking on your gear, such as a magazine. There are many methods and variations on how to dye polymers; we will take a look at just one of the ways.

Dyeing polymer is not an exact science. It takes trial and error to get just the right color or tone that you are looking for. Practice makes perfect, so be prepared to re-dye parts several times to achieve your goal. Don't be shy to experiment; in most cases, the dyed polymer can be returned to about 90 percent of its original color by treating it in bleach. (See sidebar, "Want to Start Over?")

A key for successful dyeing is to start with light-colored polymers. Many begin with items colored tan, Flat Dark Earth, or Foliage Green. (In case you were wondering, black polymers cannot be turned a lighter color.) In our examples, we start with items colored tan or Flat Dark Earth. Because manufacturers use slightly different pigments and compounds when producing their parts, the dye will affect surfaces differently. The following project shows how it's done with a TangoDown Battlegrip for an AR-15. The same process was applied to the rest of the rifle's polymer furniture to create a unique camouflage look.

As always, make sure any firearms that you are working on are unloaded, that no magazine is inserted, and that there is no round in the chamber. Strip your firearm of the parts you intend on dyeing. Because the parts being dyed require submersion in water, we suggest

WARNING!

When handling firearms, always observe safety rules and the precautions set forth in the firearm's owner's guide. Practice all steps on scrap materials before proceeding. Practice makes perfect. Perform this modification at your own risk.

TOOLS

1. A pot large enough for the part being dyed
2. A measuring cup
3. Measuring spoons
4. A water-resistant or deep-fry thermometer
5. Tongs
6. A hobby knife
7. A Heat gun or a blowdrier
8. A timepiece (not shown)

MATERIALS

1. Rit Dye (liquid or powder)
2. A vinyl sticker sheet or electrical tape

EOTech
★BCM.

removing all metal parts from assembled polymer items, such as springs from magazines. When in doubt, practice with polymer scraps before you dive into it.

We will go through not only how to dye polymer, but how to apply a mask as well. Adding a graphic element as a mask gives you more options for how you want your project to look. Dyeing multicolored patterns takes some forethought and planning. Always go from light colors to dark.

01 A few square feet of vinyl sticker sheeting from a sign shop is a good option for masking material. Electrical tape is another option, but we find that using a larger vinyl sheet allows for more design options. When doing a camouflage pattern, it helps to study the design you want first. To get started, cut the camouflage shapes onto the sheet using your hobby knife.

02 Here, you can see the sheet after it has been cut with random camouflage shapes.

03 Apply the shapes to the surface of the part being dyed. In this example of a four-layered color design, we are applying the masking to cover the lightest layer we intend on keeping. In this case, it is the original tan of the polymer. On rough textures or surfaces with sharp curves, this can be challenging. The mask is only as good as your application, so take your time.

04 Use a heat gun or hair dryer to help soften the vinyl so that it can better adhere to the surface. The tighter the application, the sharper the lines will be. A loose application will show blurred lines. Depending on how you want your pattern to turn out, you can play around with this.

01

02

03

04

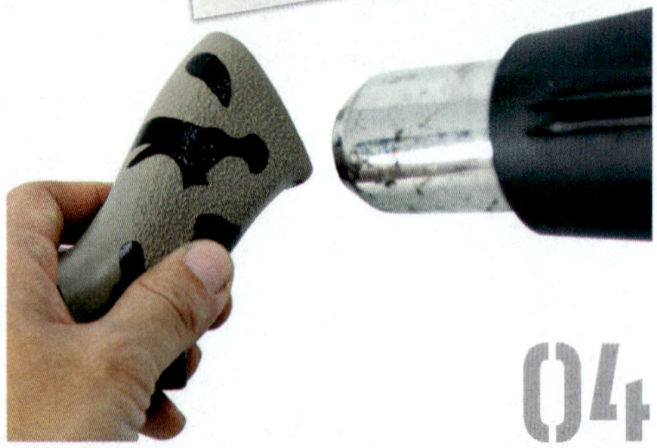

05

06

EDITOR'S TIPS

In the photo studio, we are using a portable stove. Any container and heat source able to get water up to temperature will do. A microwave to heat the water and large disposable styrofoam containers to contain the dye and water mixture is an alternative option. Although we found that by using a stove, we have the ability to heat the water to maintain a consistent temperature, which is a benefit. For larger items such as buttstocks and handguards, we used a large oven pan, like the kind you would roast a turkey in.

05 A mixture of half a teaspoon of powdered Rit Dye per cup of water worked for us for most colors in most cases. Note that Rit Dye in its liquid form is about twice as concentrated as in its powdered form, so when using liquid Rit start by trying a ratio of a quarter teaspoon to one cup of water.

06 While stirring the dye-water mixture, heat the water to a temperature between 150-175 degrees Fahrenheit. Any hotter than that isn't necessary for the dye to take effect, and excess heat can negatively affect the vinyl mask.

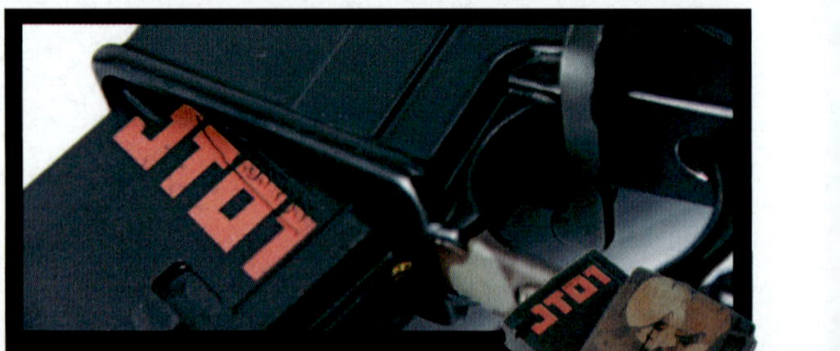

USING DYE FOR IDENTIFICATION

Dyeing identifying markers on your accessories such as magazines is a more permanent solution to paints that might chip or fade over time. In this example, we had a sign shop cut a sheet of stickers of initials and numbers to help mark magazines. For the pictured example, we took a Magpul PMAG colored Flat Dark Earth and dyed it red, applied the stickers as the mask, then dyed the magazine black. The masked areas remained red, leaving just the initials and magazine number showing on the black magazine.

07

08

09

10

11

12

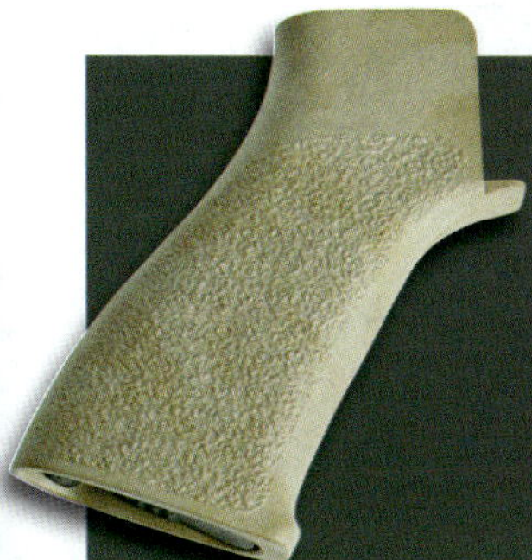

WANT TO START OVER?

If you want to start over at any point, simply soak your part in a bath of one part bleach to two parts water. Heat the mixture to 140 degrees F and let the part sit in the bath for approximately 30 minutes to an hour. (Some parts will require more time.) Most of the dye will bleach out, leaving you with a mostly blank canvas to start over with. Remember to do this in a well-ventilated area and try to keep from breathing in the fumes. The grip on the left is the same grip that was used in the article. As you can see, the bleach has removed most of the dye. Alternately, you can always dye the part black to cover any mistakes.

07 Using the tongs, gently submerge the grip into the mixture. Have a time piece ready because you will want to pull the part out of the mixuture every minute or so, to check the color. Once you get the color you want, pull the part out and pat it dry with a clean cloth. Be careful, the polymer surface will be VERY HOT. Also pay attention not to splash the mixture; you're working with very hot water that can injure or permanently stain surfaces and clothing.

08 This is the result after three minutes at 175 degrees F using Rit Pearl Grey Dye.

09 Leave the previous masking layer on the surface; once the item has cooled to a temperature safe enough to handle, apply a new layer of masking to cover the portions of color that you want to keep.

10 Repeat steps 5 through 7, alternating to darker colors as you go along. Continue doing this for every layer in your pattern. In this case, we followed the initial, three-minute, Pearl Grey layer with two layers of Dark Brown at one minute each, capped with a final dip in Pearl Grey for two minutes.

11 When you're done, the part should be covered in multiple layers of masking.

12 After the part has cooled to a safe temperature, peel off the masking to reveal the results. It's simple and effective, and best of all, the dyes are resistant to surface scuffs and light scratches. Now try your hand at dyeing other polymer accessories.

The Ergo F93 stock was disassembled and dyed multiple colors.

Details such as the Magpul MOE Trigger Guard complete the custom look.

Most polymer parts can be dyed, even this Magpul MBUS Gen 2 Rear Folding Sight.

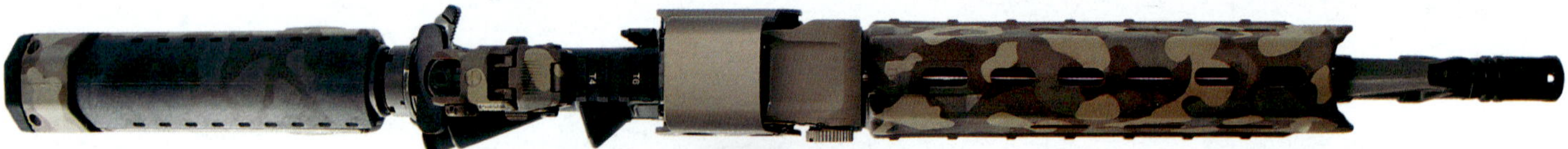

THE HOT-GLUE MASKING METHOD

An alternative to masking with vinyl tape is using a hot glue gun. The hot glue is easily applied and covers small crevices and sharp angles that might be very difficult or impossible for traditional masking. The best thing about it is that it peels right off most polymer materials with ease. One disadvantage is that it isn't very good for detailed work because the glue can be difficult to control. Pictured here are two examples of polymer-bodied Surefire G2X Pro flashlights in which hot glue was used for masking. Of course, first test a small spot on your material prior to trying this. R

STOP LOSING DETENTS AND SPRINGS

AR Rear Takedown Detent Modification

By Jonathon Perez

The AR-15 and its variants are hands down some of the most popular rifles seen at ranges. That's quite an accomplishment for a rifle that was invented in 1957. Around the time the USSR launched Sputnik, Eugene Stoner was tinkering with a platform that would eventually become the main battle rifle for the United States into the 21st century. It goes without saying that technology has progressed by leaps and bounds since the '50s, but there's no reason why we can't upgrade the rifle that we currently have while the government figures out the next rifle our forces will wield.

The AR platform, as great as it is, can be a pain to work on due to the many detents and springs that are used in its assembly. One such detent-and-spring combination is used to install the lower receiver's rear takedown pin. The usual way of installing these parts is to do a sort of intricate tango dance with your 15 fingers. You install the takedown pin, insert a detent, follow the detent with a spring, then smash it all into the lower receiver with the rear latch plate, all while screwing in the buffer tube and castle nut. One false move and *spoing*! The spring goes flying across the room with the detent in hot pursuit.

We are here to show you how to turn that freewheeling spring into a captured spring that will make losing parts as passé as doing atomic bomb duck-and-cover drills. As an added benefit, this modification makes swapping out buffer tubes and installing rear sling plates a breeze, too.

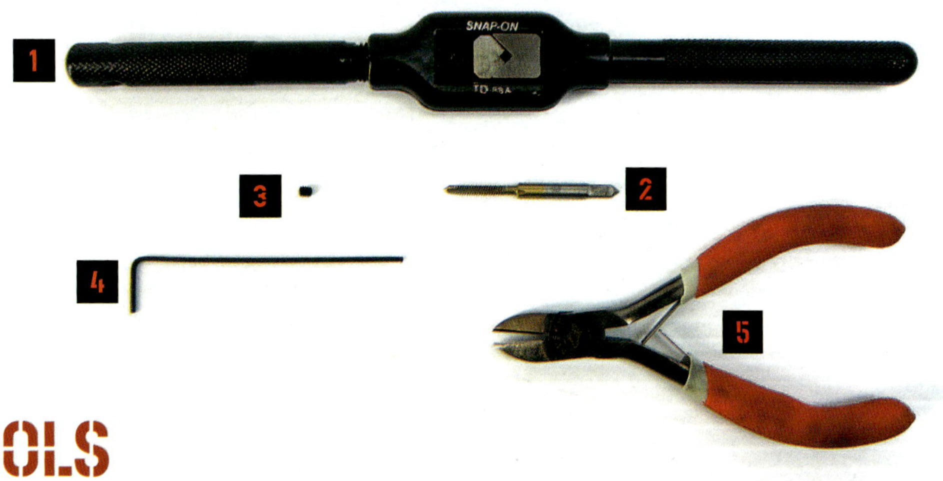

WARNING! When handing firearms, always observe safety rules and the precautions set forth in the firearm's owner's guide. Practice all steps on scrap materials before proceeding. Practice makes perfect. Perform this modification at your own risk.

TOOLS

You'll need the following items for the job: 1 tap wrench, 2 4-40 tap, 3 4-40x1⁄8 set screw, 4 0.050 Allen wrench, 5 a pair of wire cutters. Most hardware stores will have these items available. If your lower receiver is assembled, you will also need a stock wrench. We suggest that you perform this modification in an area that is clear so that you can find your spring and detent in the event that you do launch them.

01

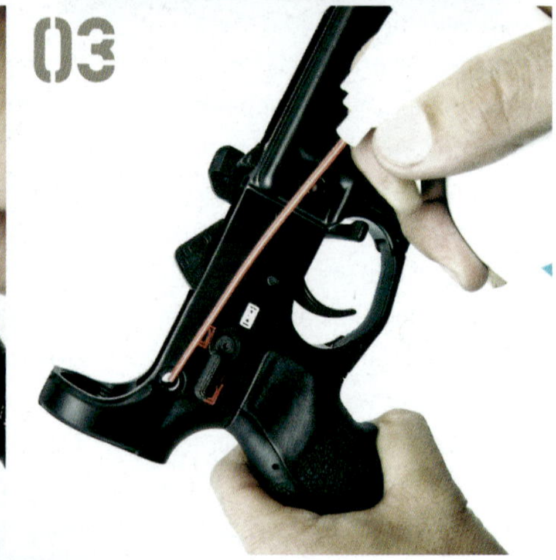

01 Start by ensuring that the gun is unloaded, that there is no magazine inserted and that there is no round in the chamber. This modification can be done to stripped or prebuilt AR lowers. If your lower is already assembled, start by uninstalling the buffer tube assembly. With the buffer assembly removed, you will find the hole for the detent and spring that help keep the rear takedown pin in place. This is also where the spring wants to shoot out from when you're not paying attention.

02 To keep that from happening again, thread the hole with the 4-40 tap. Go slow and straight in. No drilling is required. The existing hole size is perfect for the 4-40 tap to thread. Thread down to a depth of a little over $\frac{1}{8}$-inch.

03 We used a can of compressed air to blow out excess metal from the detent side of the hole. If you have an air compressor, use it. If you don't have either, just pucker up and blow a good burst of air into that hole.

04 Here's the newly threaded hole.

05 Aluminum doesn't rust, but it can corrode, so we dabbed some aluminum bluing fluid into the hole. We recommend Birchwood Casey's Aluminum Black for this job.

06 Now let's move on to the spring. Using the wire cutters, snip 1/8-inch off the spring.

07 The cut compensates for the length of the set screw. If the cut isn't made, the spring will over-compress and the rear takedown pin will be very difficult to push.

08 Now it's time for assembly. Install the rear takedown pin into its place, ensuring that the orientation of the pin has the detent groove pointed rearward toward the detent. Insert the detent followed by the cut spring into the now-threaded hole. This next part can be a little tricky, so this is why you perform this modification where you can easily find parts if they go flying: With the 4-40x⅛ set screw on the 0.050 Allen wrench, carefully push the spring in with the set screw while screwing it in.

09 Keep screwing the set screw in until it is flush with the surface of the receiver. If it isn't flush when you have the screw all the way in, you will need to add a few more turns of threading. If the rear takedown pin is really hard to push in and out, you will need to cut another coil or two off the spring.

10 Here is the finished product. Now you can change out rear sling plates and other buffer tube accessories without worrying about losing your rear takedown pin spring and detent. **R**

05

07

08

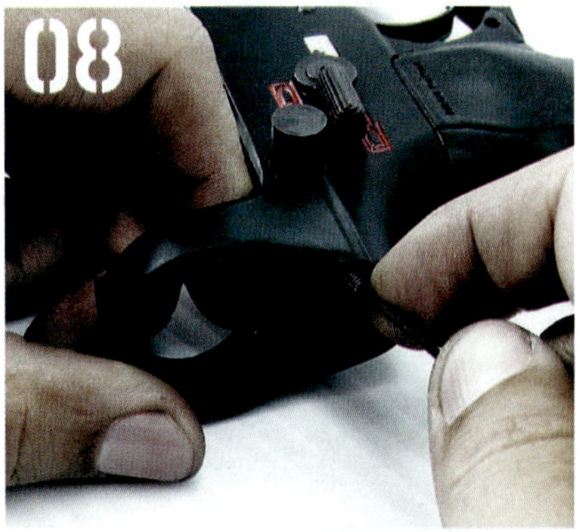

06

09

10

POLYMER STIPPLING

Better Grips for Polymer Weapons

By RECOIL Staff

Polymer has quickly become the standard material from which firearms and their accessories are made, and for good reason. Polymer is lightweight, impervious to corrosion, durable and can take almost any shape. The stippling of polymer handguns is gaining popularity these days. This modification provides a more positive grip over the slick feel of factory-stock surfaces.

The stippling effect is similar to that of traditional checkering on semiautomatic 1911 handguns. If you're shooting in soggy, wet weather or sweating it out in the heat, an aggressive stipple job may help you keep a firm grip on your potentially life-saving hardware.

Custom stipple jobs can be expensive when the gun is sent out to a specialist. With the right know-how, though, you can learn how to do it yourself on the cheap. This guide lays out the basics of stippling polymer.

Before you begin, make sure you understand that there's no going back after you start. You will be actually melting the polymer on your gun, so it's a good idea to practice on some scrap polymer before you go to work on your own weapon. Rail panels are great pieces to practice on.

WARNING!

When handling firearms, always observe safety rules and the precautions set forth in the firearm's owner's guide. Practice all steps on scrap materials before proceeding. Practice makes perfect. Perform this modification at your own risk.

TOOLS

To melt polymer, pretty much any soldering iron will do. We used an inexpensive 1 soldering iron with a pointed tip. It was a 40-watt model that can be picked up for about $15 at any hardware store. The size of the tip will dictate how fine the stippling will be. You'll also need a 2 Sharpie fine-point marker, a 3 file or sandpaper, a 4 clean cloth and 5 rubbing alcohol.

SUREFIRE
STAINLESS
M&P 45

01 Ensure that you perform the stippling procedure in a well-ventilated area. Melting polymer gives off fumes you won't want to breathe. Also ensure that the gun is unloaded, no magazine is inserted, and there is no round in the chamber. Grip the gun and figure out which areas you want to stipple, then field strip the gun to remove the slide.

02 You will be working with only the frame from this point on. Use the marker to trace the areas you intend to stipple. On this Smith & Wesson M&P 9, we went with an area around the entire grip as well as part of the area under the trigger guard and the pads forward of the slide release to act as reference points. In this example, the ink is black and the gun is black, but you can still see the outline under the light.

03 With the soldering iron already heated up, run the tip over the marker outlines you just drew. This will give you the borders in which to stipple. Take extra care to make these lines deliberately. The slower you go over the lines, the deeper the outline will be. How deep you want this outline is a judgment call you have to make.

04 Here you can see the outline made with the iron.

05 After you are satisfied with the outline, begin stippling. The art of stippling is to press the tip of the soldering iron just deep enough to melt the surface and then move on to the next spot and repeat... and repeat and repeat. Keep stippling until you have the entire surface covered. What you are accomplishing by doing this is "mushrooming" the area around the point of penetration. It is this mushroomed polymer that will give you the extra grip you seek. The deeper you burn into the polymer, the more grip

you're going to have. But be careful not to go deep enough to burn through to the other side.

06 When all of the surfaces are covered, try out your new, more grippy grip. You can fine-tune areas that are too aggressively stippled by reducing the sharp edges with a file or sandpaper.

07 If you still see marker residue, you can take if off with some rubbing alcohol and a clean, soft cloth. And then, congratulations. You just gave your handgun a drastically more aggressive gripping surface while saving hundreds of dollars by stippling it yourself. Now that you're a stippling pro, try your skills on other surfaces. The same premise can be used on all sorts of polymer parts such as rail panels and rifle grips. **R**

TIPS ABOUT TIPS

Different soldering iron tips will give you differing results. Here are a few that we played with:

FINE TIP / FINE TEXTURE
The most common style of stippling you see around is this fine texture. By using a fine-tipped soldering iron, it is easier to control what areas are stippled, especially in tight spaces.

ROUND TIP / ROUGH TEXTURE
This texture is more aggressive and gives a lot more grip. It is easier to do since one hit of the round tip covers a lot more area than the fine tips does. This tip is actually the fine tip pictured above, flipped around on its end.

CHISEL TIP / SHARKSKIN TEXTURE
This style is commonly called "sharkskin" because (you guessed it) it resembles a shark's skin. To get this treatment, we ran a chisel tip on its thin side across the panel. We closely followed up one line with the other. This effect is easier to get than it looks.

IF YOU DON'T DO IT YOURSELF

If you're not up to burning holes in your gun and accessories, let a pro do it for you. Here are a few places we Googled that offer stippling services or already stippled items.

Boresight Solutions	Nash Gunworx
www.boresightsolutions.com	www.stippledgrip.com

4-MINUTE FIXES

Make Simple Improvements to Your AK in the Time it Takes to Poach an Egg

Story and Photos By Dave Merrill

Though the AK and her variants may be the most prolifically used rifle in the world, in some (alright, many) aspects it can be a bit long in the tooth. About a decade ago, the lower prices of the AK made for a cheap and fun project gun to tinker with in the garage or workshop — and you could do so on a tight budget. This article hearkens back to that time, and though the base models may have increased in cost, the modifications haven't.

The goal was to highlight easy projects that gift clear improvements for the modern shooter that also keep the cost as low as possible.

For this article I used a vanilla homebuild pieced together from a Romanian "G" kit. There's absolutely nothing special with this rifle whatsoever, and if you've seen one you've seen them all.

In the true spirit of the Kalashnikov, every one of these mods is robust and readily performed if you're comfortable with common tools. You can do all of them in one go for under $20. If you're not comfortable with tools, each project also has *at least* one commercially available alternative — some no doubt even featured inside these very pages. None of the projects require permanent modification on irreplaceable parts, and the parts you modify are easy to swap out if you end up hating it or simply dick it up too bad. Almost everything shown can be completed with handtools so you can pretend you're in a cave in the Khyber Pass, though the vaunted Dremel is highly recommend for some.

Whether you're just trying to save a little money, want a fun weekend project, or simply desire to make your rifle a bit easier to rock — this one's for you. So suit up and prepare to do things that will make the purists purple with indignation.

In no particular order ...

WARNING!

When handling firearms, always observe safety rules and the precautions set forth in the firearm's owner's guide. Be certain that your firearm is unloaded and made safe before proceeding with this DIY.

SIGHTS

One issue with standard AK iron sights is that the slot in the rear sight is just too darn tiny. Lining up your front sight within that small notch may be fine at a static slow-fire range, but at close dynamic ranges where speed is key it can be detrimental. While AKs from some nations have wider rear notches than others, all can be improved. Widening the rear slot not only makes for faster target acquisition, it also won't affect accuracy too much — we are talking AK here after all. While you're at it, you can take off some of the sharp corners if you want.

Opening up the rear with a triangle file is easy. Go slow, keep it square, then touch it up with some cold blue, or my favorite: spray paint. Shoot for a notch between 0.135 and 0.145 inch.

HANDGUARDS

You shoot, the handguards get hot. Science. While some AK handguards have heat shields, not all of them do. Even if you wear gloves, heat can be an issue. One solution is to make your own cheap heat shield, and all that's required is a can of cheap beer. It's actually important that you go with a cheap brand, because some of the nicer ones have plastic liners on the inside.

Drink the beer. Cut out a handguard-ish section, and shove it into the lower handguard. Some guys like to use epoxy to keep it in place, but I don't bother.

While you're at it, if you have a Romanian donkey dick handguard and you're tired of it mucking up your magazine changes, give it the old chop-chop. Play your cards carefully and you'll have an Afghan AFG.

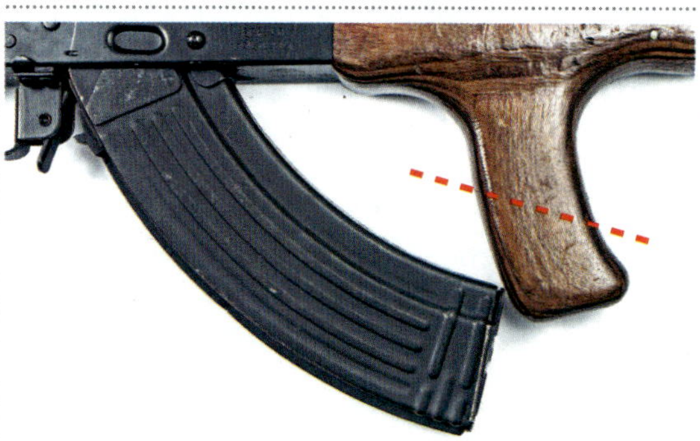

LIGHT MOUNT

There've been a lot of different ways to mount lights on AKs over the years — from handguard replacements to barrel mounts and more. For this you need a piece of add-on Picatinny rail (I used a polymer Magpul MOE section, because I had one laying around), four large washers, and two longer machine screws of the appropriate pitch for your rail.

Assemble the rail with two washers and a nut on each screw. Remove the top handguard, place a washer on either side of the lower handguard, and reinstall the top rail. Tighten. You may have to make a small relief cut to make the rail even, depending on your handguard and rail placement.

No, you wouldn't want to put a laser on it, but now you can mount any rail-compatible light easily to your rifle.

SLING MOUNTS

Mounting anything other than a factory carrying strap can be difficult. You have that small front attachment point and further complicating matters the rear attach point is sub-optimal for modern slings.

Make a better front point with 8 inches of 1/16 coated steel cable and a pair of ferrule stops. My cable happened to be ugly green; maybe you'll have better luck. Use the ferrule stops to make a loop with the coated cable, ensuring to remove the plastic that goes inside the stops. Clamp it down. If you want to get fancy, put some heat-shrink tape on it. Loop it through the factory front point and attach your sling. Using two ferrule stops allows for a rigid section that's a perfect place for your sling.

There are two ways to take care of the rear mount: You can either remove the buttplate and loop some 550 cord around the top screw and replace the plate, or simply move the rear sling mount from the bottom of the buttstock to the outside-top. If you're doing the latter, ensure that you don't inadvertently screw through the channel for the cleaning kit.

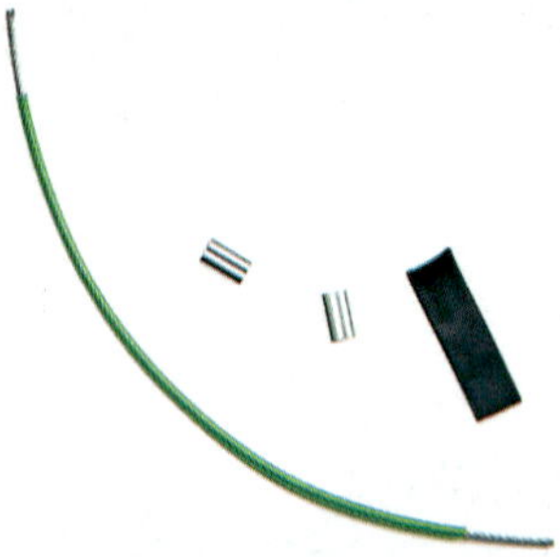

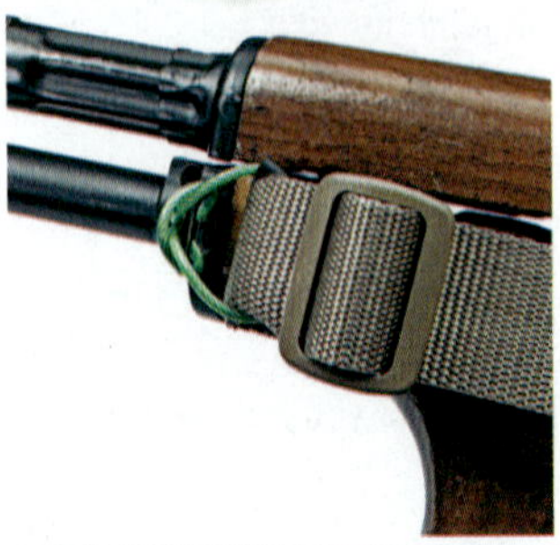

SAFETY

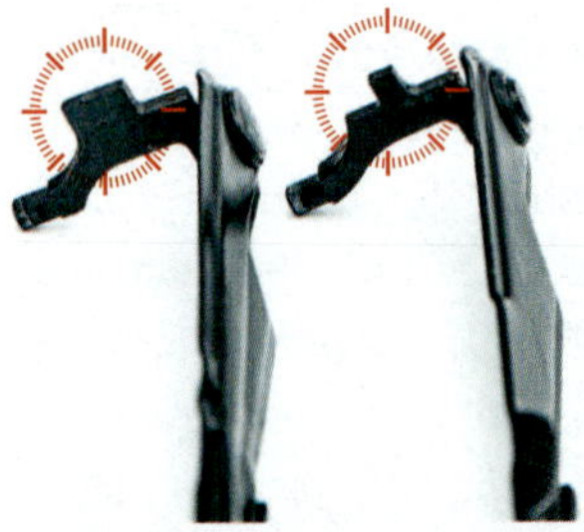

There are several beneficial mods you can perform on the AK safety. Many guns, especially ones built from kits in America, feature a full-auto selector. No, it doesn't do anything cool for you like make it go bang faster, and can actually be a hazard. The "auto" portion of the safety pushes against your disconnector and drops the hammer during removal, and sometimes does this even if you just snap it on "safe" too hard. No bueno. Simply contour that tab to match a semi-auto safety and not only will it be easier to remove and reinstall, it's much safer as well.

Unless you're using some very particular models, your AK doesn't have the ability to reliably lock the bolt to the rear. Bust out that Dremel and cut a charging-handle-sized notch in it. Now you can pull the bolt back and sweep up the safety to lock the carrier in place, which is especially handy on ranges that require a bolt locked back when the line is cold.

If you have fingers like Les Claypool, the AK safety is probably easy to employ. For those of us with human-sized phalanges it can be a pain in the ass. I used some scrap metal to add a tab to the safety for easy manipulation with the trigger finger. I happened to spot-weld the tab in place, but I've also seen it done with rivets and other forms of welding. Deburr and spray paint when complete.

GRIP

While currently grips on AR-15s are becoming more and more upright and AK-esque, there are some out there that want an AR grip on an AK. While you can buy an adapter, it's not needed for most of them. Obtain your desired grip that doesn't have a high backstrap. Hog out the front and top with a Dremel to match an AK grip. Drill a new grip screw hole, and insert a shorter M6 screw. A2 grips make perfect project grips for this. Remove the nubbin on the front or stipple to your heart's content. The inside will be rough, but no one sees that anyway.

FLUFF AND BUFF

No, don't use sandpaper. Just a little polish on the rails and bearing surfaces to smooth it out. You'll be pleased with the results. R

FEEL THE BURN

A Rough Guide to DIY Stippling

Story by Nathan Murr
Photos by Nathan Murr & Rob Curtis

There a few things that you really, really don't want to drop.

A few that come to mind include babies, live grenades, urns, beakers of acid, Fabergé eggs, and loaded handguns. Ever since the birth of handguns, their creators have sought different means to improve a shooter's grip on a gun through surface textures. From the hand-cut checkering of old flintlocks to the finger grooves of the latest German-made über-blaster, points of traction have been a must. Speak with any firearms instructor, and they'll tell you shooting a pistol starts with a good, secure grip. The shooter shouldn't need to adopt a white-knuckled death grip on their pistol to keep it under control. Not only is that bad form, it's also completely exhausting.

There's also environmental influences that can leave the grippiest gun feeling as slick as goose snot. Imagine — you're pushed to the muddy ground by a mugger before you can even start your draw. Or, you're a cop who's spent the day standing in the icy rain grabbing a slick-slabbed service weapon with a wet, numb hand. Maybe you're a competition shooter with sweaty palms on a hot August day. Or the worst of all, you're trying to fight for your life with hands covered in blood.

What's the solution to a slick sidearm? Firearms manufacturers and accessory companies have offered overmolded rubber grips, textured tape, and molded finger groves paired with light texturing. But, these and most other solutions are half measures that do little to improve traction, particularly on polymer framed pistols. Luckily, there's an inexpensive way to get the perfect grip — and with a little testicular fortitude, you can do it yourself.

Stippling isn't anything new, but in the roughly 35 years since polymer-framed handguns have been in wide scale production, you'd think Glock, Smith & Wesson, HK, and others would have the factory grip texture down solid. But, that's never going to happen. What one shooter loves, another will hate. And keep in mind, far too many shooters base their purchase on what feels good in their hand. So, don't expect substantial factory texture changes anytime soon.

Stippling your favorite gun is easy if you possess patience, and doing it yourself allows you to indulge your own preferences. While stippling costs little in way of tools, it will cost you heavily in time and attention. If you can't sit still for more than five minutes and complete a detailed, repetitive task in a slow and focused way, you're probably going to screw your gun up, or possibly destroy it.

This is the part of the article where you pause for a moment and do some soul searching, because no one knows you like you know yourself.

Let's be honest here, if you think your first stipple job is going to turn out looking like a high-end custom Glock, or one of the many cool-guy blasters flooding Instagram, you're kidding yourself. Quality comes with experience, which only comes with time. That's why the big boys charge a premium, and local amateurs are willing to spend three-plus hours with your frame for $100 and offer a two-day turn around. The novices are using your gun to gain skill and experience, and make a little money in the process. Not that there's anything wrong with that, but how about getting your own soldering iron and doing it yourself?

Stippling is a growing hobby in Gun Culture 2.0, with many younger shooters mastering the craft and pushing the envelope of style. You can join their ranks by putting in some practice and building your skill level. That doesn't mean you need a dozen 9mms sitting in the safe ready to feel the burn; quite the opposite. You likely already own what you need to get started.

TOOLS

First you need to gather your tools; if you don't already have them, you can find them locally (and cheaply). One stop at a nearby big box store will net you a soldering iron (or alternatively a wood burning kit), sandpaper, tape, razor knife, and optional Dremel tool.

Many opt for a wood burning kit over a soldering iron because they usually include an assortment of burning tips. Small tips create pin-sized holes, requiring more passes over a small area to achieve a uniform appearance. Larger tips leave larger holes burned into the polymer, resulting in faster stippling progress on the same sized area. The larger tips also create a more aggressive gripping surface, which you may or may not want on your gun. Aggressive textures will wear your cover garments out faster and may not be comfortable for everyday carry against your skin. Try doing a circle pattern, a box pattern, and lines back and forth. Find a system that works for you, and avoid the temptation to just randomly stab away.

You'll need the right ingredients before you bake your first Kydex pie. Photo by Patrick Vuong

Start on something low risk. Grab a PMAG and get to work. A magazine body offers an excellent canvas to experiment with different tools. Use the recesses as boundaries and practice staying within the lines.

Once tools and supplies are gathered, it's time to get started. First, take the polymer frame you intend on stippling and put it in the other room. Don't even look at it. You won't be touching it with your new tools anytime soon.

DONOR POLYMER

Next, assemble a collection of plastic objects of various sizes and shapes to practice on — polymer AR-15 magazines, spare A2 pistol grips, an old travel mug, a cast away holster, or an ex-wife's Tupperware collection. Various polymers react in different ways when sanded, cut, or put in contact with heat. Softer, thinner polymers are less forgiving to a heavy hand wielding a soldering iron. You absolutely need to practice, even if it feels funny to stipple a broken dollhouse or wiffle ball bat. Use this time to learn your tools.

JUST THE TIP

Press tip to plastic, noting the time it takes to burn to different depths. When changing to a different polymer as you advance in your practice sessions, do the same to learn the feel of that particular medium. Some polymers melt faster and easier than others, necessitating concentrated focus on the task. Patterns are created by the choice of tip you select, and the consistency in which you apply it. You can achieve a sandpaper-like fine pattern with a needle sharp tip, a golf ball pattern with a larger round tip, or a cross-hatch pattern with a chisel tip.

Don't be afraid to create your own tip by filing a factory tip to your preferred size and shape. If you examine different patterns on finished guns, you can figure out what size and type of tip was used purely by its appearance. This is the exciting part of stippling, otherwise a tedious endeavor. You get to choose what style to use on your pistol, resulting in a truly unique (and permanent) personal touch. Experiment with stippling a logo, unit crest, emblem, or anything more detailed than just standard rows of tiny holes. You might decide to incorporate a similar design

Begin with a light touch and a low temperature setting on your soldering iron. Make some mistakes on purpose so you learn the feeling, sound, and smell of an error in the making.

Drum sanding attachments on a rotary tool are a great way to remove material from recessed areas, such as trigger guard undercuts and beavertails.

into your gun, such as your initials or something meaningful to you. If done well, expect your friends to ask you to work your new magic on their guns.

DREMEL UP

Once you feel like you have the plastic burning skills down solid, it's time to grab your Dremel and run through the same familiarization process with whatever polymer detritus you've got left. Get a feel for sanding drums, buffing wheels, and tiny router bits. These are the most commonly used attachments for the Dremel-wielding stippler.

Once you feel up to it, grab a pistol frame and pop your stipple cherry. Sanding drums are used for removing or reducing finger groove ridges, which many dislike on stock Glocks. They're also used to remove the light factory texturing on an area to be stippled. And, with a skilled hand, a sanding drum can leave a uniform appearance across the frame. Remember that different grits and speed settings will remove polymer at different rates, so consistency is extremely important here.

Many shooters prefer to remove material from the trigger guard where it meets the lower frame, allowing a more choked-up grip on the gun. Slow and careful removal of material with the sanding drum will accomplish this undercut.

Additionally, a small notch on the trigger guard for the support hand to lock into when using a two-hand grip can be added in the same fashion. But remember, whatever material is removed is gone forever. Enthusiastic sanding has been known to weaken a frame or leave a paper-thin trigger guard. If you're worried that you're going too deep,

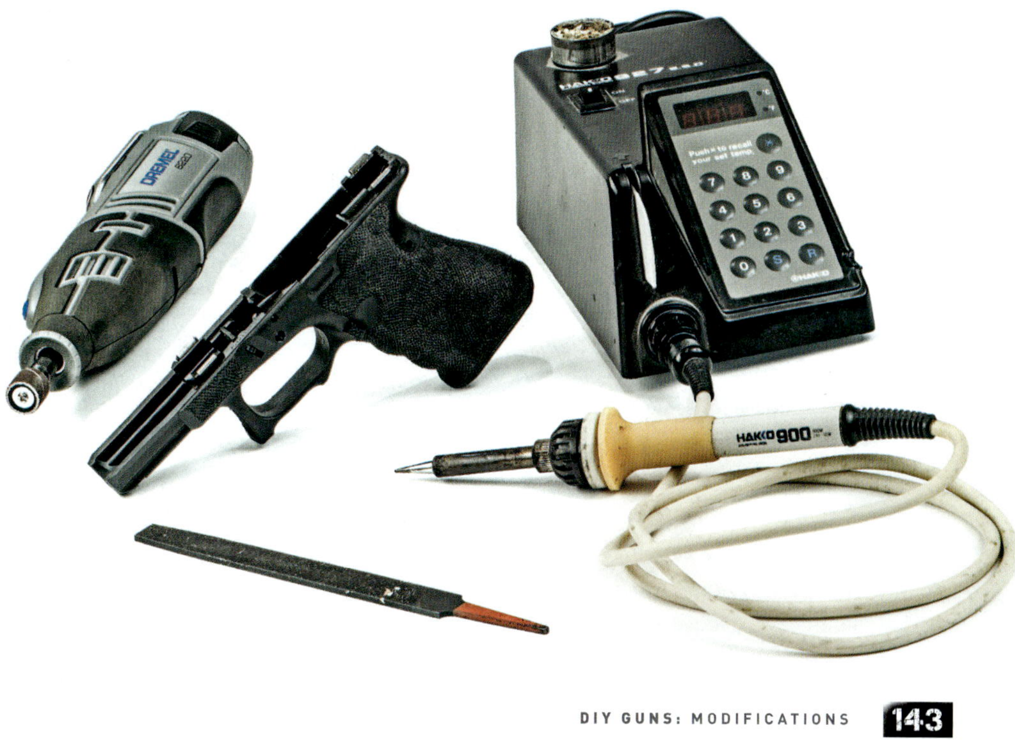

Adjustable temperature soldering stations are more versatile and more forgiving than $10 toys. This once, high-end Hakko soldering station was found on eBay for less than $50.

you probably are. So go slow and check your work often.

After rough blending with the Dremel, you should smooth the frame out by hand with sandpaper. A uniform appearance matters and will show through the stippling if heavy ridges or inconsistencies are left. At this point, some will bust out the buffing wheel and turtle wax. This'll leave any area touched by a tool smooth as glass. It's a consideration for some areas where comfort is key, such as the trigger guard area or beavertail section of the frame.

GET GOING

Once you've prepped the battlefield, it's time to determine the stippling layout. This means deciding where you want your stipple lines to start and end by marking the frame. Use tape to create straight lines, a marker for the curves, and a razor knife to lightly score the outer border of the area to be stippled. If you're going to add a logo or unique design, place that on the frame as well.

Now that all your prep work is done, it's time to melt plastic. Expect this phase to take you several hours and for the smell of burning polymer to surround you like bad cologne. (We recommend using a small fan, and possibly goggles if you'd like, to keep the smoke out of your eyes and respiratory system.) It sounds silly, but don't allow yourself to become distracted. Don't talk on the phone, watch YouTube, check Facebook, or do anything other than stipple. If you need to take a break, that's fine. You don't have to do the whole thing in one sitting. But, it's a good idea to finish the work as soon as possible to retain continuity in the depth and style of the pattern.

If you become bored out of your mind in the process, good. That means you're doing it right! The boredom of burning a million tiny pock marks by hand into a $500 firearm is likely another contributing factor to the cost of professional services. Once finished with the iron, a small round-head routing bit can be used to clean up the edges and blend the stipple work to the untouched frame. Blending isn't easy, so remember to practice this step beforehand, as well.

None of this is rocket science, but there's certainly the possibility for things to go terribly wrong if you don't invest the practice time needed to development the basic skills that'll prevent you from turning a pistol grip into a molten mess. Slow and cautious is the mantra, and it absolutely needs to be followed from start to finish. If you compromise on this point, it'll definitely show in your work.

Stippling's a fun hobby, and a growing one among younger shooters. The benefit of doing your own work is money saved and bragging rights when executed well. If you're a crafty do-it-yourselfer, don't be scared to give stippling a try. ■

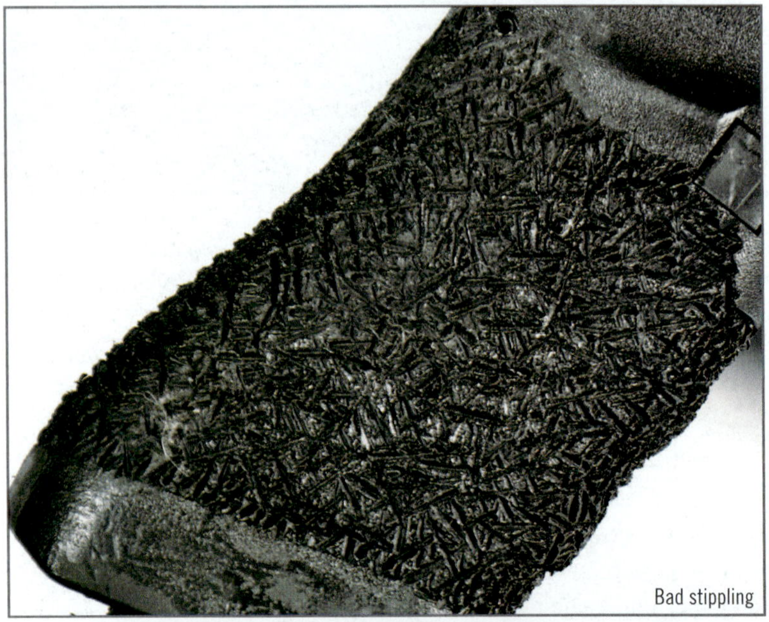

Bad stippling

Good stippling

INK AND IRON

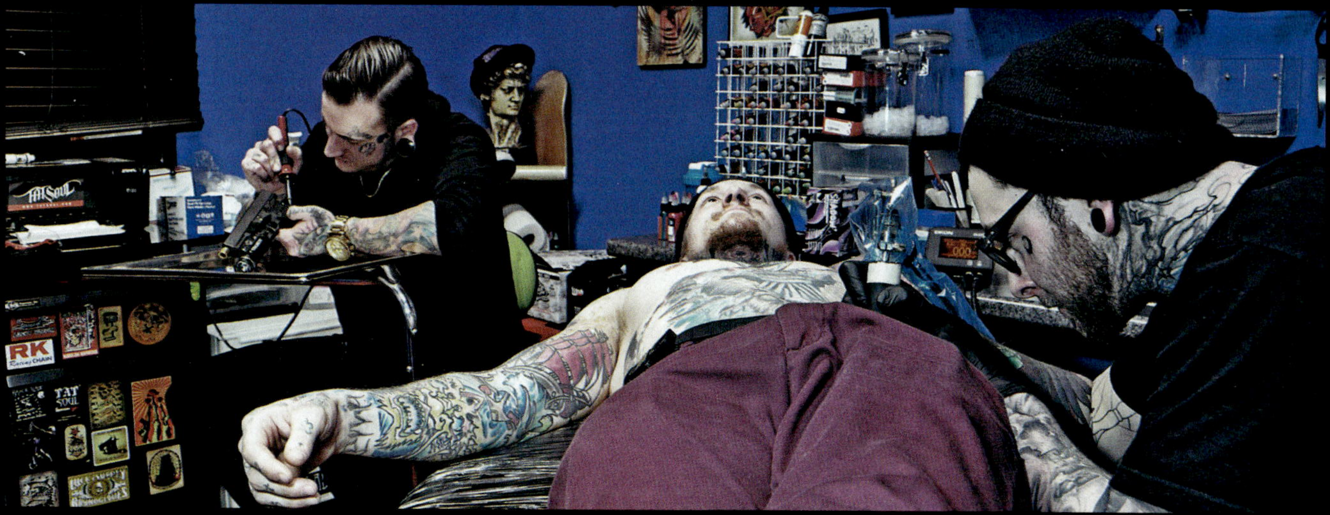

Instead of showing the work of a master stippler, we decided to show the work of a semi-pro so you don't cry when you compare your first, second, and third attempts to the likes of a Boresight, Zev, or Salient grip job. Mike Bortner is an accomplished tattoo artist, fierce supporter of the Second Amendment, and experienced stippler at Black Locust Tattoo studio in York, Pennsylvania. Like many, Mike wanted to try his hand at texturing his own Glock and eased into stippling as a hobby before offering his services to friends and customers. Being a tattoo artist requires attention to detail, a steady hand, and the ability to envision the completed work. The fine line work of the tattoo machine transfers directly over to the soldering iron.

Mike Bortner, Frank Deacon, Ryan Bushemi, and the other artists at this unique studio all hold CCW permits, and carry daily. They're also all into stippling pistols and at various stages of experience. Their tattoo shop is arguably one of the best defended, and perhaps the only one where you can get your pistol stippled as you also get inked.

"Trial and error is how you discover new patterns and styles," Bortner told us as his fellow artist Frank received a Glock logo exactly where his G19 sits in his appendix holster. "All of us here love guns, carry religiously, and appreciate the art that goes into a good stipple job. Just like when you see a bad tattoo, no one wants to be told their work sucks!"

Bortner admitted his work on polymer isn't as good as his work on skin, but both required a lot of time to get it down. "Just like a tattoo artist will continuously improve with time," says Bortner, "so will your stippling work." He says once you get the basics down, you need to push past your comfort zone to really improve.

Deacon agreed, as he's just starting to learn stippling himself. "Mike has given me a lot of pointers," says Deacon, "but ultimately it's up to you to get a feel for your work, and develop an eye for what looks good." They both agreed that stippling your first pistol is as nerve-racking as tattooing your first customer.

CH. 4

Gun Laws

HOLD THE LINE

What Assault Weapon Bans Look Like and How People in Ban States Are Building Legal Guns

By Steven Kuo

It's said that one of the hallmarks of wisdom is learning from past mistakes. If that were the case, then you'd think so-called assault weapon bans would themselves be banished. In 1994, the Public Safety and Recreational Firearms Use Protection Act, known as the Federal Assault Weapons Ban (AWB), was passed, with a subsection banning weapon systems that looked scary to those who didn't know better. It expired in 2004 due to a sunset provision, thus providing ample evidence in follow-up studies that AWBs do nothing to reduce crime.

Yet enemies of liberty persist, ever pushing for more gun control. Biden's 2020 presidential platform stated it in plain English: "Get weapons of war off our streets. Ban the manufacture and sale of assault weapons and high-capacity magazines. Regulate possession of existing assault weapons under the National Firearms Act." And it goes on and on from there. Several states have long-standing AWBs. Make no mistake, by legislation or executive order fiat, more battles over the Second Amendment are coming. Where governments have turned their back on their citizens, such as California, the grassroots fight tirelessly continues in courtrooms, thanks to organizations like the Firearms Policy Coalition, Second Amendment Foundation, and others.

Please note we use the term "assault weapon" frequently in this only because it's used in AWB legislation not because it's a meaningful term.

Let's break down how AWBs are typically structured and how the residents of those states are coping. Let's look at how they have to configure their guns and some clever products enterprising Americans have developed to deal with the laws. Those in occupied land may not realize they can still own reasonably capable guns. For those in free America, look what's happened to your brothers and sisters. Don't let it spread. The will and spirit of Americans can't be diminished — hold the line against infringement of the Second Amendment.

Featureless grips make it harder to work your safety selector, so look for ambidextrous and/or extended selectors like this one from Resurgent Arms.

Juggernaut Tactical's Hellfighter kit locks the magazine release in place until you crack the upper open a bit using a quick-release rear takedown pin. They also offer a lower with their rear pin and a pre-installed Hogue Freedom Fighter mag lock for minimal impact on the exterior of the gun. You can loosen the lock when in free America to use the mag release normally.

COMPONENTS OF AWBs

Despite the expiration of the Federal AWB, DC and a number of states currently have AWBs, including California, Connecticut, Hawaii (pistols only), Maryland, Massachusetts, New Jersey, and New York.

Some typical elements:

» Bans on named firearms: This is like playing gun control whack-a-mole, but most AWBs ban a list of specific guns.

» Bans on particular features: These laws rely heavily on classifying certain functional features of a firearm as particularly heinous. For instance, a semi-automatic rifle or pistol may be classified as an illegal assault weapon if it can accept a detachable magazine and also has one (or two, depending on the state) of the prohibited features listed in the law. This may include items such as a pistol grip, folding or telescoping stock, forward pistol grip, flash suppressor, thumbhole stock, bayonet mount, and grenade/flare launcher. So in areas with a one-feature prohibition, such as California, Connecticut, District of Columbia, and New York, you can't have any prohibited features if your gun also takes detachable magazines. In two-feature areas, you can have just one evil feature in addition to an open magwell. Maryland has a short list with two required to be illegal, so many rifles are legal there, including heavy barreled builds since the Colt HBAR was originally exempted by name. On the other hand, New York lists everything, plus also threaded barrels. Configuring a weapon to avoid being classified as an AW has become known as a "featureless" build.

» Magazine restrictions: All ban states plus Colorado and Vermont have magazine capacity restrictions. Some prohibit possession entirely, while others carve out grandfathered magazines.

» Registration and transfer restrictions: Most ban states required guns classified as assault weapons to be registered prior to a certain date. Some also prohibit transfer of the registered guns, so your children can't even inherit them.

MAGAZINE-FED GUNS

The initial test in these laws is if the gun is semi-automatic and capable of accepting a detachable magazine. To avoid the AW classification, you can either disable semi-automatic operation or fix a 10-round magazine in place (numerous companies offer kits for this) — but both routes severely restrict functionality. But if you do so, you can go to town kitting out your rifle, since it won't be classified as an AW.

California has a specific definition of a fixed magazine — it can't be removed without disassembling the firearm action. So companies have designed products to quickly pop open the upper to release the magazine. For example, Juggernaut Tactical's popular Hellfighter system for ARs locks the mag release until you crack the upper just slightly open

A SELECTION OF FEATURELESS GRIP OPTIONS:
1. Thorsden Customs FRS-15 Gen II
2. Thorsden Customs FRS-15 Gen III
3. Strike Industries AK Simple Featureless Grip
4. Strike Industries Megafin Featureless AR Grip
5. Resurgent Arms Featureless AR-15 Rifle Grip
6. Strike Industries AR Simple Featureless Grip
7. Solar Tactical Kydex Grip Wrap
8. Juggernaut Tactical Featureless Grip

with a quick release rear takedown pin. Then, you can drop the empty mag, while the action is disassembled and no longer functions as a semi-auto.

Mag changes are actually pretty quick with this type of system. However, California law restricts you to 10 rounds in a fixed mag, unlike a featureless build, and malfunctions that require dropping the mag to clear are a pain in the ass. Other companies offer fixed mag systems too, notably one from AR Maglock that solves the malfunction clearance problem.

States like New York, New Jersey, and Hawaii don't define "capable of accepting a detachable magazine" any further, so it's not clear how these devices might play out in court. Tread carefully in these jurisdictions.

Of course, you can opt for guns with traditional rifle stocks like the M1A and Ruger PC carbine, which require little to no effort to make ban-state legal.

Instead of flash hiders, you can run muzzle brakes and compensators that provide the benefit of reduced recoil and muzzle rise, at the expense of blast and concussion.

PISTOL GRIPS

One commonality in gun control mythology is the pistol grip, which apparently makes guns look more evil, like the military's M16. The absurdity is clear when you compare the ubiquitous AR-15, vilified as an unconscionable weapon of war, with the M1A — considered to be completely fine, since it looks more like a traditional hunting rifle, even though the M14 was also military-issue and is chambered in a more powerful caliber.

Several states, such as Massachusetts and New York, only stipulate that a prohibited pistol grip "protrudes conspicuously beneath the action of the weapon" with no further details, while California goes on to define it as a "grip that allows for a pistol-style grasp in which the web of the trigger hand (between the thumb and index finger) can be placed beneath or below the top of the exposed portion of the trigger while firing." On the other hand, Connecticut defines it as "a grip, such as a pistol grip, a thumbhole stock, or other stock, the use of which would allow an individual to grip the weapon, resulting in any finger on the trigger hand in addition to the trigger finger being directly below any portion of the action of the weapon when firing." Naturally, these unhelpful government agencies don't provide official determinations on prospective designs, but with the detailed description outlined in California law, various companies have released so-called featureless grips designed specifically to avoid such categorization.

One of the simplest types are Kydex grip wraps, such as those made by Solar Tactical, which are pieces of Kydex formed around standard pistol grips to prevent the shooter from grasping it normally. If you're handy and have some Kydex and a heat gun, you can make your own. Advantages include consistency in grip angle, simple installation and removal (e.g. when taking your gun to and from free America), support for a wide assortment of rifles, and no need to replace your original grip. Easy removal is also a downside if you're worried about unscrupulous individuals messing with your grip wrap, leaving behind an "illegal" pistol grip.

Next are one-piece fin grips, which install in place of the regular grip. They function similarly to grip wraps without the worry of being easily modified into an illegal state, though they're also not as easy to swap out. They're only available for the most popular platforms; examples shown here for ARs and AKs are from Strike Industries.

Ironically, precision rifle shooters often pull their thumb over to the strong side of their weapon to break a shot. The idea is to isolate the trigger press and minimize movement of the weapon. While it's not ideal for dynamic movement, you can still run a rifle effectively without wrapping your thumb around the grip.

Finally, purpose-built grips contort themselves specifically to comply with the "web of the trigger hand" language while still allowing you to grasp them to retain positive control of the weapon. They feel and look unconventional and aren't as comfortable with their long, raked grips. But they're surprisingly workable, as long as you don't have stubby fingers for the long reach to the trigger or big meat puppets that get pinched. Thorsden Customs offers two versions of a grip/stock assembly that are popular and work well, goofy looks aside. If you want a regular buttstock, Resurgent Arms has a contoured stubby little grip that you can grab on to. Both Thorsden and Resurgent offer re-contoured end plates for more comfort when rubbing on your hand. And Juggernaut pushes further with a featureless grip they describe as the "most aggressive yet." Unlike the others, it's aluminum and noticeably heavier. There are additional options from other companies too. When moving to and from free America, these grips also take more effort to swap out — don't lose that spring and detent.

BUTTSTOCKS

Almost all AWBs require you to run a fixed stock for a featureless configuration. You could simply install your favorite fixed buttstock; just make sure there's enough clearance for your featureless grip. Or secure an adjustable stock in place — modify it yourself or use a stock designed for this purpose. For example, Mission First Tactical makes a version of their BattleLink stock with a roll pin to lock it. You can also insert Strike's AR stock stop into the underside of your receiver extension to cover up all but one hole, so your stock locks in one position.

Not being able to adjust your stock is quite an annoyance, but practically speaking it doesn't detract from your ability to run your gun.

OTHER EVIL FEATURES

Other prohibited features can be excluded without materially impacting your ability to deliver shots on target. When's the last time you affixed a bayonet? Many like forward grips, but would you trade your mag well for one?

It's infuriating though, as these restrictions all infringe on the Second Amendment. They're nonsensical, confusing gun owners on how to remain legal. But so long as the laws stand and law-abiding citizens wish to remain so, there are many ways to configure a compliant rifle. An important point on semantics: Some refer to these products as loopholes; that's a misnomer — if something is compliant with the law, it's not a loophole, it's simply legal.

Unfortunately, gun control like AWBs is more about control than public safety. If you're in free America, stand firm and hold the line against more laws. You don't want to defile your rifles like the ones shown here. If you're behind enemy lines, get yourself more modern sporting rifles. However weird they may look, train with them, and you can still use them effectively to protect yourself, hunt, compete, and have fun. We've met prospective new gun owners in ban states with no idea how silly the AWB laws actually are, but boy do they light up behind an AR. Help educate them and open their eyes. Finally, support the organizations that are fighting hard for the Second Amendment. R

Cut and insert Strike Industries' AR stock stop into the underside of your receiver extension to cover up all but one hole, ensuring that your stock can only lock in one position.

DEATH AND TAXES

A Walk-Through of the ATF eFile Process

By Tom Marshall

With the focus of this issue being on how to navigate the legal side of the firearms world yourself, we figured it only appropriate to spend a little time talking about the legal process to obtain your very own not-so-long gun.

Despite the undisputed sex appeal of SBRs, the extraneous red tape required to get one is more than a little intimidating at times. In a rare stroke of common sense, the ATF has established an electronic system for filing some of their NFA-related forms, including the Form 1 "Application to Make and Register a Firearm." This is the form to use when you are making an SBR yourself. If you're purchasing a factory-configured SBR, you'll need to file a Form 4, which is a separate process. It should be noted that "making" a firearm doesn't require a professional assembly line or QC shop. Slapping a store-bought 10.5-inch AR upper on a lower you already own counts as "making" a firearm. In this case, you'd simply file a Form 1 to get the lower approved to take shorty uppers.

The screenshots you see here were for the registration of the Project JSOK short AK seen in RECOIL issue 44. In that case, we obtained a receiver from Petronov Armament on a standard Form 4473. Then, we e-filed a Form 1 for that receiver before building it out. The e-file process itself is surprisingly user-friendly. The website does run a little slow and sometimes requires multiple clicks on something to get it to work. But it does run. Here's what we had to do to get our Form 1 into the pipeline:

1. Create user profile on the ATF eForms portal

Enter your basic contact info like name, address, and phone number. Then, choose a password. Your username is automatically generated by the system. Once all the required fields are filled in and submitted, you'll receive an email from ATF with your information.

2. Log in to eForms and select the appropriate form to file

In this case, we're filing a Form 1, as an individual. There are "special instructions" for individuals, mostly to do with fingerprints.

3. Fill out the digital version of the Form 1

This consists of several pages' worth of information, including:

Eligibility Questions: Just like the 4473, there's a list of questions about your criminal history and citizenship in order to determine eligibility.

List of Responsible Persons: All of the individuals covered by this application. Current policy requires each person being listed to fill out a questionnaire and submit fingerprints. You must also upload a photo of yourself to include with the application.

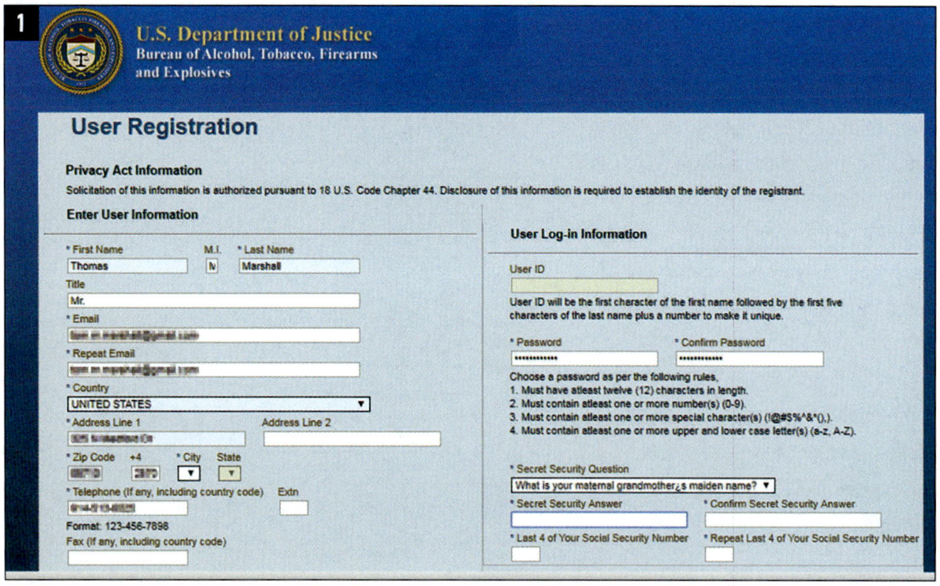

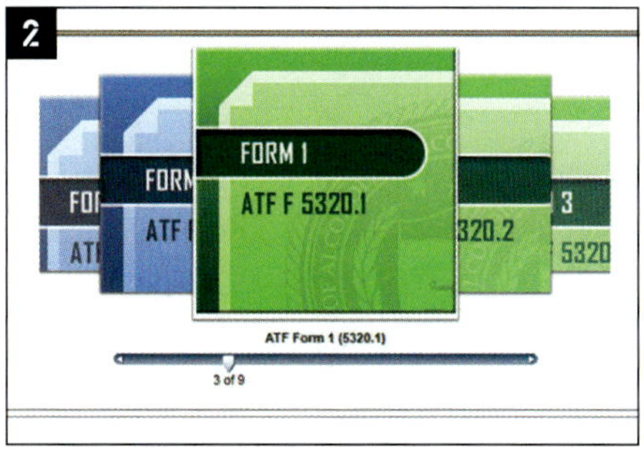

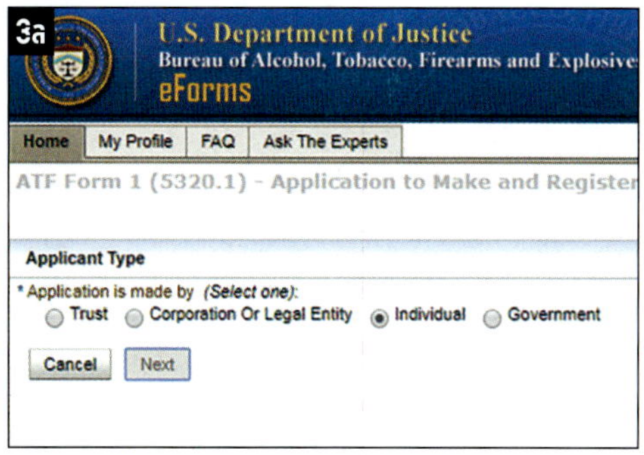

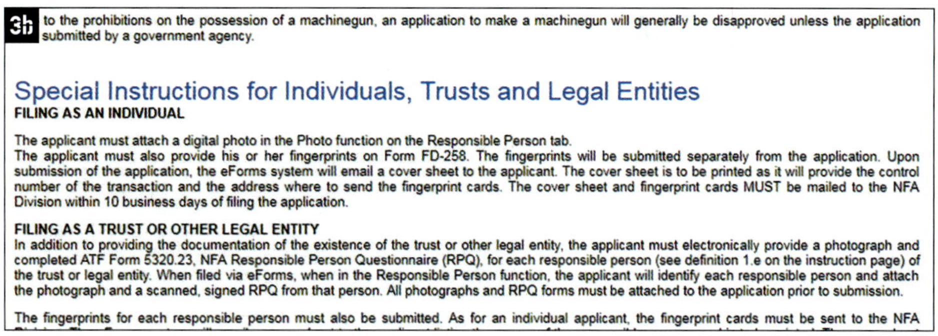

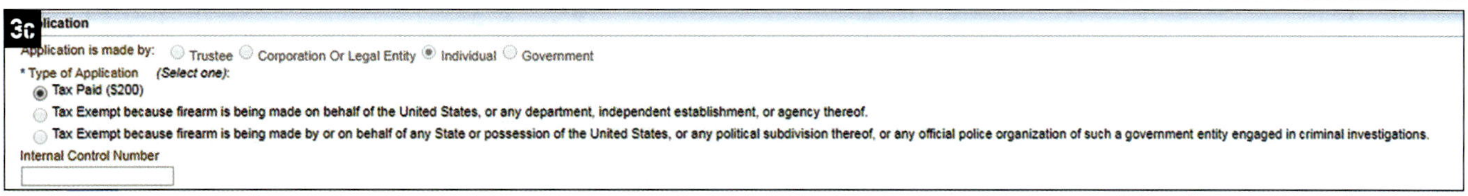

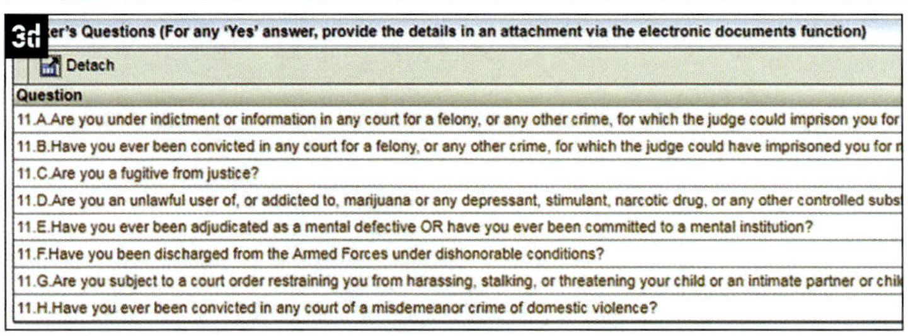

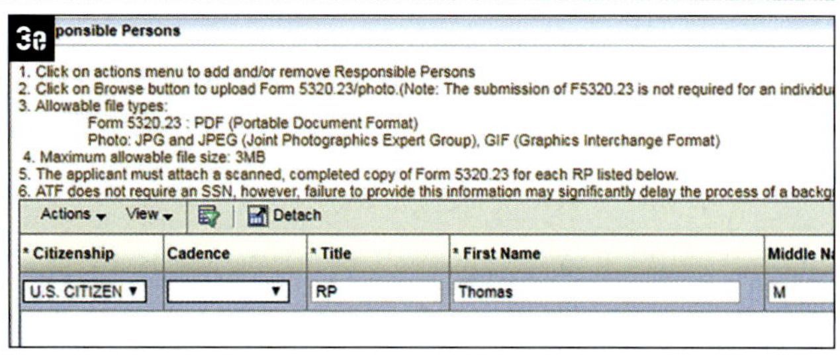

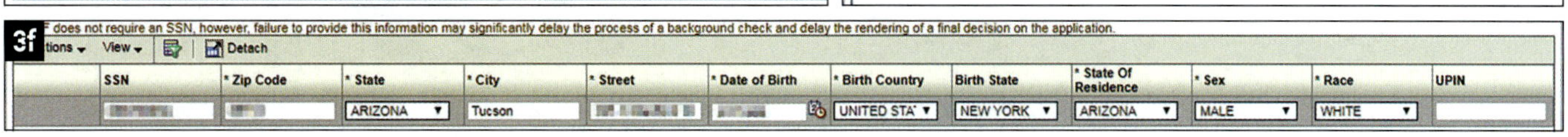

CLEO Notification: There's no longer a requirement for the applicant to get permission from their respective chief law enforcement officer (CLEO), but you do have to include contact information for your CLEO for the purpose of notification. This will typically be your Chief of Police or Sheriff, depending on where you live. It should be noted you're not required to personally contact your CLEO separately to notify them.

Line Item: You're required to select or manually enter the make, model, and description of the firearm you intend to register.

Electronic Documents: Any supporting photos or documents.

4. Pay Tax Stamp

Break out your wallet and fork over your allowance money. Fortunately, the eForms systems allows you to pay with a debit or credit card.

5. Submit Fingerprint Card

Unfortunately, the eForms system cannot accept fingerprint cards. Once you submit your Form 1, the eForms automated system will generate a cover letter for you to mail in with a standard print card. The cover letter includes your application number so that when you mail everything in to NFA Branch, they know who to match your application up to.

6. Wait

The worst part. At time of writing, the author has heard everything from six days to seven months in terms of Form 1 lead times. Once the fee is paid and the paperwork is in, your best bet is to find a few other guns to occupy your time until the ATF can run through your application and grant your tax stamp.

That's it. All in all, the process is a little tedious but, for this first-timer, much less intimidating than initially expected. See the included sidebar for a rundown on engraving requirements. R

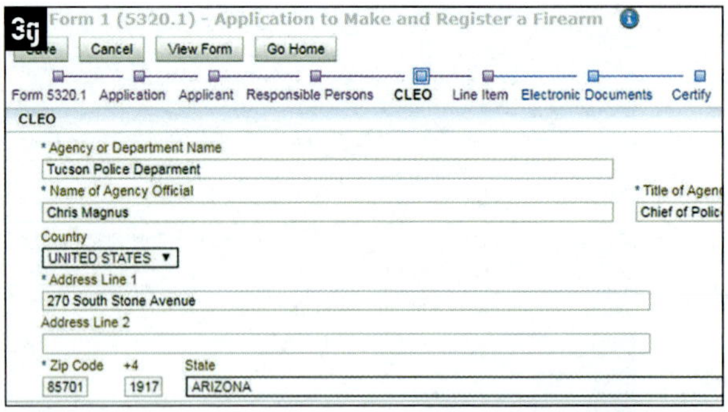

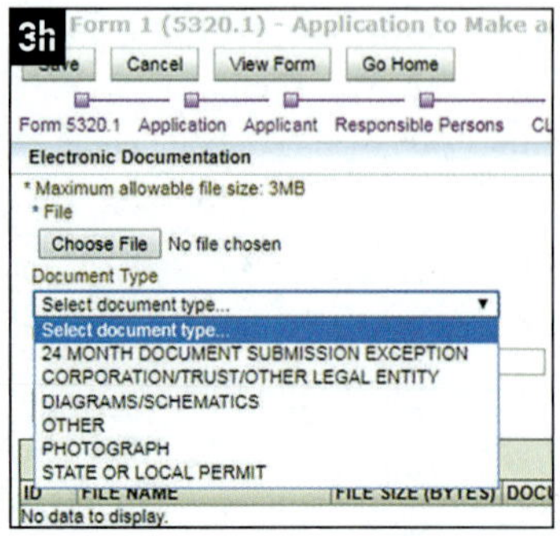

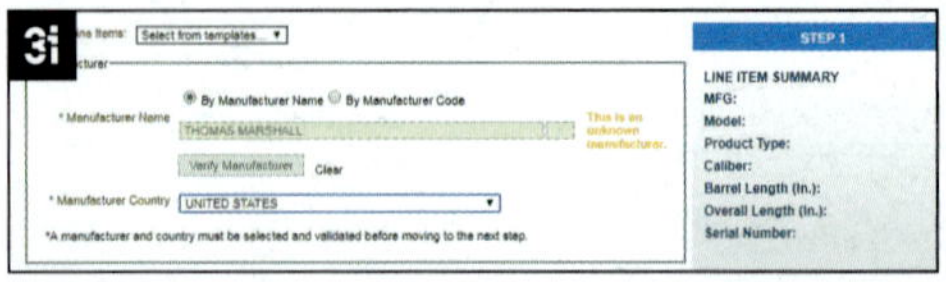

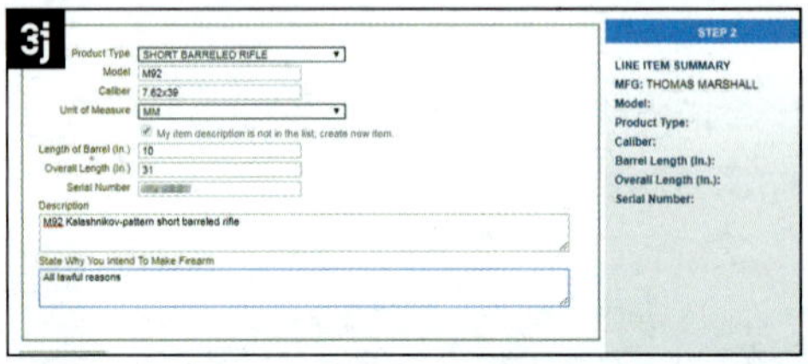

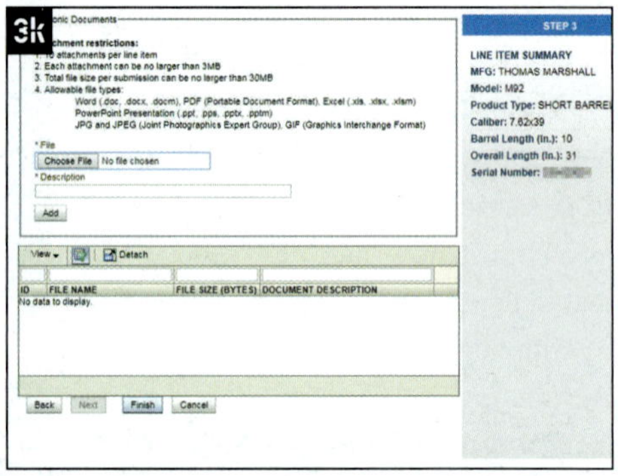

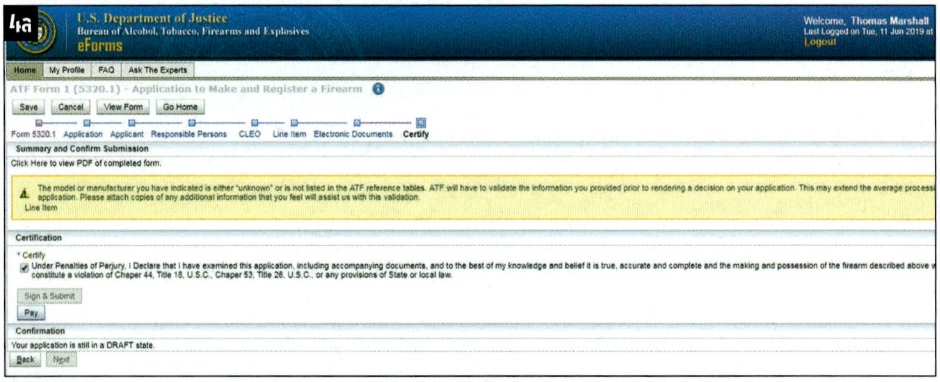

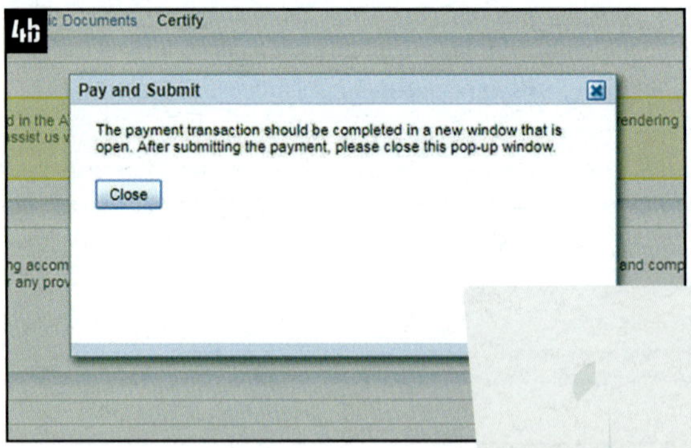

FRIGGIN' LASER BEAMS

The Internet Was Wrong — You Have to Engrave Form 1 Items

By Dave Merrill

WHY

If you Form 1 an NFA item, disregard whatever purported "truth" a low-information trash weblog told you and listen to what we've heard directly from the BATFE: You have to mark any personally manufactured NFA item. The BATFE has gotten so many questions due to this garbage rumor they've even been stamping some forms reminding folks it needs to take place.

MINIMUM REQUIREMENTS

The depth must be no more shallow than 0.003 inch with a text size no smaller than 1/16th of an inch. But we actually recommend going deeper and larger. Here's why: If you only engrave to the minimum size and depth, it's incredibly easy to inadvertently cover the marking with paint or something like Cerakote. And so long as you place it someplace inconspicuous, the marking can be more than twice the depth and size without being anything close to obvious.

WHAT

Per 27 CFR § 479.102:

"In the case of a domestically made firearm, the city and State (or recognized abbreviation thereof) where you as the manufacturer maintain your place of business, or where you, as the maker, made the firearm"

There are two ways to go here. If you're making an NFA item from another established item, such as an AR receiver, all you need is the name of the manufacturer and city of manufacturing. However, if you're making a complete NFA item from scratch, in addition to the above you'll need to add a serial number, caliber, and model.

This means if a trust was used to manufacture the firearm, the trust name must be listed, and if you did it as an individual, your name must be listed. Along with the other information required.

Let's say a standard AR is used to manufacture the firearm (as is common), such a marking would look like this:

RECOIL TRUST
PHOENIX, AZ

If an NFA item is built from scratch, such a marking would look like this:

RECOIL TRUST
PHOENIX, AZ
SN: 0006969
CAL: 7.62 MM
MODEL: YEET CANNON

While serial numbers themselves are required to be engraved/stamped/pressed onto metal (à la the metal tab on a Glock), all other information can be on plastic. So don't freak out that you don't have a piece of metal to engrave on with a Glock.

WHERE

While if you're creating a firearm from scratch all makings need to take place in a conspicuous place on the receiver, with an NFA item it's not so straightforward. You're allowed to mark on the receiver, frame, or barrel.

While marking the barrel itself may seem strange, it doesn't when you leave the AR world and start looking at non-ARs like SCARs, AKs, and Thompsons. Note that it says "conspicuous place," and generally that means a place that can be accessed without requiring tools. The easiest place to hide something on an AR is the upper inside of the trigger guard, but you can't put it someplace underneath the pistol grip.

HOW

Our favorite place to get this performed by a large margin is Tarheel State Firearms. They charge $30 for metal receivers with $15 return shipping for any number of firearms sent. However, if you can find someplace local, even at a higher cost, you don't have to bother with shipping. We hit up CNS Engraving in Powell, Ohio, for the photos you see here.

WHEN

The standard route is to send the receiver at the same time as the tax stamp. While it makes a certain sort of sense to only engrave after an NFA item is approved, it's also a helluva lot easier to replace a receiver lost in the mail before it becomes an official Title II item. If there's a place that can perform the service locally, we recommend rolling with one of those, even at an additional cost, to save on time and shipping.

Basically: Roll the dice when you want to do it.

LAW ABIDING

Yep, 80-Percent Lowers Are 100-Percent Kosher

By RECOIL Staff

As the old axiom goes, knowledge is power. So, empower your arsenal by building your own firearm from an 80-percent lower receiver. Make your own gun? "Isn't that illegal?" you might ask. Well, actually, it's not. According to federal law, it's completely legal to make your own firearm.

Under the Gun Control Act of 1968 (GCA), an individual can make a firearm for personal use. Surprised? If you are, you need to read on and find out more about what you can and can't do with an 80-percent lower. Because, as Duke taught us in the '80s, knowing is half the battle.

DISCLAIMER
While we always strive to report the most accurate information, laws can change and vary from state to state. Always double-check the laws in your region and seek out professional legal counsel when necessary.

WHAT'S AN 80-PERCENT LOWER?

Think of it as an incomplete sculpture that still needs some stone to be chipped away. Essentially, it's a partially formed lower receiver that requires the owner to tool in the missing 20 percent — hence the name. This usually means you'll have to drill or machine out the fire-control pocket, as well as drill holes for the trigger, safety selector, hammer pin, and trigger pin. Once you've done that, you have a working lower that you can use to assemble a firearm.

HOW IS THIS LEGAL?

Because the federal government says so. The Bureau of Alcohol, Tobacco, Firearms and Explosives (ATF) has determined that 80-percent lowers are not firearms. In the eyes of the law, it's just a blank piece of material, for all intents and purposes.

SO AN 80-PERCENT LOWER IS NOT A FIREARM, BUT CAN I BUILD A FIREARM FROM IT?

Yep — at least from the federal government's point of view. (Check your state laws, as several have passed restrictive legislation.) The GCA states that an unlicensed individual can make a firearm as defined in the GCA for personal use, but not for sale or distribution. That means it's kosher to complete an 80-percent lower and assemble a working firearm from it, so long as:

- **It's for personal use:** No gifting, transferring, or willing it to relatives.
- **You build it yourself:** You can't bribe Tim "The Tool Man" Taylor to do the dirty work for you.
- **You're not prohibited from owning a firearm:** Sorry, felons, but there's no loophole here for you to exploit. If you're already prohibited from owning a regular firearm, the same penalties apply to possessing one that you made yourself.

DO I NEED TO PUT A SERIAL NUMBER ON IT?

Nope. Serial numbers are only required on guns sold and distributed by licensed manufacturers. "Individuals manufacturing sporting-type firearms for their own use need not hold federal firearms licenses," says the ATF, though officials there suggest that you "at least identify the firearm with a serial number as a safeguard in the event that the firearm is lost or stolen."

Another reason why you might want a unique marking on your lower is for future service or repair. It'll be a lot easier for a gunsmith or coating company to keep track of your receiver, and some may be uncomfortable working on an unmarked lower.

CAN I SELL MY 80-PERCENT LOWER?

Sure. If it's still in 80-percent form, then it's not a firearm and you're free to sell it just like you would smartphones or silver bullion — or like that vintage Walkman you had hoped would become a collector's item...but didn't.

CAN I SELL MY FINISHED 80-PERCENT LOWER?

This question has been a source of much haranguing on the Interwebs. The ATF says that "a firearm may be made by a non-licensee provided it is not for sale and the maker is not prohibited from possessing firearms." The key phrase being "provided it is not for sale." The question here is not whether you can sell it, but whether you had the *intent* to sell it before finishing it.

If a court can be persuaded that you did indeed have that intent, then you can kiss your liberty goodbye for a while. While this may be a tough one to prove, we are not about to add to the body of case law on this subject, so none of our personally made firearms will ever end up on GunBroker. However, if you do decide to sell a homemade lower at some point, and if your state allows it, you will then need to engrave on the receiver your name (or recognized abbreviation), a serial number, the caliber of the firearm, and the city and state where you made it.

DOES THIS SALE NEED TO GO THROUGH AN FFL?

No, not under federal law, so long as you're selling it to someone who lives in your state. "When a transaction takes place between private (unlicensed) persons who reside in the same state, the GCA does not require any recordkeeping," the ATF says. "It is not necessary under federal law for a federal firearms licensee (FFL) to assist in the sale or transfer when the buyer and seller are 'same-state' residents." Your state may have additional restrictions or requirements, so make sure you research them.

CAN I USE MY 80-PERCENT LOWER TO BUILD A FULL-AUTO, SHORT-BARRELED RIFLE WITHOUT ANY BIG BROTHER INTERFERENCE?

Negative. All other firearm laws — i.e. the National Firearms Act and regulations in your state, county, and municipality — still apply.

Making a full-auto firearm can score you an unpaid vacation to Club Fed. If you want to make an SBR from your 80-percent receiver, you'll need to get the ATF's approval and pay the $200 tax stamp for building an NFA gun. Always check the laws in your area first.

LOOKING FOR YOUR OWN 80-PERCENT LOWER?

Quick, go to our buyer's guide in RECOIL issue 12 for more.

HOW DO I FINISH AN 80-PERCENT LOWER?

Funny you should ask. Check out this issue's "DIY" elsewhere in this book for a step-by-step walkthrough on how to complete your own.

SOURCES

Bureau of Alcohol, Tobacco, Firearms, and Explosives
www.atf.gov

United States Code (Title 18, Chapter 44)
uscode.house.gov

NFA BASICS

Understanding the National Firearms Act

By Steve Kuo

Photos by Jeff Chen

If you've been around guns for a little while, you've probably heard of the NFA. But, what does NFA really stand for? No Fun Allowed? No F***ing Around? Since it regulates silencers, it can't be Noise Free America. It's actually the National Firearms Act, and in addition to silencers, it tightly regulates machine guns, short-barreled rifles and shotguns, so-called destructive devices, and a whole host of other types of firearms.

Now for the legal disclaimers we got from the guys at the office with expensive suits, fancy German cars, and a hint of self-loathing: This article is intended as a primer on the NFA to give you a basic understanding, and it should not be construed as legal advice. You should seek competent counsel and understand state and local restrictions before proceeding with acquiring any NFA items.

Like most firearms-related legislation, the National Firearms Act (NFA) was enacted in response to high-profile criminal activity at that time – namely gangland crimes like the St. Valentine's Day Massacre – and intended to curtail possession of "gangster weapons," such as machine guns and short-barreled shotguns. Unable to ban them outright, the NFA achieved its objectives via federal registration procedures and associated taxes that were intended to make transactions in NFA firearms totally cost prohibitive. The typical $200 fee was a hefty sum back when the NFA was passed in 1934 – in today's dollars, that $200 fee would represent about $4,000! But, the $200 fee has not changed since 1934, so these days it presents less of an obstacle to law-abiding citizens. A little fine print for students of the law out there: The original NFA of 1934 was essentially nullified by a Supreme Court ruling in 1968 and subsequently amended that same year to the NFA currently in effect – known as Title II of the Gun Control Act of 1968.

1 AA-12 SLEDGEHAMMER
Boasting an incredible 300 rounds per minute rate of fire, this fully automatic 12 gauge shotgun has an effective range of 100 meters with slug ammunition. Although the AA-12 can be classified as a Destructive Device under NFA classifications, the fact that it is capable of firing fully automatic pushes it into the Machine Gun class.

2 M60 MACHINE GUN
Introduced in 1957, the M60 was made famous during the war in Vietnam. The 7.62x51mm caliber belt-fed fully automatic machine gun fires from an open bolt position is gas-operated and air-cooled. While many think it is impossible for civilians to own such a gun, they are mistaken. Dependant on location, the transfer of machine guns to civilians is legal, but can be very expensive.

MACHINE GUNS

Machine guns are defined as any firearm that discharges more than one round per trigger press, encompassing both fully automatic and burst fire modes. The receiver itself is considered the regulated firearm, and any parts that are designed to convert a weapon to a machine gun are also covered, such as a Drop in Auto Sear for an AR15. In 1986, the Firearm Owner's Protection Act prohibited ordinary citizens from transferring or possessing machine guns, except those already in legal circulation at that time – essentially capping the supply of civilian-legal machine guns and subsequently driving up their value. Select-fire weapons manufactured after this date can only be sold to government agencies and law enforcement or held by licensed manufacturers and dealers, commonly referred to as "dealer samples."

SHORT BARREL RIFLES AND SHOTGUNS

Short barrel rifles (SBR) have a buttstock and a rifled barrel less than 16 inches long or an overall length less than 26 inches. The overall length is measured with any folding or collapsing buttstock extended. Similarly, short barrel shotguns (SBS) have a buttstock and a smoothbore barrel less than 18 inches long or an overall length less than 26 inches. These types of firearms could be manufactured in this configuration or assembled by a private owner.

SILENCERS/SOUND SUPPRESSORS

Any device designed to silence, muffle, or diminish the report of a portable firearm is considered a silencer and is regulated under the NFA. Many refer to them as suppressors, but the NFA terms them "silencers," even if they don't actually result in complete silence. This also includes associated parts, not just completely assembled units. It does not, however, include devices designed to silence chattering spouses – but unfortunately nothing really can, so the NFA can't help you there.

3

4

5

3 SIG SAUER SIG551-2 SP SWAT (SIG551)
The SIG551 is a semi-automatic gas piston carbine that is chambered in 5.56x45mm. This example is fitted with a 14.3 inch length barrel, which puts it squarely into short barrel rifle territory.

4 M26 MODULAR ACCESSORY SHOTGUN SYSTEM (MASS)
In this configuration, the 12 gauge MASS has an overall length of just under 20 inches with its stock extended and a 7.75 inch length barrel. Because it is fitted with a buttstock, this shotgun does not qualify as an AOW.

5 RUGER MK.II [INTEGRALLY SUPPRESSED]
This integrally suppressed version of Ruger's Mk.II .22LR pistol features a barrel with a sound suppressor built around it and is whisper quiet. This particular pistol was custom built by S&H Arms.

6 CUSTOM ASSEMBLED 300BLK SBR
This 300 AAC Blackout PDW has a compact 9 inch barrel, qualifying it as a Short Barrel Rifle.

MEASURING BARREL LENGTH

The official ATF procedure for measuring barrel length is to measure from the closed bolt or breech face to the farthest end of the barrel or permanently attached muzzle device. Close the bolt, insert a dowel all the way into the barrel, then mark and measure it. Per the ATF, permanent methods of attaching muzzle devices include full-fusion gas or electric steel-seam welding, high-temperature (1,100°F) silver soldering, or blind pinning with the pin head welded over.

ASSEMBLING A SHORT BARREL RIFLE

If you want to build a short barrel rifle as a private individual, you must submit the paperwork to register it as an NFA firearm before you acquire the necessary parts. So, for instance, if you have a typical AR and you wish to buy a new 11-inch upper for it, you should register the AR as an SBR before you even take possession of the new upper to remain in full compliance with the law, unless you own an AR pistol lower. Some may debate this point, but *better safe than sorry* is always a good plan when dealing with the NFA. Once this is done, you could use multiple different short uppers on the same registered lower.

6

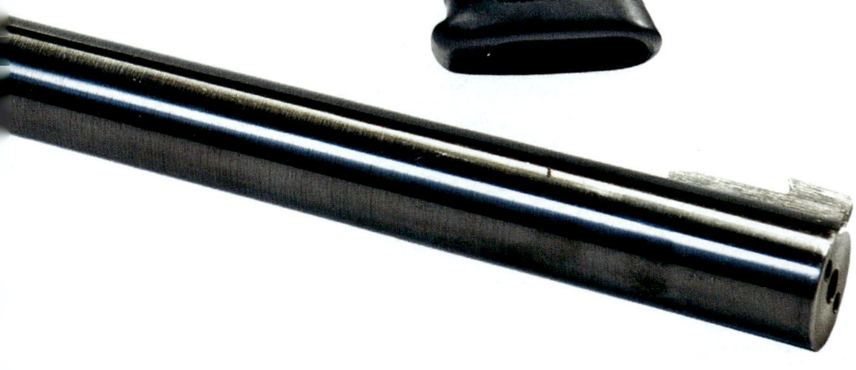

DESTRUCTIVE DEVICES

The term destructive device (DD) refers primarily to explosive devices such as grenades, bombs, explosive missiles, and large-bore weapons. The latter includes any firearm with a bore more than a half inch, except for most shotguns and some large hunting calibers, which are deemed to be for "sporting purposes," and covers such weapons as rocket launchers, mortars, and cannons. It should be noted that unlike the other NFA definitions, large-bore weapons using non-explosive propellants, such as compressed air, are included.

7

THE FORTY MIKE MIKE

Many readers may be familiar with the venerable M203 40mm grenade launcher, which clings beneath many a serviceman's M16. If you've always wanted one of your own as a civilian, it is possible by acquiring one pursuant to the NFA as a destructive device (subject to local laws). Actual explosive rounds for it, with more than ¼ ounce of explosive, however, would each be considered a separate DD and each incur the $200 tax, not to mention potential storage requirements. A variety of other types of non-explosive projectiles not covered by the NFA are available. For those who cannot or do not wish to follow the DD route, several companies manufacture 37mm launchers that are not regulated and can launch flares, smoke, and other signaling rounds. You can register such a launcher as a DD if you wish, but if you haven't, it is critical that you do not have any cartridges that might be considered anti-personnel in nature, even wood/rubber pellets or bean bags — or else the mere possession along with your launcher will be classified as a DD. Also, your signaling rounds must not contain more than ¼ ounce of explosive material in order to avoid DD status.

TRUSTS AND CORPORATIONS

Rather than owning them personally, some folks use trusts or corporations to acquire NFA items. This is permitted by the NFA, and was often used in the past by those whose Chief Law Enforcement Officer (CLEO) wouldn't sign off on NFA items. However, the CLEO requirement was eliminated in 2016. Additionally, fingerprints or photographs used to be unnecessary for trusts, but that's also no longer the case. These were two of the most popular benefits of using an NFA trust prior to 2016. There are other potential pros and cons to using this method, which is best discussed with competent counsel that specializes in firearms law. The criminal penalties for violating the NFA are quite severe, so do it right!

7 MILKOR MK14. MOD 0 LAUNCHER, GRENADE

Essentially a giant revolver, the MK14 can lob six rounds of 40mm grenades well beyond 400 meters per load. A civilian version designated the M32 is classified as a Destructive Device because with a 40mm bore, it is considered a large-bore weapon.

ANY OTHER WEAPON

The "any other weapon" (AOW) designation is a bucket containing various types of devices capable of being concealed on a person that can discharge a projectile, such as disguised guns, pen guns, cane guns, and so on, smoothbore pistols, short barrel shotguns manufactured without a buttstock, combination shotgun/rifle barrels between 12 and 18 inches in length — how about that for specificity! — and pistols with a second vertical grip.

8 SERBU SUPER SHORTY
This Super Shorty is built on a Mossberg Maverick Model 88 12 gauge shotgun. Because it doesn't and never was fitted with a buttstock, this pistol-gripped 16.5-inch overall shotgun is considered an AOW. Its 6.5-inch barrel and folding foregrip make this a popular gun.

DEALING WITH DEALERS

Private individuals can register NFA items (subject to state and local laws); you don't have to be an NFA dealer. However, many manufacturers only distribute to licensed dealers who are able to deal in NFA items. An importer, manufacturer, or dealer in NFA firearms must pay a special (occupational) tax each year, hence the commonly heard term "SOT." Then, SOTs can make and transfer NFA firearms among themselves without paying additional taxes, but a subsequent transfer to you would incur an NFA transfer tax. You may have heard of Class 1, 2, or 3 — this refers to the special occupational taxes paid to be an importer, manufacturer, or dealer, respectively, of NFA firearms. Thus, a so-called Class 3 dealer is typically what you would look for to shop for NFA items.

8

9

9 ITHACA AUTO & BURGLAR "MODEL A"
This double-barreled, smooth-bore pistol was manufactured in different variations between 1921 and 1934. It fires 20 gauge shotgun shells and has a 10- or 12.2-inch barrel, depending on model (10-inch shown here). Originally intended for self defense, this gun became subject to NFA laws enacted over a decade after its introduction to the public. The 1934 NFA laws effectively killed the production of this gun, which has now become a prized collector's piece.

At the federal level, law-abiding private individuals may own one of these wondrous devices, but state and local laws may impose additional restrictions. Any NFA item has to be registered with the Bureau of Alcohol, Tobacco, Firearms, and Explosives (ATF), and generally speaking, to acquire one you must seek approval from the ATF, submit to a detailed background check, and pay a tax. You must be at least 21 years old, a U.S. resident, and have no felony record. The process can take a long time, depending on backlogs at the ATF. You must also get permission from the ATF whenever you move it across state lines; note this isn't required to transport silencers or AOWs. To privately make any type of NFA firearm, such as installing an 11-inch barrel on your AR to create an SBR, the making tax is $200. To transfer ownership of an NFA firearm incurs a tax of $200, except for AOWs, which is just $5. Until and unless you transfer the item again, there are no other taxes to be paid. Approved paperwork returned from the ATF bears a stamp confirming that the tax has been paid, hence the commonly used term "tax stamp" when referring to registered NFA items.

So, the NFA requires you to jump through some hoops, but it's a straightforward and manageable process and the reward can be well worth it. Whether for the joy of rockin'-and-rollin' full auto fire, while your wallet is rapidly depletin', the handling benefits of an SBR, or saving your eardrums with a suppressor, you should definitely consider taking a closer look. **R**

URL
www.atf.gov/firearms/nfa/

CH. 5

Gun Maintenance

THE GLUE THAT BINDS US

Everything You Ever Wanted to Know About Threads and Threadlocker

By John Brooks

There's a saying in Formula One that a 15-cent screw can ruin a race just as fast as a blown motor. The little things matter, especially in a 13:1 compression engine spinning at 18,000 rpm. Guns don't exactly run on gentle unicorn toots either; they're an instrument of intense heat, pressure, and extreme acceleration in an area no bigger than a Pringles can. If your life depends on your firearm, or a major match is looming on the horizon, you'll want to prep your gun just like any F1 pit crew would, and that means securing all those 15-cent screws.

I was recently at a pistol match with my Sig XCarry, my slide freshly hacked up to accommodate an old Insight MRDS I had bouncing around the bottom of the parts bin. The fitment of the sight was rock solid. However, in the middle of my second stage — *BOOM* — no dot. I did my best to salvage the remainder of the stage, but blindly shooting your way through a plate rack and a Texas star is a tricky business. I thought maybe my early GWOT-era dot had sh*t the bed, but after a secondary inspection, it seemed the two screws that kept the battery door shut had simply come loose — a small oversight that cost me a decent stage score. Had it been a defensive shoot, it could've cost me my life. I thought to myself, "What a boner move! A dab of blue Loctite would've prevented all that." Any of that sound familiar? Well, then read on — I've got some primo nuggets of knowledge for your ass-pack of facts.

THREAD BASICS

Before we jump straight into the ins and outs of threadlocker, let's do a quick review of threads and screws. There are various types of fasteners, but the most prevalent of these used in guns would be those with an isosceles "V" thread form, between 6-32 to ¼-28, and come with either a head or as a set screw (a screw with no head and a

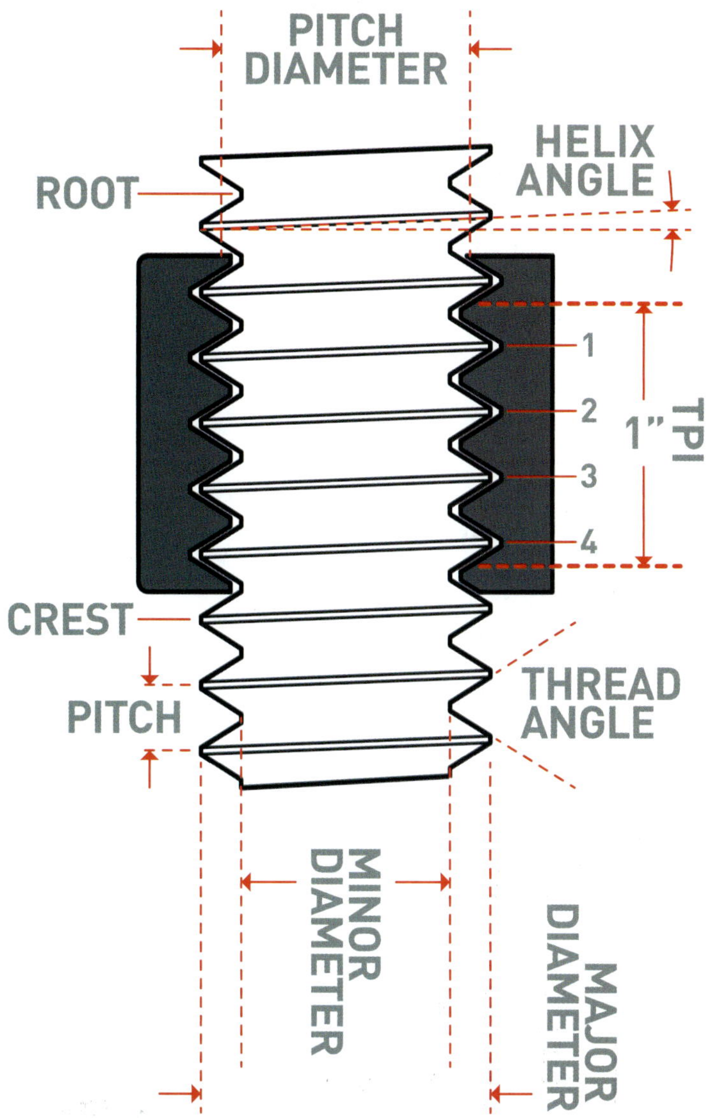

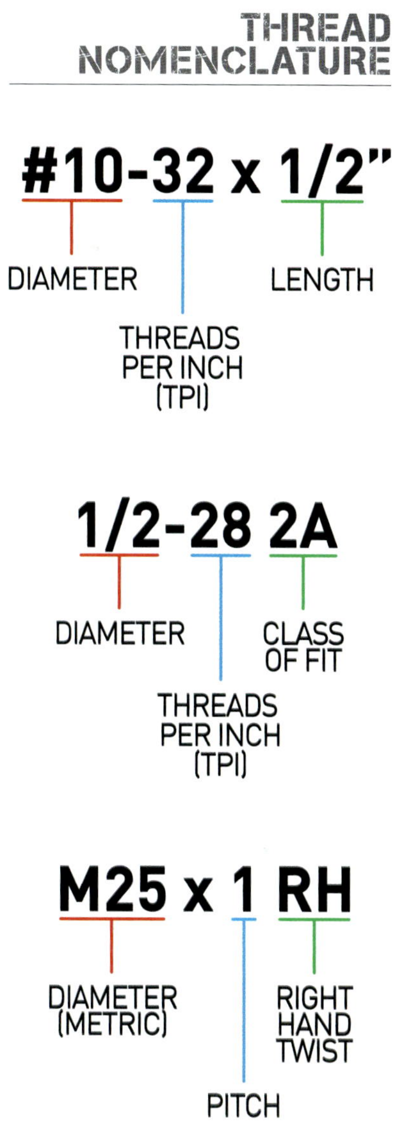

machined or broached drive integral to the body). If you're not quite sure what all the numbers mean, let's decode it.

For Imperial threads, the first number is the theoretical major diameter of the thread body. I say *theoretical* because the sharp points of V threads are truncated to varying degrees in order to allow clearance between the tip, or crest of the male threads, and the root, or bottom of female threads. The second number is the threads per inch, or TPI. That is, counting from crest to crest, how many rotations are formed about the body in one linear inch.

Metric threads diverge slightly, as the second number is the pitch of the threads, or the distance between one crest to the next. You might also see some thread callouts with a tolerance attached in certain examples. These classes of fit range from loose to tight as a 1, 2, or 3 and A or B (External or Internal). The thread helixes in either a clockwise or counterclockwise direction denoting a right- or left-hand thread. The existence or the absence of the head on the screw distinguishes which direction torque is being applied. Headed screws are designed to fasten a part externally to another, while set screws fasten one object inside another. Prime examples of this would be attaching a scope base to the top of a bolt-action receiver with headed screws or anchoring a gas block to an AR barrel.

Set screws exert compressive force while headed screws stretch, much like a spring, to exert linear tension. Torque is the rotational force we apply to a screw based on its design parameters to achieve the desired effect. The optimal torque setting directly correlates to the material and finish of the screw — the diameter and the pitch. While it's common knowledge that too little torque can be detrimental for obvious reasons, did you know that too much torque can be even worse?

Screws on rifles and threadlocker go hand in hand — just make sure you use it properly.

Stretching a screw past its nominal yield strength (commonly about 75 percent) can weaken the integrity of the screw body, reducing the clamping strength over time and exposing it to repetitive lateral shear forces, which, coincidentally, occur quite often perpendicular to screws in firearm applications. So the next time you think it best to white knuckle a screw with your hex wrench, think twice. "Well, how much torque should I be using then, John?" you might ask. Before we delve into that, first and foremost, get an adjustable torque wrench in inch-pounds and a set of bit drivers for every screw head type you plan on using.

Torque wrenches are relatively inexpensive, so if you don't have one, there are plenty of brands and models available. On the topic of specific torque values, I'll always consult the manufacturer that specified the use for that particular fastener. Their design intent might differ from the typical torque value you might expect from a certain screw size, such as securing a polymer body MRDS to a pistol. Now, if you're some poor degenerate with no torque wrench, outside the realm of any form of internet or mail-order system and just need to figure it out, I find a prudent practice is to thread the screw in with two fingers until resistance is met, then go another quarter to half of a turn and evaluate for tightness. Then, pray to the machine gods for forgiveness, heathen.

ON THREADLOCKER

You've probably started to wonder to yourself, "If I'm applying the proper torque to a screw, what's the purpose of threadlocker?" Short answer: insurance.

There are a few factors that contribute to a screw coming loose, and it just so happens most of these occur with firearms. Shock, vibration, and thermal expansion are all colluding to undo your meticulous efforts with every round fired.

I wouldn't say the best way of putting threadlocker on screws is necessarily tribal knowledge, but from where I'm sitting, it sure seems elusive to most. I can't tell you exactly how many thousands of guns I've handled or disassembled, but I can confidently comment on the American firearm owner's love affair with the subtle art of way overdoing it. What we should strive for in a gun build is the careful balance of assembled rigidity and ease of disassembly. At some point, that screw you thread in might need to come out, so it'd be great if we could accomplish that without the associated blood, sweat, and tears.

So how do we achieve this delicate equanimity as gun builders? Enter Loctite Threadlocker. Manufactured by the Henkel Chemical Company, Loctite produces an array of phenomenal products that are surprisingly more extensive than the few options you likely see in the glue aisle at your local home improvement store. These single component, anaerobic adhesives are comprised of liquid resins that cure to a tough solid when exposed to metal ions in the absence of air. The anaerobic cure mechanism allows adhesive to flow along threads to fill the grooves of fasteners without curing prematurely. Any excess threadlocker that overflows the threads during application will remain liquid and can be easily wiped away.

As Loctite cures, it forms polymer chains that find their way into all the empty space between threads. Thread friction increases with the application of threadlockers as they bond to imperfections in metal's surface. A fully cured application of threadlocker prevents lateral movement and protects the joint from corrosion that can result from moisture, gases, and fluids. Using threadlocker, much like bedding a rifle, provides 100-percent contact between metal surfaces, whereas a typical nut and bolt assembly may have as little as 15-percent metal-to-metal contact. Available in different strengths for different applications, threadlockers will hold critical clamp load pressures even in the most severe environments.

WHAT WE SHOULD STRIVE FOR IN A GUN BUILD IS THE CAREFUL BALANCE OF ASSEMBLED RIGIDITY AND EASE OF DISASSEMBLY.

A soldering iron can be used to break down threadlocker on individual screws without spreading the heat around too much.

SELECTION

While Loctite isn't the only threadlocking compound available, it's the most widely used in all the shops I've seen. For the purpose of simplicity, I chose to focus on that product. If you prefer to use Permatex or another brand I'm unaware of, be sure to follow manufacturer guidelines. Loctite Threadlockers are available in low-strength formulations (Purple) for easy removal, medium-strength grades (Blue and Green) that can be removed using common hand tools, and high-strength formulations (Red) that offer the highest holding abilities. While none of these could be considered truly permanent, Loctite Red requires direct exposure to temperatures of up to 450 degrees F to remove, which might limit its use near anything that tends to melt or burst into flames.

Loctite advises that the strength and viscosity of the adhesive needed for an application are directly related to the size of the fastener used. Purple is used on screws up to ¼ inch in diameter, for tiny screws or ones that might need future adjustment. Blue is typically meant for fasteners up to ¾ inch in diameter, but I've found through experience that it's a good general-use threadlocker for just about any

Automotive stores will typically have more options than a big-box hardware store.

Improper application or selection of threadlocker can lead to situations like this: three of four screws broken off in this slide!

threads you'd typically have in a gun. Red Loctite is ideally used on larger thread sizes up to an inch, but since we typically deal in the smaller sizes, I find it best to reserve its use to only the most extreme examples. How extreme you might be wondering? All of the fasteners in the miniguns we build at my day job get Red. It doesn't get more extreme than 3,000 rounds per minute, and I haven't heard any first-hand account of any screw or bolt coming loose.

During my research for this article, I came across an interesting bit of information that perfectly explained some frustration from my early time as a gunsmith that you may have experienced as well. Apparently, the type of metal used on the fastener is critical to the performance of the adhesive. If the surfaces to be bonded are made of two inactive metals such as stainless steel, magnesium, black oxide, anodized aluminum, or passivated titanium, a primer may be required to facilitate cure. I'd wager a guess that, like me, if you've ever been underwhelmed by Loctite, this might be the root cause.

APPLICATION

While some of the newer offerings of Loctite offer some oil and solvent resistance, I find that thoroughly degreasing all the threads to be bonded using a strong solvent is an absolute must. My favored method is removing all the screws I needed to coat at once and dipping them in my small steel bench ramekin full of MEK. I let them sit for a few minutes, then remove and blow dry with compressed air. I strongly advise the use of rubber gloves and that this be done in a well-ventilated area. This stuff is super gnarly and will for sure give you face cancer.

For the most effective bond, the adhesive must coat the complete length of the thread engagement area. Loctite is a thixotropic formula, which means it is thick and viscous under static conditions, and as it spreads along a surface under capillary action, it becomes thinner and will seem almost self-guiding. You'll find a modest drop on a section of threads will find its way around the circumference and "creep" up and down the helix. When threaded in, this action compounds and the adhesive will work its way up and down the length of the threads quite easily.

For large parts, applying to both faces provides the most adequate adhesive application as well as for blind-hole assemblies such as cap screws — threadlocker should be applied to both the bolt and the mating threads. If applied only to the screw going into a blind hole, air pressure tends to force the liquid threadlocker out of the hole as the bolt is torqued down. The anaerobic process hardens in minutes at room temperature and fully cures within 24 hours.

When determining the amount to use, it's important to remember that less is more, and dipping the screw to the head will not make it "more better." The space between threads allowed by tolerances is actually quite a bit smaller than you think, and the location that the adhesive is applied is much more important than the amount. If you're trying to determine which level of adhesion fits your specific needs, start on the low end and work your way up gradually and try experimenting a bit if you have the time and resources.

If it comes time to remove a screw that has been secured using Loctite, first ensure you're using the best-fitting bit for the job. Any play between your screwdriver bit and screw allows space for deformation and/or breakage of one or the other under the increased torque load you're likely to encounter with the use of threadlocking adhesive. If it's still being stubborn, I alternate between using a heat source, and a soak of either solvent or Kroil. If you're going to resort to heat, don't go straight for the MAPP gas torch. Try to keep heat localized around the female threads — start with a heat gun and gradually work up to higher heat. Some advocate using a soldering iron to heat the screw directly. You'd be surprised what the thermal expansion coefficient in some metals are.

ROCKSETT

Contrary to popular belief, some parts of a gun can get really hot. I know; shocker, right? If your range sessions strangely resemble the shootout scene from the movie *Heat*, it's a good bet your gun will be a tad on the warm side. Suppressed rifles can easily achieve temperatures in excess of 800 degrees around the muzzle, which would cook any application of Loctite Red.

If you've ever bought a SureFire can or a muzzle device, you're well aware of the little translucent tube that accompanies it, which states simply and in bold red "ROCKSETT." Rocksett is a water-based cement containing silica manufactured by Flexbar Machine Corp. It originally started out as a clear version of an adhesive used for bonding machine components in high-temperature environments. Its proprietary formula is both nontoxic and impervious to most chemicals. At its inception there were a few gunsmiths who were quietly using it, but with the rise of the internet and search tools, Rocksett really caught on among gun users.

Unlike Loctite, the structure of Rocksett can withstand temperatures from -350 to over 2,000 degrees F. This makes it ideal for use around the muzzle, but the Director of Sales at Flexbar, Lou Valenti, assured me it can be used effectively in any threadlocking application. This is made possible by the somewhat counterintuitive way it's removed: a soak in hot water.

The correct application of Rocksett uses a thin or sparing amount, because it tends to dry from the outside in somewhat like a skin hardening. Once application is complete, the water in the formula needs to evaporate in order to leave only the silica behind. This is a physical process rather than a chemical one. The drying/curing process is benefited by forced warm/hot air movement across the surface. Rocksett recommends that for a light application, 24 hours at room temperature will suffice, followed by 20 minutes of heat at 175 degrees for a full cure.

If the application requires a fair amount, better results can be achieved by a gradual bake from room temperature to 300 degrees F in linear steps over the course of six hours. Excessive application is problematic for Rocksett. While excess Loctite can be wiped away, too thick an application of Rocksett won't cure throughout, encapsulating uncured adhesive. When any additional heat is applied, the trapped water will boil, forcing its way out and destroying the bond. As is expected, ensure all parts to be bonded are free of oil and contaminants.

LOOSE ROUNDS

Having a strong familiarity with bolts and screws can serve all aspects of your tinkering lifestyle, and since fasteners are everywhere in our daily lives, we can better understand how they hold things together. You may even feel the urge to go put threadlocker on everything now, but don't think you absolutely have to. When I'm putting a gun together, I often don't use Loctite, unless it's a part I suspect might be subject to excessive vibration or heat, or has shown a predisposition to come loose in the past. I trust my torque wrench implicitly when I've prepped my fasteners and surfaces correctly.

If you plan on using Loctite or Rocksett on your next build but are uncertain if it might be the right choice, don't hesitate to contact either company. I found both to be enthusiastically forthcoming with product information, and if you can get an engineer on the phone at Henkel, they'll actually help you determine the exact color and grade that'll work best for whatever you're doing. Once you have everything dialed in, enjoy the confidence that comes with a well-built gun. ■

If a little is OK, then more is better, right? Wrong.

CHEAP DIY

DIY Upgrades for Your AR and Glock

By Sean Murphy

In a world of custom, John Wick-inspired, lightened, and Cerakoted rifles and pistols, price tags can exceed a few mortgage payments. For those who work for a living or live with a spouse's watchful eye, we decided to make budget-friendly upgrades to an affordably priced DPMS AR-15 carbine and a second-hand Glock pistol. Rather than making Gucci guns, we worked to enhance the performance of both platforms without breaking the bank.

The AR in AR-15 originally meant Armalite Rifle; some would say it now stands for America's Rifle due to the millions of AR-15 based rifles now in circulation. The proliferation of the AR-15 design has resulted in rifles ranging from inexpensive to several grand. We chose a brand-new, value-priced DPMS to upgrade.

And if the AR-15 is America's Rifle, Glock has created the universal pistol. Found in every corner of the world, Glocks are known for their reliability and ability to perform when needed for duty, competitive, or defensive uses.

BASELINES

Before making changes, we performed baseline testing on both guns to determine the effectiveness of our upgrades. We tested accuracy by shooting different types of ammunition at known distances. Shooting performance was tested with a variety of drills incorporating multiple targets of different sizes, different target distances, movement, and reloads.

The DPMS delivered 3 to 5 MOA performance from a variety of ammo, with some heavier match ammunition performing a little better. The Glock ran well and could be shot decently, but the sights held the pistol back. With prior use breaking in the gun, the trigger was smooth, although a little heavy, in stock configuration. Both rifle and pistol showed promise but could use some upgrades.

The upgrades we applied to our basic-bitch AR-15 saw group sizes shrink by nearly half from 3- down to 5-MOA groups.

THE CARBINE

Starting as a weapons parts manufacturer in the mid 1980s, Defense Procurement Manufacturing Services (DPMS) became a large player in the AR-15 world. Our base carbine for this project was the LCAR model, a simple Mil-spec-type carbine that looks similar to a military M4.

The AR-15 platform is highly modular, enabling a variety of upgrades and modifications. We made changes to help a shooter deliver more accurate rounds on target quickly. To do this, we upgraded the handguard, switched to an adjustable gas block, installed a nicer trigger, added a muzzle brake, and changed the furniture.

Handguard and Gas Block

Replacing a stock handguard with an extended rail offers numerous benefits, including better ergonomics for shooting the rifle, better airflow for cooling, the ability to bolt on accessories, and a free-floated barrel for some models. For the tightest budgets, a drop-in handguard replacement doesn't require changing the barrel nut but might not free-float the barrel. If you're on a budget and/or don't have the tools to do a whole free-float rail upgrade, there are handguards that use the GI barrel nut. The only challenge is you'll have to hacksaw or Dremel off the front sight to provide clearance for the rail.

We wanted all the benefits, particularly free-floating the barrel to increase our accuracy potential. Free-floating removes contact with the handguard to eliminate interference with barrel harmonics. This increases consistency as well as accuracy and reduces point of impact shift when you place the handguard against something like a barricade.

We selected a 14-inch Midwest Industries Slim Line rail. To install it, we had to either cut down the front sight block to fit under the rail or replace it with a low-profile gas block. Doing the latter with an adjustable piece would also allow us to reduce the gas flow and tune the gun for reliability. With short gas tube lengths on carbines, too much gas pressure increases the bolt carrier speed, harming reliability and increasing recoil unnecessarily. Thus, we installed an adjustable gas block from Seekins Precision in Idaho.

The rail installation involves removing the muzzle device, front sight base, and original barrel nut. An armorer's wrench such Magpul's has the appropriately sized

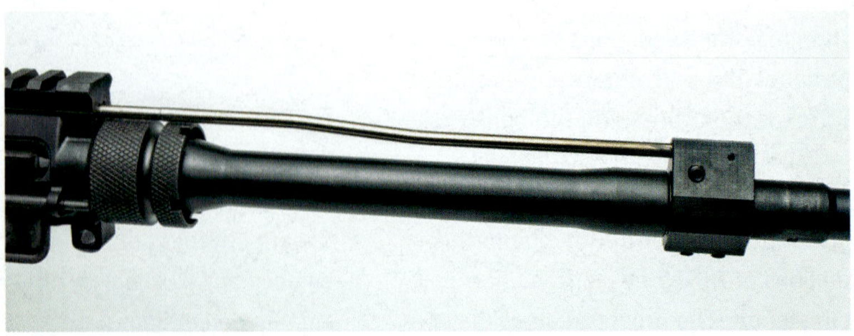

Getting rid of the M4-style gas block with its integrated sight tower and replacing it with a low-profile gas block allows the use of a slim, free-float handguard that improves handling and accuracy.

cuts and surfaces to do the job correctly. Installing and torqueing the rail-specific barrel nut is next, followed by installing the new gas block. Take care to ensure the gas block and tube are installed and aligned correctly.

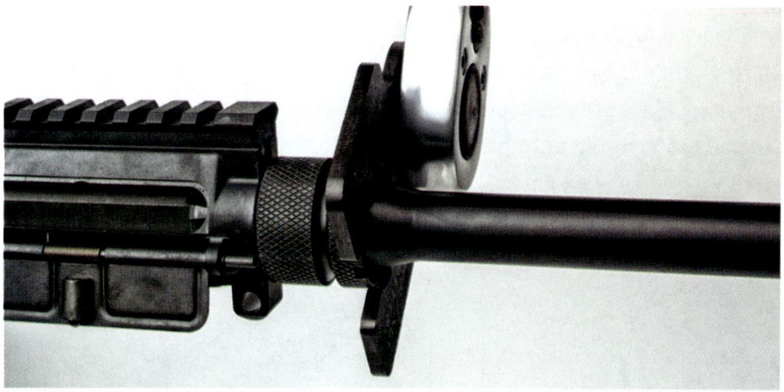

M4-style handguards that have a delta ring barrel nut (top) require a common armorers wrench, but your replacement handguard may use a proprietary barrel nut and require the use of a matching, proprietary tool (bottom).

Trigger

Mil-spec AR triggers were designed for reliability and functionality on a combat rifle. While reliable, they're usually handicapped by a heavy and gritty trigger pull. Our DPMS's stock trigger was no different, breaking inconsistently between six and seven pounds. Even though the AR's better than some other military rifles, a lighter trigger with a clean break can allow for faster and more precise work. Rock River Arms has been offering solid rifles and parts at affordable prices for many years. Their national match trigger is an affordable and reliable two-stage unit and has been a popular choice within the service rifle community.

Installation is very simple, starting with removing the original trigger. With a punch, first drift out the hammer pin and remove the hammer assembly. Then, remove the trigger pin, trigger, and disconnector. Install the Rock River trigger by following these steps in reverse. Beware to install the springs in the correct orientation for proper function, and note that some lowers might require removal of the safety. Our new trigger broke at approximately four pounds, with a defined first and second stage and a crisp reset.

If you're lucky, swapping your AR's trigger only involves a couple pins and a little swearing while lining up the internals. Some trigger swaps require the removal of the safety.

Muzzle Brake

Muzzle brakes are used to tame recoil on weapons platforms from cannons to rifles. The DPMS came adorned with a standard A2 flash suppressor. While effective at disrupting gases to reduce flash during firing, it doesn't redirect these gases for recoil control. A 5.56mm rifle isn't a heavy-recoiling rifle by any means, but a muzzle brake aids in getting rounds on target faster.

Within their parts catalog, DPMS offers the Miculek brake, originally designed for competitive shooting. While an older design, it's very effective at reducing recoil without breaking the bank. Designed with Jerry Miculek, one of the fastest action shooters in the world, the brake has three ports per side to direct

gases for reducing recoil. The consequence of this is increased noise and concussion. However, for going fast without muzzle climb, a muzzle brake is the solution.

Furniture

Our rifle came with very simple Mil-spec-type furniture. Unless you've been living under a rock, you've probably heard of a little company called Magpul. Originally started with a magazine pull to easily remove magazines from pouches, Magpul has grown to be an industry leader with parts, accessories, and lifestyle products for shooters. While we didn't change anything mechanical, upgrading the furniture made the rifle more ergonomic for shooting and manipulations.

We replaced the original buttstock and pistol grip with MOE-SL replacements, and the trigger guard with an MOE piece. Additionally, we added the Type 2 M-LOK rail covers to the handguard. All these accessories are made of polymer to save weight while remaining durable. The grip and panels are textured to help hold onto the rifle in any condition, while the grip angle and stock shape offer greater comfort and better fit with the carbine.

Upgrading the most basic handguards generally require swapping the barrel nut, but the new part makes a huge difference in handling, modularity, and possibly accuracy.

THE PISTOL

For the past several decades, Glock pistols have walked the beat with more law enforcement agencies than any other. As pistols retire from service, they're generally traded-in and resold to the commercial market. The average street price of a new Glock can range between $450 to $600, but many used models can be found for as low as $250 to $300. As former police guns, most have been carried a lot and shot a little. This means most second-hand Glocks still have plenty of life left in them.

The base gun for our project was a well-used Gen 3 Glock 17 in 9mm. We chose to breathe new life into the old gun with upgraded sights, trigger, barrel, magwell, and new magazines.

Sights

Replacing the sights should be the first order of business for any Glock owner, as the OEM sights are made of plastic and suck. For a defensive-oriented pistol, consider at least a front night sight or "straight-8" design with tritium vials to provide visibility in low-light situations. For competitive shooting or just having fun on the range, a fiber-optic front sight transmits ambient light to the

Generally, every stock Glock will see a performance benefit when upgrading several common parts, namely the trigger, barrel, and magazine well.

end of the fiber for fast and easy sight acquisition. My preference in sights for over the last 10 years has been Warren Tactical Sights.

Designed by Scott Warren, a former FBI Hostage Rescue Team member and successful competitive shooter, his sights have a unique wave shape to the rear sight. The center is a U-shaped notch with 90-degree angles at the top-inside edges for alignment with the front sight, while concave corners open up the field of view of the target and surrounding area The lack of hard angles allows your eye to quickly find and focus on the front sight, yet deliver the precision needed for proper sight alignment. Finally, the front face of the sight is cut to allow for one-handed manipulations.

Changing the sights on a Glock is straightforward. The front sight is held in place with a screw; be sure that the new front sight is straight in the cutout when you install it. The rear sight requires a sight tool or use of a hammer and punch to remove the old sight and install the new one. When installing it, center the sight in the dovetail, and if necessary, make slight adjustments after you shoot it.

Trigger Kit

Stock Glock triggers are advertised at 5.5 pounds of pull weight, although most tend to have a heavier trigger. Without going crazy for some of the full kits available, we picked up a ZEV Professional Starter Spring Kit for $29. This kit includes a replacement striker spring, firing pin safety spring, trigger return spring, and connector. The lighter springs reduce resistance on the firing components, while the connector alters the geometry of the trigger operation to provide better trigger feel and further lighten the trigger pull.

Replacing the trigger parts isn't hard, but if you haven't done it before it might seem complicated. A Glock armorer can replace these parts in their sleep but for the uninitiated, the process can be found online. The slide assembly is also simple to

Glock front sight posts are screwed in through the slide and easily changed with a screwdriver. Newer models require a shallow nut driver that's found with a Google search for "glock front sight tool."

There aren't many parts in a Glock to begin with, and the ones that make the most difference are easily swapped for aftermarket parts. Highly polished parts make trigger pulls smoother and small changes in trigger connector geometry reduce trigger pull weights.

take-down once you've seen it done, just beware to not lose any parts and ensure installation is done correctly. We've also outlined these steps in Issue 14's DIY column.

Since our second-hand Glock had been used heavily, the parts were worn in nicely. The trigger had smooth take-up and a positive wall, breaking at 6 pounds. The upgraded trigger was still smooth, while the new connector dropped the weight to 4.5 pounds. The wall before the break was also lighter and crisper.

Barrel

Factory Glock barrels are hammer-forged and known for being a good barrel capable of shooting a variety of ammo with a long life. Replacing the barrel isn't always necessary if the pistol wasn't shot a lot. However, if you're interested in chasing accuracy or adding a suppressor or compensator, a barrel replacement might be right for you. Additionally, if you happen upon a screaming deal for a .40 S&W but wanted a 9mm, there are drop-in caliber conversion barrels.

The only thing easier than upgrading a Glock's barrel is changing the magazine.

Adding a magwell funnel to your pistol decreases the accuracy needed when slamming a mag home and increases the speed you can reload. They're generally cheap and easy to install.

We obtained one from SilencerCo, which offers high-quality drop-in barrels for a variety of pistols, including Glocks. These barrels are made of 416R stainless steel (common in many match barrels) and finished in black nitride for extra hardness. The muzzle is extended and threaded for mounting a suppressor or muzzle brake with U.S. and metric threads, depending on your attachment of choice. Replacing the barrel is as easy as field stripping the pistol and swapping in the new part.

Magwell and Magazines

In 2015, Magpul announced their first Glock accessories, including a magwell and 9mm magazines. When shooting pistols, being able to rapidly reload allows a shooter to stay in a fight, in the game, or get shooting faster on the range. A magwell aids this by providing an expanded opening at the bottom of the frame. This increased diameter is more forgiving, especially when going fast, and a magazine might not always be perfectly aligned. Some competition magwells can be quite large, but Magpul created a streamlined unit that can be used on a carry or competition gun. Installation was as simple as placing the unit over the magwell opening and tightening down the supplied screw.

Magazines are often the weak point to any firearm, whether it's a good magazine at high cost or a magazine that simply doesn't work reliably. Glock offers a good polymer molded magazine with a metal liner. However, if on a tight budget, Magpul offers a variety of polymer magazines that are lighter in weight and inexpensive. Retail price is around $40 for OEM Glock mags, while the Magpul options start at just $12.50. We picked up three flush-fitting 17-rd and three extended 21-rd magazines for the equivalent of two OEM magazines.

FINAL TESTING

Once we completed our swapping of parts and garage gunsmithing, we wanted to see how much of a difference we actually made. Running the same drills as our baseline tests, we

PARTS	MSRP
Midwest Industries Slim Line – 14-inch Rail	$155
Seekins Precision Adjustable Gas Block	$59
Rock River Arms Trigger	$120
DPMS Miculek Muzzle Brake	$40
Magpul MOE-SL Stock	$60
Magpul MOE-SL Grip	$20
Magpul MOE Trigger Guard	$9
Magpul M-LOK Rail Covers – Type 2 (quantity 3)	$39
Total:	$502

PARTS	MSRP
Warren Tactical Sights – Fiber Optic Front	$77
Zev Professional Starter Spring Kit	$29
SilencerCo Barrel	$190
Magpul Magwell	$25
Magpul Magazines (quantity 6)	$75
Total:	$396

$502 invested in our DPMS AR netted a 2-MOA increase in accuracy and noticeably faster split times. Optics are outside the scope of this article, but we used the pictured Nightforce to evaluate performance before and after our upgrades.

RATHER THAN MAKING GUCCI GUNS, **WE WORKED TO ENHANCE THE PERFORMANCE OF BOTH PLATFORMS WITHOUT BREAKING THE BANK.**

found both guns had changed. The rifle saw noticeable performance changes for the better, with group sizes shrinking by nearly half from its original 3 to 5 MOA groups. Pistol accuracy improved with the fresh barrel, but it wasn't as drastic. With an assortment of FMJ practice and JHP duty ammo, the OEM barrel delivered 4- to 5-inch 10-shot groups at 15 yards, while the SilencerCo replacement was tighter at 2.5 to 3.5 inches on average.

As far as the ability to apply accurate fire under speed, both guns showed big changes on the timer and by feel. With the rifle, we ran El Presidente drills from low-ready as well as a 100-yard drill consisting of 5-shots from standing, kneeling, and prone. El Presidente times improved by 25 to 30 percent, while the 100-yard drill went from A- and C-zone hits in almost 25 seconds to almost all A-zone hits in 15 seconds. While louder to the shooter, the muzzle stayed flat on the target rather than bouncing around and requiring more time to re-aim. The trigger was significantly nicer than the stock unit, allowing faster splits between shots and greater accuracy at high speed with the crisper break. Finally, the longer handguard increased leverage by allowing the support hand to grasp it further forward, while the more ergonomic furniture made the gun handle that much better as well.

The pistol performance gain wasn't as great as with the rifle, but the changes brought the heavily used gun back to life. We ran several iterations of the El Presidente and other similar drills with draw, multiple shots, and reloads. We found the most noticeable change came with the new sights. The factory sights were hard to shoot accurately and obscured much of the target. Going to a thinner front sight with a fiber dot and having more of the target in view made target acquisition faster, while also allowing greater precision. The trigger upgrade was noticeable, also helping with precision and speed. On a double-stack gun, the magwell was a nice addition but not as noticeable of a change on the timer. On average, the upgrades resulted in tighter accuracy and shaved 10- to 20-percent off of the recorded drill times, with noticeable increases in speed for first round hits and reloads, indicative of the better sights and larger magwell. The original Glock factory barrel wasn't quite ready to retire quite yet … but the SilencerCo barrel tightened up the groups. We didn't accuracy test with a suppressor, but the ability to go quiet put a big smile on our face. **R**

$396 invested in our stock Glock netted less drastic performance improvements, but we did notice reduced split times and faster reloads.

AR ARMORER TOOLS

The Right Tool Makes All the Difference

Story and Photos by Rob Curtis

Some would have you think the modern AR-15 is a mystical device and working on one requires an eon of apprenticeship and a degree in mechanical engineering. Not so. All you really need is a willingness to learn by breaking stuff, enough cash to replace those broken parts, and a good set of tools.

We spoke to a handful of armorers to get a feel for the best tools to get you through the most common jobs. Robb Jensen is a commercial armorer with Virginia Arms in Manassas, Virginia, Corporal Sean Dougherty is a SWAT officer and team armorer for Union County Police Department in New Jersey, and Nathan Murr is a prior service Marine and lay armorer who's leveraged his firearm experience into a career working on government guns and designing his own firearm accessories.

Far and away the most common task performed by our armorers is swapping a G.I. handguard for a free-floating handguard. This is a pretty intense job because it generally involves pulling the barrel and replacing the gas block. Not the simplest of jobs, but one that can be done by anyone with the proper mindset and the right tools.

The second most common job is installing an aftermarket trigger. After that, the array of work includes installing ambidextrous fire selectors, new grips, and new muzzle devices. Some of these jobs are simple, some require taking other components off the gun to complete the primary objective. Building a gun from parts is less common for these guys, and most professional shooters will tell you it takes a lot of experience to choose the correct components to build a reliable and safe Frankengun. That being said, assembling a stripped lower and combining it with a factory-built complete upper receiver is a very popular and viable option for less experienced builders who don't want to deal with next level headaches such as headspace issues or gas system troubleshooting.

All of our armorers agree that wrenching on your own guns is the best way to learn. Get good, and it won't be long before you're installing handguards and triggers for your less mechanically inclined friends. Just be aware that you're responsible for everything you do to a firearm. There are some things armorers don't do — and we're suggesting you steer clear of things like stoning triggers. It's too easy to bone up a trigger and get someone hurt when a gun decides to dump a mag on its own.

Lastly, before anyone whines about spending money on tools, please realize that none of these tools are necessary for working on an AR. Our intention is to point out tools that make any job faster, easier and, therefore, way more pleasant.

THE BASICS

Whether building, repairing, or servicing just about any firearm, it's certain you'll encounter roll pins and screws. To deal with them effectively, you'll want at least a set of roll pin punches, a lightweight hammer, and a set of screwdrivers.

A **Brownells Roll Pin Punch Set** will last dozens or more projects, with individual punches needing replacement when a shaft bends or a tip begins to mushroom. To make life easy, protect your fingers and increase tool life, consider adding a **Brownells Roll Pin Holder Set** and **Roll Pin Starter Kit** to your toolbox. A roll pin holder looks like a hollow-ended punch. You drop a pin in the end and it holds it on-axis while you tap it in. Without the holder, you're left holding the pin with your fingers as you tap it. Roll pin starter punches help get tough pins moving when they require forces that'll bend longer, thinner roll pin punches. Once the pin is knocked loose and it's moving, switch from the stubby starter to a regular roll pin punch to push the pin all the way out the other side.

Starting with a **Brownells Premium 1-Inch Delrin Tipped 8-ounce Brass Hammer** will keep you from bending punches and marring finishes. At some point, you're going to want a bit more smack and something like a **Williams 20543 12-ounce Ball Pein Hammer** will let you apply more force exactly where you need it.

There aren't many screws on a stock AR-15; you might find one holding the grip on the lower receiver. But, updated guns, and updating guns, will call for a driver to install and work on many aftermarket handguards. A ¼-inch bit holder with a spread of quality bits will offer a lot of versatility without taking up a lot of room in your toolbox. The shorter the bitholder handle, the easier it is to control. We have a **Brownells Magna-Tip Standard Set No. 1 with LE Handle**. The LE handle is longer than a stubby, but shorter than a regular length handle. Some bit holders retain bits in the head with a magnet, some use a mechanical clip. Preference is up to the user; magnetic bits are fast to swap, but they tend to stick in the bolt heads. Clip holders retain the bit using a thin wire that catches a detent on the side of the bit. We find this system a little more secure, but the holding wire will eventually wear out. You'll want an array of ¼-inch shank Torx, Phillips head, slotted head, and hex head bits to cover all your bases. While a bit holder will offer the most versatility, hex keys are fast and the short end of these tools are great for breaking loose, stuck bolts. Just be careful; poor-fitting driver heads made from cheap metal will strip out, or worse, strip out a screw before you realize how screwed you are. If you're working on a gun with lots of bolts, or you just like nice things, have a look at a **Wera 816 RA ratcheting bit driver**. It's a European company known for technician's tools of exceptional quality. Think of the kinds of tools Formula 1 and MotoGP technicians use and you're there. The smooth action, short handle, and autolocking collar combine to make a lustworthy tool that'll find you looking for things to fix just so you can use it.

1 Brownells Roll Pin Punch Set
MSRP:
$20
URL:
www.brownells.com

2 Brownells Premium Roll Pin Holder Set
MSRP:
$32
URL:
www.brownells.com

3 Brownells Roll Pin Starter Kit
MSRP:
$35
URL:
www.brownells.com

4 Brownells Premium 1-inch Delrin Tipped Brass Hammer 8 oz
MSRP:
$22
URL:
www.brownells.com

5 Williams 20543 12 oz Ball Pein Hammer
MSRP:
$20
URL:
www.snaponindustrialbrands.com

6 Brownells Magna-Tip Standard Set No. 1 w/ Clip Tip LE Handle
MSRP:
$50
URL:
www.brownells.com

7 Wera 816 RA Ratchet Screwdriver
MSRP:
$40
URL:
www-us.wera.de

HOLDING IT DOWN

The backbone of an armorer's den is the bench and vise. A workbench of nearly any size will accommodate an AR. Something as small as two 2x3-foot sheets of ¾-inch plywood glued together and mounted to sawhorse legs will work, so long as it's stable enough to hold a vise. The vise is your brawny, third hand and its value is in its ability to hold resolutely, move smoothly, and last forever. The vise you'll want will have at least 4-inch jaws. Six inches will give you better purchase (more clamping area), but we've found 4-inch jaws work well enough holding common AR fixtures that it's hard to justify the added size, weight, and expense of a much larger vise.

You've got three routes to a vise. You can get a decent quality bench vise from Harbor Freight or any big-box hardware store. You're likely looking at a cheap cast-steel body that'll hold up for a while before something chips or bends. Another option is to pay beaucoup dollars for a brand-new, heirloom quality Wilton machinist vise. A thousand bucks will get you a great ductile iron vise with a shrouded main screw that will outlive your grandkids' grandkids. But, the best option is to hit Craigslist and find an older, high-quality bench vise that's of no more use to its liver-spotted owner. Look for older Wilton, Columbian, Ridgid, Record, Reed, and Yost vises. Don't worry about surface rust; pay attention to the jaws and the action. Look for jaws that are square to one another and don't wiggle. Movement of the main screw should be smooth, and jaw pads should be without deep scratches or burrs. Just be aware there are vise geeks out there collecting the good ones to refurbish and resell. When you bring your vise home, clean it up, lube the main screw with plenty of grease, and mount it with a few ⅜-inch bolts.

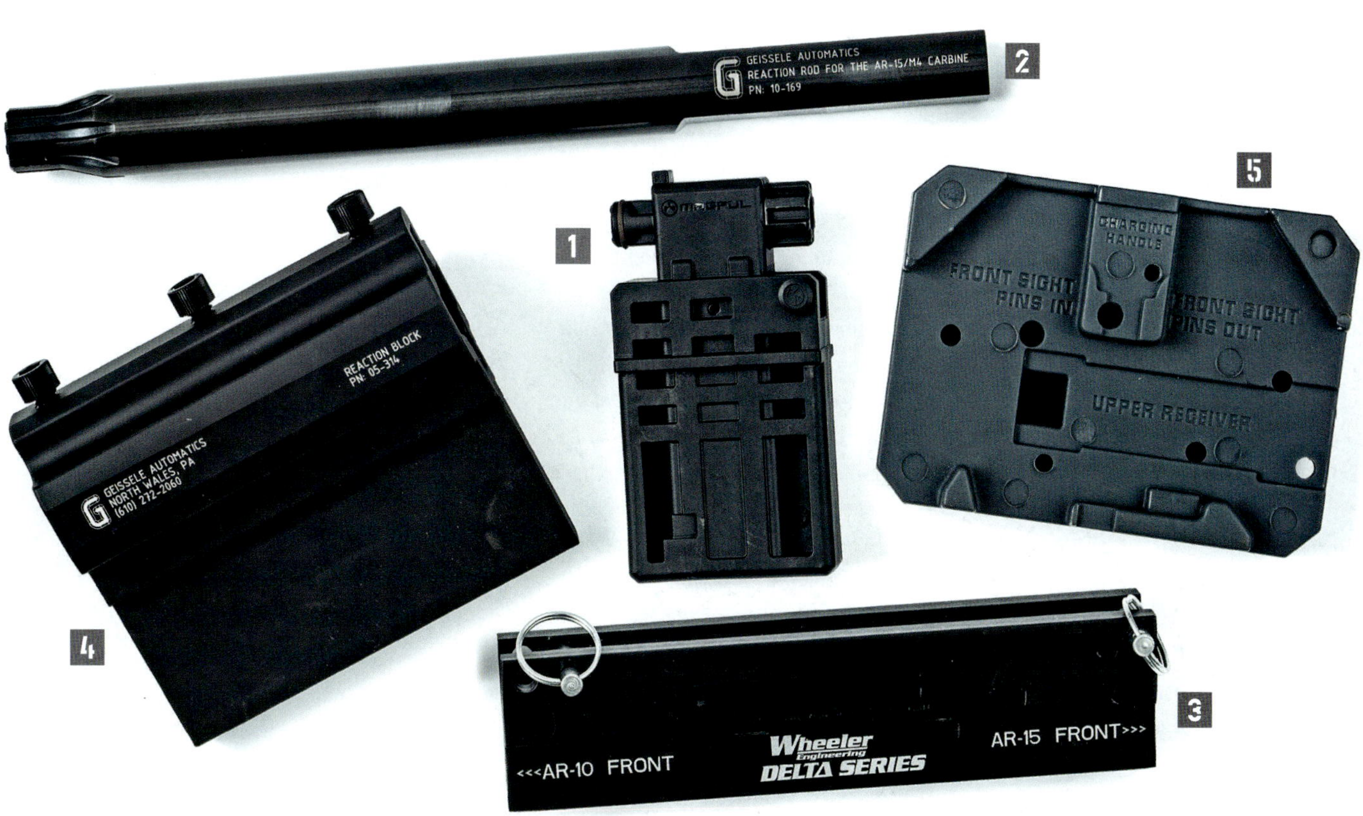

A vise alone is great for holding anything solid with a two parallel faces, of which there are zero on an AR. Rather than crush aluminum receivers in vise jaws or curse the odd, unhandlable shapes of gun parts, we look to fixtures as the savior of our sanity. The first fixture you'll likely encounter in the AR world is the simple lower receiver vise block. Slip one half of the block into the magwell of an AR lower and tighten a vise on the other, exposed portion, and the receiver is going nowhere while you install the receiver extension and most of the lower parts kit. The basic Brownells version has been around forever, but take a look at the **Magpul BEV Block**. It's a couple of fixtures in one. Not only is it a great lower receiver vise block, its primary duty is holding an upper receiver while preventing the barrel extension from turning when installing a new barrel, muzzle device, or handguard. It does this by engaging the chamber lugs on one end, and snapping the bolt carrier into the other for a wobble-free mount. While the BEV is versatile, the **Geissele Reaction Rod** is the most elegant and bombproof way to work on an upper receiver yet invented. Set the flats of the rod in a vise and slip the upper on it like a mandrel. The splines on the fixture engage the barrel extension and hold it immobile while installing a barrel, gas block, handguard, or muzzle device. The best feature is its speed. There's nothing to lock, snap, or twist to lock the barreled upper receiver. Want to rotate the upper for a better angle? Slip it out an inch, rotate, and slide it back on so the splines reengage the barrel extension. The **Wheeler Engineering Delta Series AR Upper/Pic Rail Vise Block,** comes in handy when lining up the top rail of a handguard with the upper receiver rail. It's a great value since it also does the job of an upper receiver block on both AR-15 and AR-10 platforms. The **Geissele Reaction Block** is an alternative to the standard lower receiver vise block. It captures the receiver extension in one of four positions, allowing you to work with the sides of the lower facing up in addition to the straight up and down orientation you get with a regular vise block. It also takes the stress off the receiver when tightening the receiver extension nut; handy for polymer receivers, but overkill for aluminum. The **Wheeler Engineering AR Armorer's Bench Block** is a multi-use fixture that holds receivers, charging handle, and front sight post steady while driving pins. It's a great way to save some frustration while working on a standard receiver set; just be aware that it's not going to work with thick polymer receivers.

1 Magpul Bev Block
MSRP: $50
URL: www.magpul.com

2 Geissele AR15/M4 Reaction Rod
MSRP: $99
URL: www.geissele.com

3 Wheeler Engineering Delta Series AR Upper / Pic Rail Vise Block
MSRP: $54
URL: www.btibrands.com

4 Geissele Reaction Block
MSRP: $99
URL: www.geissele.com

5 Wheeler Engineering AR Armorer's Bench Block
MSRP: $60
URL: www.btibrands.com

Adding a free-float handguard is probably the most common job performed on the AR platform. In nearly all cases, this means taking the front sight base/gas block off the gun and replacing it with a low-profile block that fits under the new handguard. It's a frightening job that seems like a heart transplant. But, with the proper tools, it's about as stressful as performing an oil change. Getting a well-seated taper pin moving can bend a standard punch. For this, use a **Starrett Taper Pin Starter** and finish with a regular punch. Once the FSB is off the barrel, use the **Geissele Gas Block Pin Punch Set** to get the tiny, 0.078-inch gas tube roll pin out. To get it back in, use the **Precision Reflex, Inc Gas Block Assembly Fixture**. It comes with a roll pin holder sized for the gas tube roll pin. Lay your low-profile gas block in the fixture and tap away. Without a fixture like this, the gas block sits at an odd angle, and it's very easy for punches to slide off as you try to tap the gas tube roll pin in. Occasionally, you may run into a taper pin that doesn't respond to your gentle taps while trying to drive it home. You might use a **Drillco 4550E Series 2/0 HS Steel Taper Pin Reamer** to clean up tapered gas block and barrel holes before resorting to the Big Hammer. More than likely, though, you're replacing a FSB with a low-profile gas block that uses set screws to hold it in place. The easiest way to install a set screw gas block is using a fixture like the **BRD Gas Block Dimple Jig**. Slide the jig on your barrel, use the locator pin up top to find the gas port, then use the two guides on the bottom to drill dimples in the barrel. Replace the jig with a gas block and tighten the screws. With the gas block screws in the dimples, the gas block isn't moving. **HB Industries Gas Block Alignment Pins** are a handy way for the anal-retentive to make sure the gas block hole is lined up with the gas port. With the gas block off the gun, drop one of the pins into the port, slide the gas block on and turn the upper assembly upside down and wiggle the gas block 'til the pin drops and you're aligned. Tighten the gas block retaining screws, use a cleaning rod to snap the plastic pin, and drive out both halves.

Turning to friction, and the chemical control of said phenomenon, let's talk installation lube and threadlock. While there are many opinions on what lubricant to use when installing a barrel nut, everybody agrees the barrel nut should have some lube to help it achieve the proper torque and protect the threads of the dissimilar metals from galling. The longtime standby is **AeroShell Grease 33MS** because it meets the Mil-spec standard the DoD applies to barrel nut lubricants. Buying it in a commercial tube will give you enough grease to last five lifetimes, so look for resellers that repackage the stuff and sell a few ounces for a few bucks. It won't hurt to dab a thin layer on the barrel extension before sliding it into the upper. If you ever have to take that barrel out, you'll thank us for reminding you. An alternative to AeroShell recommended by Robb Jensen is **Weapon Shield Grease**. The lithium-based grease won't hold up to the high temps of the AR chamber as well as the Moly-based Aeroshell, but it'll get you torqued and set just as well. You just might not have it as easy pulling the barrel apart a few years down the road than if you used the Aeroshell.

On the other end of the spectrum, some AR jobs call for thread-lock to keep screws and bolts from backing out. It's mostly used to keep the screws in aftermarket free-float handguards from falling out. If you're still using the squeeze bottle, drop applicator, you're in for a treat. Check out the new **Permatex 27010 High Strength Threadlocker Red and 24010 Medium Strength Threadlocker Blue Gel**. These gel formula thread-lockers make application so much easier and cleaner, you'll feel like a caveman going back to the liquid Loctite in the drip bottle.

1 Starrett Taper Pin Starter
MSRP: $13
URL: www.brownells.com

2 Geissele Gas Block Pin Punch Set
MSRP: $25
URL: www.geissele.com

3 Precision Reflex, Inc Gas Block Assembly Fixture
MSRP: $30
URL: www.precisionreflex.com

4 Drillco 4550E Series HS Steel Taper Pin Reamer 2/0 size
MSRP: $20
URL: www.drillco-inc.com

5 BRD Gas Block Dimple Jig
MSRP: $65
URL: www.brdengineering.com

6 HB Industries Gas Block Alignment Pins
MSRP: $9
URL: www.brownells.com

7 Aeroshell AeroShell Grease 33MS 5% Molybdenum Disulphide Grease
MSRP: $3
URL: www.shell.com

8 Weapon Shield Grease Lithium Grease
MSRP: $9
URL: www.weaponshield.com

9 Permatex 24010 Medium Strength Threadlocker Blue Gel
MSRP: $12
URL: www.permatex.com

10 Permatex 27010 High Strength Threadlocker Red Gel
MSRP: $12
URL: www.permatex.com

LOWDOWN AND A LITTLE DIRTY

Attacking the lower AR receiver is pretty painless once you have a few tricks under your belt. Every armorer has their most hated point in the lower build. Some fear crushing the trigger guard ears with a hammer and punch and instead use a press such as the **Little Crow Gunworks Roll Pin Pusher**. This little guy slowly pushes the trigger guard roll pin in and out, making alignment and insertion a breeze. Other guys have launched so many forward takedown springs and detents from their benches that they won't build a lower without a **Brownells Forward Takedown Pin Insertion Tool** on hand. Though this tool is longer and easier to use, an alternative is a **¼-inch x 2-inch Clevis Pin** from any hardware store. Both let you press the spring and detent into the recess using something like a heavy-gauge paper clip (shown), hold it there as you press the takedown pin against the face of the tool and push it back through so the detent takes hold of the takedown pin as it passes over. Another, why-oh-why operation is driving the bolt catch pin in place. It's common to see lowers that bear the scars of a newb armorer's hammer or punch that's skipped off and gouged the lower's virgin skin. Applying some masking tape will help, but getting some space to work the hammer is the best play. **Little Crow Gunworks' Bolt Catch Tool** does just that. It's a custom, extended punch set that lets you tap the punch behind the lower without fear of skipping off and causing damage. Tapping trigger and hammer pins in and out of the lower is another easy way to mar metal. Instead of tapping, try the **Geissele Trigger Fitting Pin**. Cupping the ball end of the pin in your palm lets you apply a lot of force slowly and evenly, taking some of the stress out of the three handed trigger installation job. Lastly for the lower, staking the castle nut couldn't be faster and easier when using a spring-loaded center punch. A few dozen clicks with the **General Tools 89- Stainless Steel Automatic Center Punch** will stake a castle nut in a minute or two.

1 Little Crow Gunworks Roll Pin Pusher for AR-15
MSRP:
$40
URL:
www.littlecrowgunworks.com

2 Brownells Forward Takedown Pin Insertion Tool
MSRP:
$16
URL:
www.brownells.com

3 Home Depot Clevis Pin
MSRP:
$1.57
URL:
www.homedepot.com

4 Little Crow Gunworks Bolt Catch Tool
MSRP:
$18
URL:
www.littlecrowgunworks.com

5 Geissele Trigger Fitting Pin
MSRP:
$15
URL:
www.geissele.com

6 General Tools 89- Stainless Steel Automatic Center Punch
MSRP:
$16
URL:
www.generaltools.com

The spaces beneath armorer's benches are littered with the carcasses of once favorite armorer's wrenches. Some break, some rust, some slowly wear out. Once, when the AR was young, there was only the basic G.I. barrel nut — Eugene Stoner's perfect solution. Its toothed outer ring is both a bearing surface for a wrench and a way to lock the nut using the gas tube. Despite the explosion of free-floating handguards and their necessarily customized barrel nuts, the G.I. barrel nut is still very prevalent. There are a host of other common nuts on the AR, and there's no end to the amount of armorer's wrenches that hope to turn them all and claim the title of "last wrench you'll ever need." These three might not be the last you'll ever need, but they will last a long time and have a few unique features that are worth a look.

The **Tapco's Armorer's Wrench** is a worthy wrench that covers all the bases without breaking the bank. The coated surface helps maintain your grip during use and prevents oxidation in storage. While it isn't the master of any AR bolt, it does seem to have that combination of wrench heads that makes it one of the most versatile tools in its price range. The **Magpul Armorer's wrench** is kind of a beast compared to other armorer's wrenches. It's the wrench that will beat all other wrenches unconscious and engrave "unfair advantage" on them before curbstomping the pins from their working end. It's big and fairly thick, giving great leverage without biting your hands. The surface is purposely left a little rough feeling due to the phosphate finish, enhancing durability and grip. The barrel end works on both pin and G.I. nuts, and the other end slips over a barrel extension to give three points of contact on a castle nut. The enclosed end means it's a tiny bit slower than an open-ended buttstock wrench, but the quality of engagement seems worth the reduced productivity. The wrench isn't meant to be a 500-in-1 tool, but a super durable wrench that addresses the most common AR needs without much compromise. The **Brownells CAR-15/M4 Buttstock Wrench** pulls no punches. This wrench isn't the One Ring at all, it's an elegant approach to one of the most shortchanged tools on the AR armorers bench. With four engagement points on the castle nut, this is one wrench that won't slip off and damage the finish of a gun. It's a stout, thick little wrench that takes care of the back half of an AR like no other wrench we've handled. Sure, you could use those $5 stamped metal jobs, but we don't think you've read this far into the article if you're not a buy-once, cry-once kinda tool shopper.

1 Tapco Armorer's Wrench
MSRP: $45
URL: www.tapco.com

2 Magpul Armorer's Wrench
MSRP: $80
URL: www.magpul.com

3 Brownells CAR-15/M4 Buttstock Wrench
MSRP: $80
URL: www.brownells.com

SPECIAL NEEDS

Here's a set of tools that the garage armorer can live without, but having them around will take the uncertainty out of some jobs and allow you to do a few things that can't be done with a basic set of tools.

Spline rods are hard on barrel pins. If you have a really stubborn barrel nut and are using a splined rod fixture, there's a chance you could apply enough torque to bend a barrel pin or deform an upper receiver. The **Geissele Super Reaction Rod** uses the same splined chamber lug interface as the original Reaction Rod, but adds a removable boss that spreads the torque of a PIA barrel nut removal through the upper receiver in a way that minimizes the chance of damaging the gun.

We won't lie, every armorer we spoke with said it's a rare day when they reach for an ejector removal tool. It's pretty unlikely that you'll see an ejector related problem on an AR. But, that means when someone needs ejector help, they'll be desperate. You could be a hero if you have a **Brownells Ejector Removal Tool** on your workbench. While it's on the pricey side, this tool works on a few different guns including AR-15, AR-10, Remington 700, Winchester 70, Savage, and Sako platforms.

Judging torque isn't hard, but it takes practice and experience — things that a garage armorer might not develop. A shortcut is having a set torque wrench on hand to make sure you're not under or overdoing it with the twist. For barrel and receiver extensions, something like the **Matco Tools TRB75F 3/8-Inch Drive Flex Head Torque Wrench** and a 1/2-inch socket adapter would work as well as a less expensive beam style torque wrench from a hardware store. The advantage of a pro grade torque wrench lies in the fact that you set the torque and the wrench clicks when you hit it. No guessing. The reason for a 3/8-inch drive as opposed to a 1/2-inch is versatility. The 3/8-inch works with a common automotive socket set so it can be used for jobs other than working on guns. A torque driver such as the **Brownells Magna-Tip Adjustable Torque Handle** comes in handy for the same reasons when working on handguards and optic mounts.

Most AR uppers are perfect out of the box, but some aren't. With the AR market crash of 2014 came a dearth of cheap receiver sets. So, it's not hard to find a bargain bin upper that has a little bit of run-out on the barrel interface. The **Wheeler Engineering AR 15 Receiver Lapping Tool** is a simple way to true up the face of an upper receiver. Just apply some lapping compound, slide the mandrel into the upper, attach a drill, and slowly lap the opening into a flat and true face that will fix any barrel seating woes. This has become such a common issue that some professional armorers we've spoken with lap all their barreling jobs preemptively. **R**

1 Geissele Super Reaction Rod
MSRP: $175
URL: www.geissele.com

2 Brownells Ejector Removal Tool
MSRP: $64
URL: www.brownells.com

3 Matco Tools TRB75F 3/8-inch Drive Flex Head Torque Wrench
MSRP: $333
URL: www.matcotools.com

4 Brownells Magna-Tip Adjustable Torque Handle
MSRP: $150
URL: www.brownells.com

5 Wheeler Engineering AR 15 Receiver Lapping Tool
MSRP: $36
URL: www.btibrands.com

AR-15 BOLT REBUILD

WARNING!
When handling firearms, always observe safety rules and the precautions set forth in the firearm's owner's guide. Be certain that your firearm is unloaded and made safe before proceeding with this DIY.

It Ain't Sexy, But Preventive Maintenance Will Keep Your Bad-Ass Rifle Running Strong

By RECOIL Staff
Photography by Henry Z. De Kuyper

After 2,500 rounds or so, the bolt on your AR is probably feeling its age. As part of routine maintenance, you should know how to inspect it, tear it down, and replace any wear items to keep the system running at peak performance. Note that if you ever feel the need to become a hand model, don't try running an AK at speed before the photo shoot...

INSPECTING

To inspect your bolt for wear, first remove the complete carrier from the rifle. With the bolt in its forward position, stand the carrier up on a hard surface. If it collapses under its own weight, the gas rings are probably shot. Next, take a small punch or live round and press out on the extractor. There should be significant spring tension. If not, then a bunch of malfunctions are in in your future — addressing problems caused by bad mags, faulty gas rings, and worn-out extractors and springs will take care of the vast majority of all mechanical failures on the AR platform. Finally, while holding the bolt in place, use the same small punch or live round to press down on the ejector. This should have less spring tension than the extractor, but should still feel pretty lively.

1. If your BCG will stand on its end without collapsing, your gas rings are probably GTG.

2. If it drops to the unlocked position under its own weight, it's time for a rebuild.

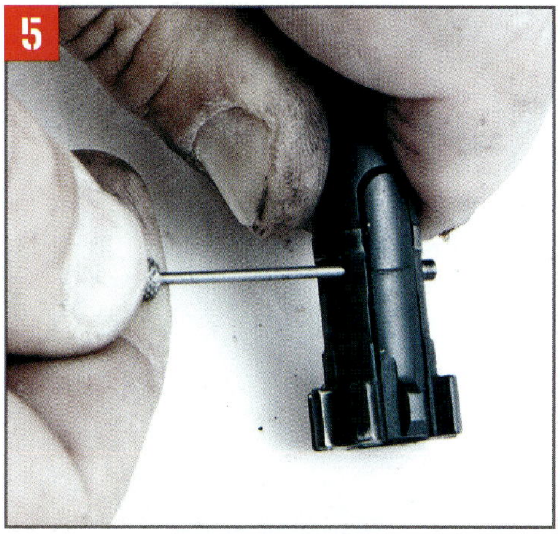

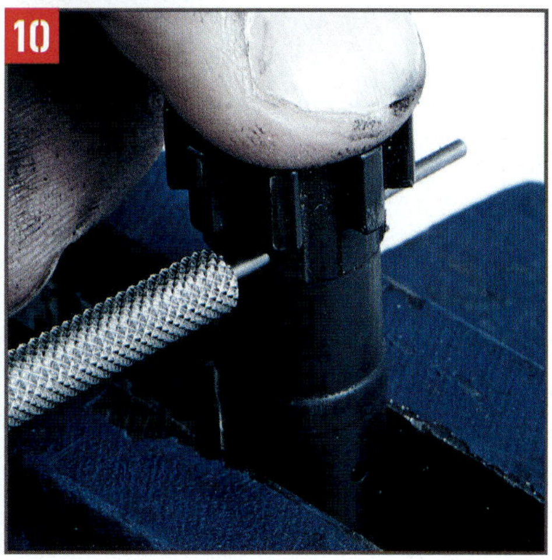

STRIPPING

To strip the bolt, remove it from the carrier as you would for normal cleaning. Check the bolt lugs for wear — watch out for rounding of the lugs. Inspect the cam pin hole for any cracks that might be developing; with high round counts, this is an area to keep your eye on. Next, remove the gas rings. These are located in a groove at the end of the bolt and can be peeled out one at a time by lifting up on one end, pushing the tail out of the groove and then running your fingernail around the bolt. Repeat until all three have been removed.

Take a small punch and press out the extractor pin, then remove the extractor from its groove in the bolt. Separate the extractor spring from the extractor, noting which way round the spring goes in the pocket. (Hint: big end toward the extractor). Replace the spring if it did not provide enough tension when you tested it earlier. Inspect the extractor claw for chips or wear — you can drag the extractor on your arm; it should feel sharp and leave a mark. If necessary, discard it and replace with a fresh one.

Place the bolt in a vise and using a roll pin punch, drive out the ejector roll pin. With the punch still in the bolt body, place your hand over the ejector and catch it as you withdraw the punch. Remove the ejector spring. Note that the ejector spring is the same one used in your takedown pin detent, should you ever need an emergency replacement.

3. Check extractor tension by pushing with this convenient tool. They come in handy, spring-loaded boxes of 30.

4. Press the ejector to check function and spring tension.

5. The extractor pin should come out easily.

6. Extractor assembly components.

7. Lift tail of gas ring up and out of groove.

8. Repeat two times. Lots of carbon buildup.

9. Use roll pin punch to remove ejector pin.

10. Capture the ejector and spring with your thumb; otherwise, prepare to spend time on your knees.

REASSEMBLING

Place a fresh ejector spring into the hole in the bolt face, then drop in the ejector. Be sure to line up the cutout on the ejector body with the hole for its roll pin. Start the roll pin, then press the ejector down while you tap the pin to engage the ejector cutout. Tap the roll pin gently into place and then finish with the roll pin punch. Test that the ejector moves freely.

Place the extractor with its new spring into the groove on the bolt and line up the pin hole. Press the pin into place and test the extractor's tension.

Wind new gas rings into the groove on the bolt tail using the reverse of the technique you used to remove them. Do not try to force them over the lip — start one end first then roll your thumbnail around the bolt tail. Stagger the gas ring gaps. We decided in this case to upgrade to a McFarland one-piece ring, which eliminates any gaps. Reassemble the bolt into the carrier and test for function. **R**

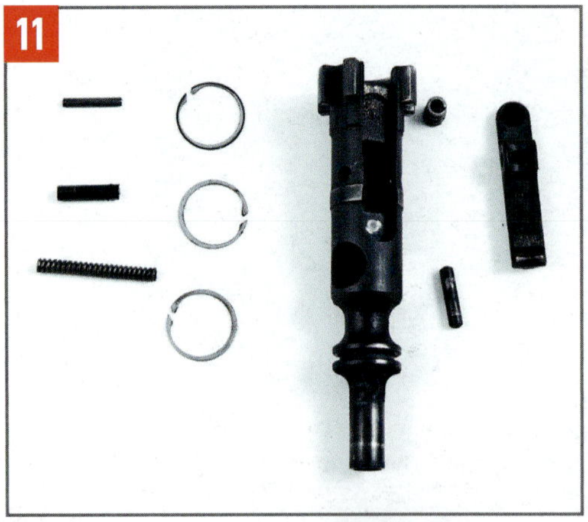

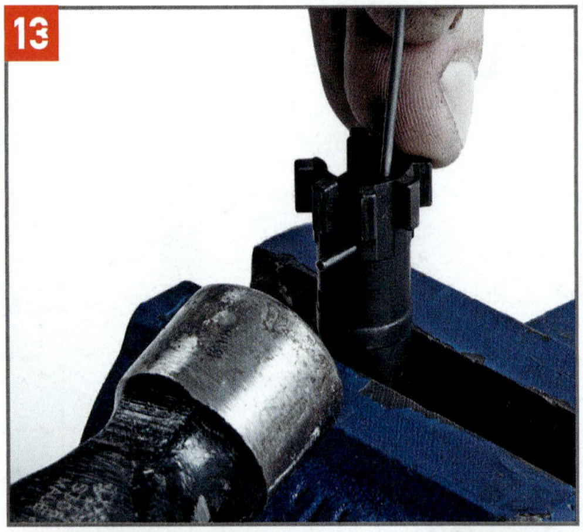

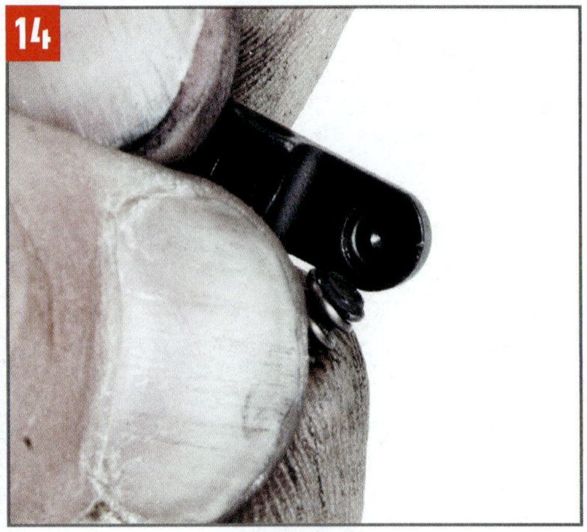

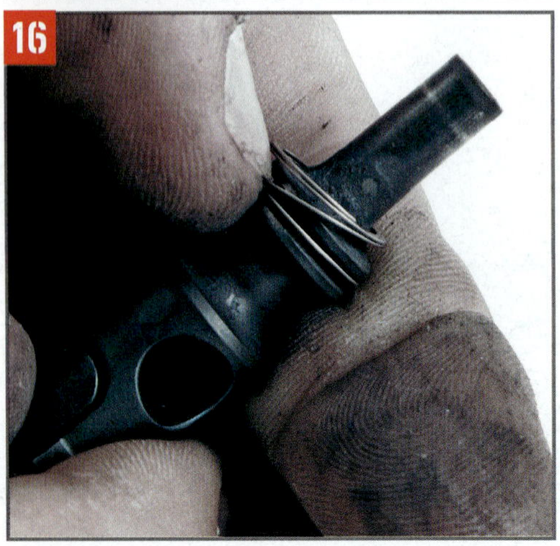

17

11. Bolt completely stripped. Now clean that filthy sumbitch.

12. Check this part of the extractor for wear or damage. If it's worn, replace it.

13. Use your punch to compress the ejector spring while you start the roll pin, then drive the pin with the punch.

14. Replace the extractor spring. Make sure it has the appropriate insert.

15. Replace extractor and line up pin hole.

16. Use thumbnail to wind McFarland gas ring onto bolt tail.

17. Good for another 2,500 rounds at least.

AR-15 UPPER RECEIVER ASSEMBLY

By John Teator

PRESENTED BY

BROWNELLS®
SELECTION. SERVICE. SATISFACTION.

WARNING!

When handling firearms, always observe safety rules and the precautions set forth in the firearm's owner's guide. Be certain that your firearm is unloaded and made safe before proceeding with this DIY.

Understanding how an AR-15's upper receiver is assembled allows you to not only assemble a fully functional upper out of parts, but also gives you the knowledge to do preventive maintenance, to swap parts to keep your upper running right, or to customize your upper as you see fit.

With the current market being as hot as it is, many AR-15 enthusiasts are finding it increasingly difficult to find factory-assembled uppers available at the store. Although buying a complete upper from a reputable company comes with assurances that the quality and workmanship are all there — along with a great warranty in many cases — there is something to be said about building your own upper.

One of the advantages of rolling your own is selecting the exact specs and components you want. Another is gaining the experience and better understanding of how each part interacts with one another. Knowing how your gun operates can be as important as how to shoot it, as this insight can help you diagnose problems on your own, rather than relying on an armorer.

The AR-15's upper receiver is where the bulk of the action occurs within the rifle. It is in the upper that accuracy and reliability are mostly determined. While there are many methods of assembling uppers that can make it quite an art form, we'll leave all the tricks of the trade to the master gunsmiths. Within the scope of this article, our goal is to focus on the basics with this build.

AR-15 UPPER RECEIVER ASSEMBLY

To keep things easier to follow, let's go about this in sections. Parts for each of the sections will be called out at the beginning of each section.

I. Stripped Upper Receiver Assembly
II. Barrel, Gas System, and Handguard Assembly
III. Muzzle Device Timing
IV. Installing Bolt Carrier Group, Charging Handle, and Accessories
V. Function and Safety Inspection

Featured parts and specialized tools used in this article are sourced from Brownells. Part numbers and prices are highlighted in brown so you can easily find the items we used at Brownells' website: www.brownells.com.

TOOLS

There are a few basic specialized tools that you will need to properly assemble an upper receiver.

1. DPMS AR-15 Upper Receiver Block **231-000-157 $45.99**
2. Smith Enterprise .223 Armorer's Wrench **851-000-068 $49.99**
3. Loctite C5-A Antiseize Lubricant **532-000-012 $7.99**
4. Brownells #3 Starter Roll Pin Punch **080-000-244 $6.99**
5. Brownells Bench Block **080-000-493 $18.99**
6. Break Free CLP **102-100-004 $6.99**
7. ½-inch Torque Wrench
8. Vise
9. Hammer

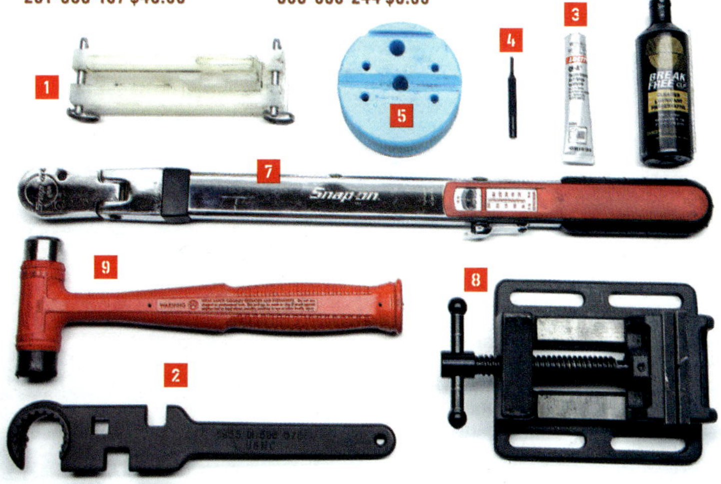

START

We want you to get as much out of this guide as you can, so we went ahead and disassembled the Daniel Defense A4 Upper Receiver Assembly (100-005-851 / $139.99) that the upper will be built off of. We're doing this in case you have a stripped receiver and upper parts kit that you need to put together.

I. STRIPPED UPPER RECEIVER ASSEMBLY

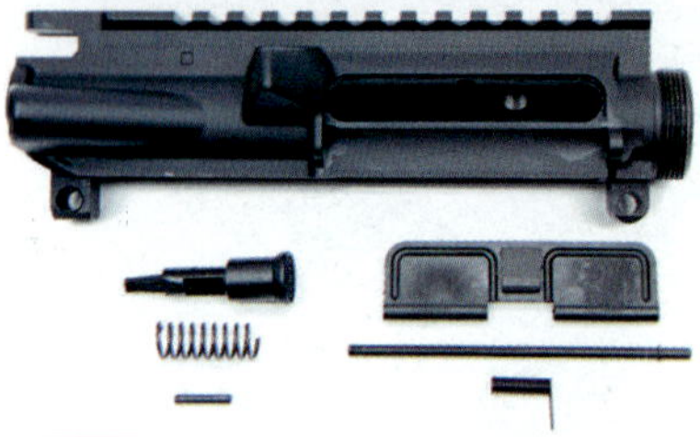

PARTS

AR-15 Stripped Upper Receiver
Forward Assist Assembly
Ejection Port Cover Assembly

1. We'll begin with installing the Ejection Port Cover Assembly into the receiver. The hinge pin has a tiny snap ring installed on it. If your pin doesn't have it installed, go ahead and do that now.

2. With the snap ring portion of the hinge pin to your right, feed the pin into the receiver's right side ejection port cover hole going from right to left. While doing that, thread the pin through the port cover and spring.

3 & 4. Be especially careful with how the spring is aligned. The part with the longer leg should be on the right. The shorter leg needs to be twisted clockwise one full turn before the hinge pin is fully inserted into the receiver. Don't let it unravel before the pin is fully inserted. It'll take a bit of finesse to get this done, but it's doable.

5 & 6. After the hinge pin is fully inserted, test that the port cover snaps closed and springs open. If it just flops open, the spring wasn't installed correctly. If it looks good, let's move on.

7. Now we'll install the Forward Assist Assembly. Place the upper receiver block securely into a vise. Place the upper receiver onto the block. It'll be easier to accomplish the installation by starting the roll pin into the receiver. Just get it started, don't drive it too deep, otherwise it'll interfere with the installation of the forward assist.

8. Now let's take a look at how the forward assist needs to be oriented. The assist works by having its leg engage the teeth on the bolt carrier to help push it into battery. We need to make sure that the engagement surface is pointed toward the carrier when we install it.

1

2

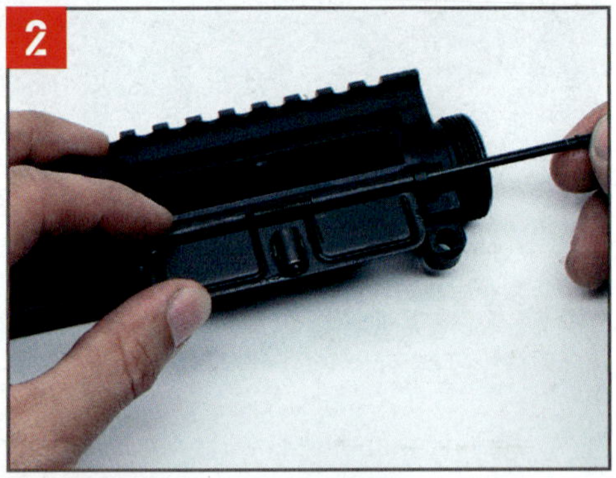

3

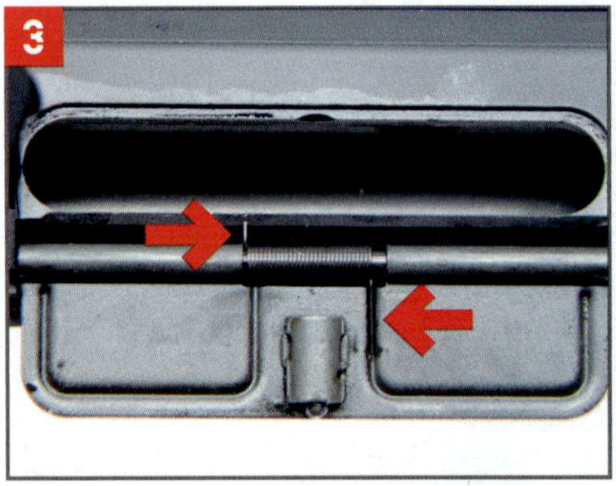

4

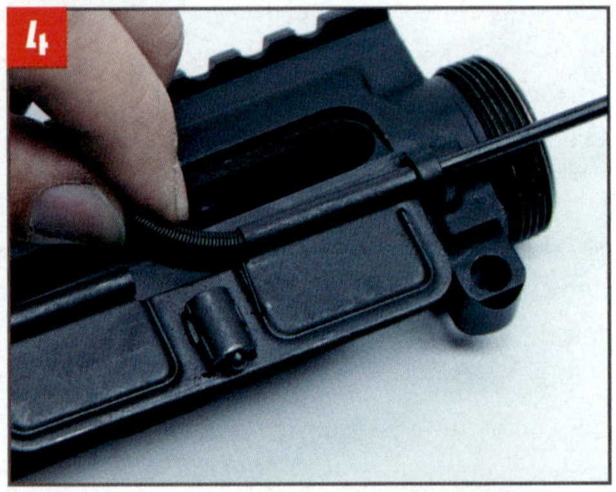

5

6

7

8

9

10

9. With the spring installed on the forward assist, slide the assembly into the receiver's forward assist opening, making sure the orientation of the assist is correct.

10. Now for some hand acrobatics. Depress the forward assist all the way in while driving in the roll pin with the roll pin punch and hammer. That spring will want to battle you, but you will win this battle.

11. Once you have the roll pin in a ways, the pin will hold the assist, making for easier hammering. Tap the pin until it is flush with the top of the receiver's surface.

12. Check that the forward assist button springs back into place after pressing it. It should move freely without binding. Here is your completed upper receiver.

11

12

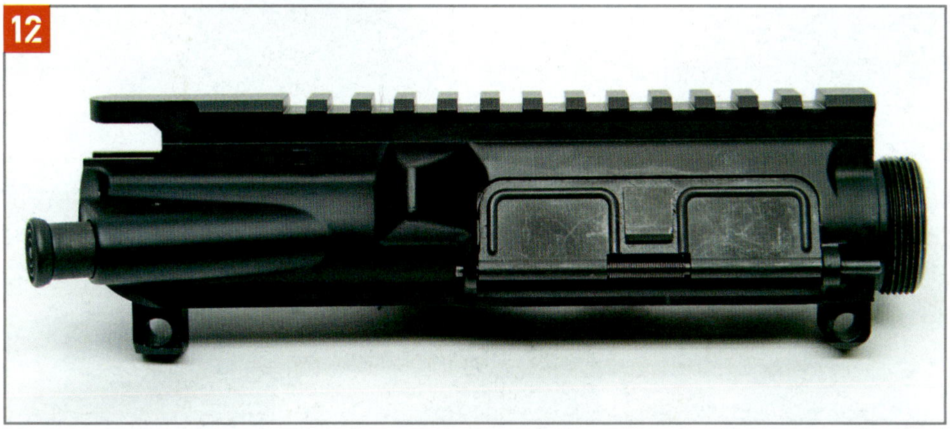

II. BARREL, GAS SYSTEM, AND HANDGUARD ASSEMBLY

BARREL AND GAS SYSTEM PARTS

Daniel Defense M4 Stripped Barrel
100-004-633 $176.95

High Standard AR-15 Barrel Nut
430-000-424 $10.23

Les Baer AR-15 Carbine Gas Tube
124-000-004 $14.99

Syrac Ordnance AR-15 Adjustable Gas Block (with Gas Tube Roll Pin)
100-008-858 $99.95

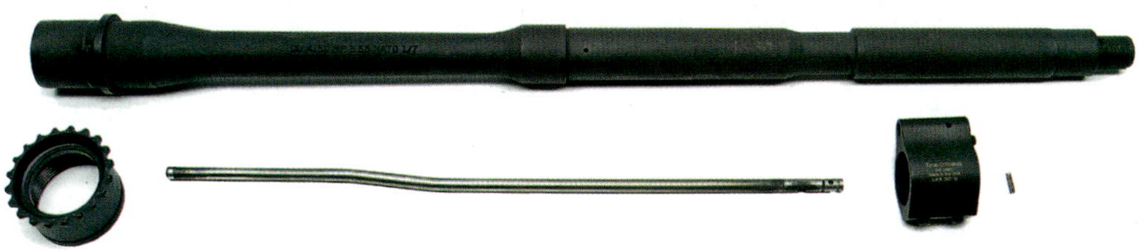

1. Take a look at the plugged end of the gas tube. You'll notice that it has three holes. The larger one lines up with the gas port and the two small ones are for the gas tube roll pin.

2. Insert the gas tube into the gas block making sure that the roll pin holes are aligned and the gas hole is pointed downward, in line with the corresponding hole of the gas block. Make sure you see daylight out the other side of the roll pin holes.

3. You'll also notice that there's a bend in the gas tube; make sure that it goes upward toward the rear.

1

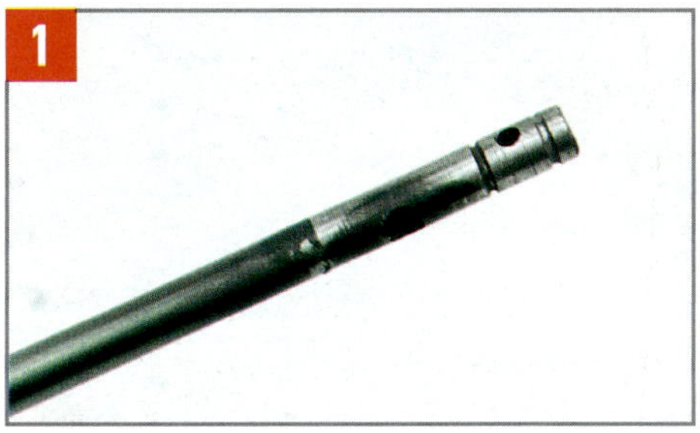

2

3

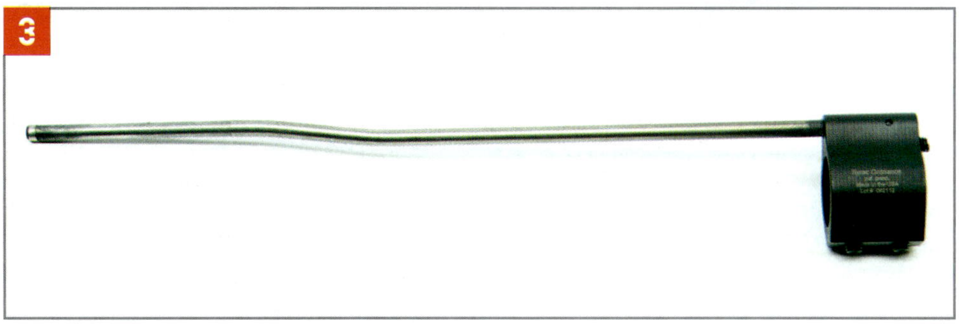

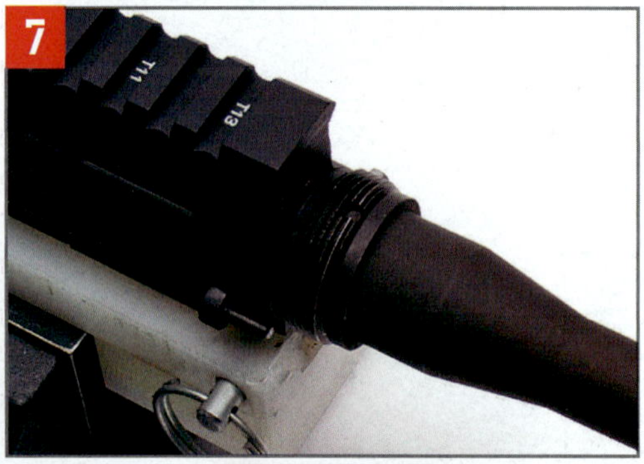

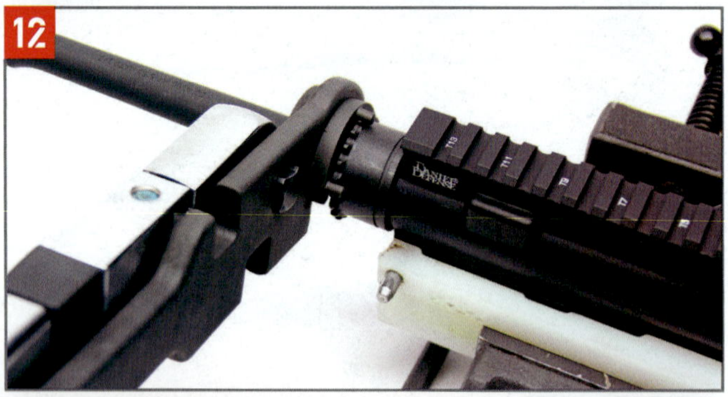

4. Using the roll pin punch and hammer, drive the roll pin into the gas block and tube while being mindful that the gas tube doesn't shift position. The gas tube will want to shift around as you work, so be careful to adjust the holes back into alignment as you go.

5. After the pin is flush with the surface of the gas block, set the assembly aside for later use.

6. With the upper receiver assembly mated to the receiver block in the vise, take the barrel and insert it into the receiver. Be sure that the pin sticking out of the barrel extension is pointed up, at the 12 o'clock position.

7. The pin corresponds to a notch cut into the front of the threaded portion of the upper receiver.

8. Let's move on to the barrel nut. The handguard we are installing uses a standard AR-15 barrel nut, so that's what we're using. Other handguards may come with proprietary barrel nuts. The basic principle of installation is the same, but follow the instructions that come with your choice of handguard. If you ever want to take the barrel off the receiver in the future, it's a good idea to lube up the threads of the barrel nut so that it doesn't seize. With a dab of antiseize on your finger, rub the compound so that it covers all the threads of the barrel nut. A light coat is fine, don't gunk it on.

9. Hand-tighten the barrel nut onto the receiver.

10. Taking a look at it dead on from the front, you'll see that the teeth of the barrel nut may not align properly with the receiver's gas tube hole. You want a gap between the teeth to align with the hole, as this is where the gas tube will go.

11. Tightening the barrel nut onto the receiver is essential to seating the barrel correctly. Take the armorer's wrench and attach the torque wrench to it.

12. Using between 30 to 80 lb-ft of torque, tighten the barrel nut so that one of the gaps between its teeth clears the gas tube hole on the receiver perfectly. You may find that it's easier to do this in stages; torque the nut, then back it off a couple of times.

13. It is recommended to brace the wrench with your hand at the barrel nut so that the wrench doesn't slip off. We also recommend using as little torque as you can because the more torque that is used, the harder it will be to remove later on and the more chance your receiver can get tweaked, but don't go any lower than 30 lb-ft.

14. Don't fudge this; the gas tube hole has to be perfectly clear of the teeth.

15. Now take the gas block assembly and slide it onto the barrel. The gas block will sit at the gas block shoulder that is cut into the barrel.

16. Some gas blocks may require a slight space between the gas block and the shoulder; others will not. You can verify this by measuring the gap between the shoulder and gas hole on the barrel and also the gas hole and end of the block on the gas block.

17. Depending on what type of gas block you get, some attach with set screws, some with pins, some require high-temperature thread treatment such as Loctite or Rocksett. Following the manufacturer's directions, we tighten the set screws to their specs.

18. Make sure the gas tube does not come in contact with the barrel nut and that the gas block is tightened in place.

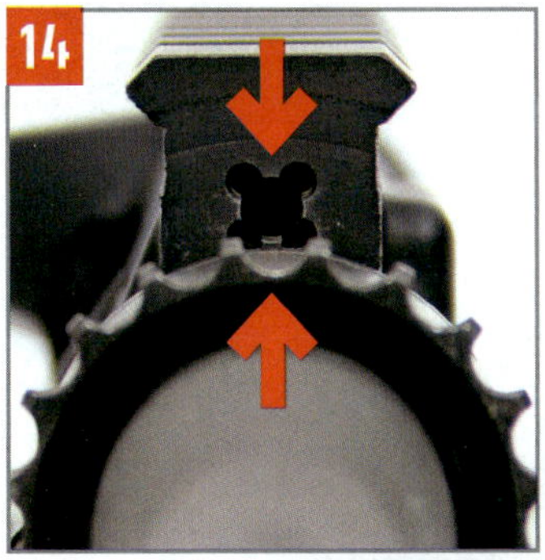

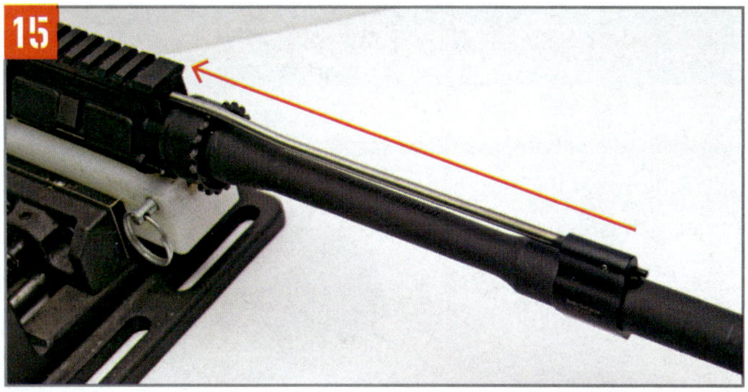

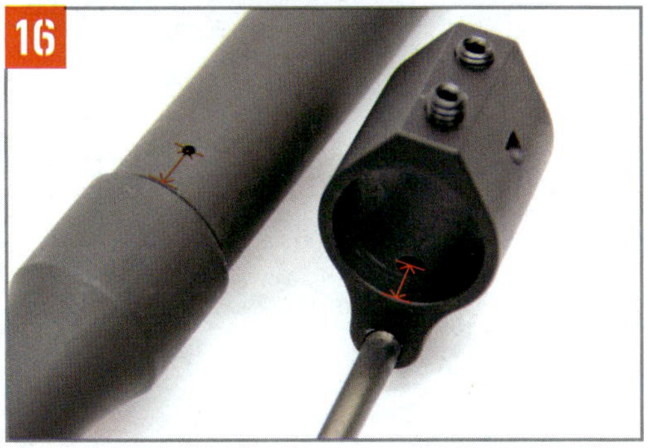

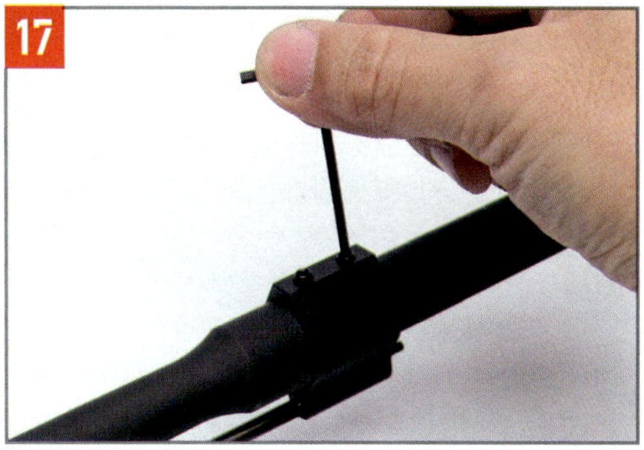

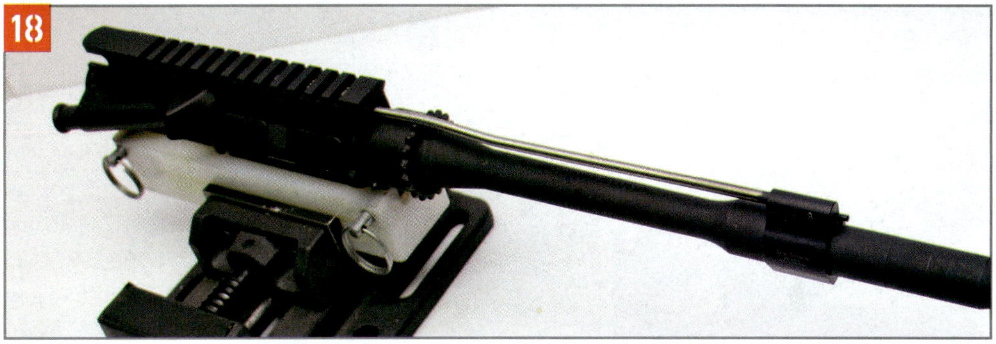

HANDGUARD PARTS

Troy Alpha Rail, 13-inch

100-009-696 $179.99

19. Following the manufacturer's instructions, we install the Troy Alpha Rail.

20. Here is the upper with the handguard installed.

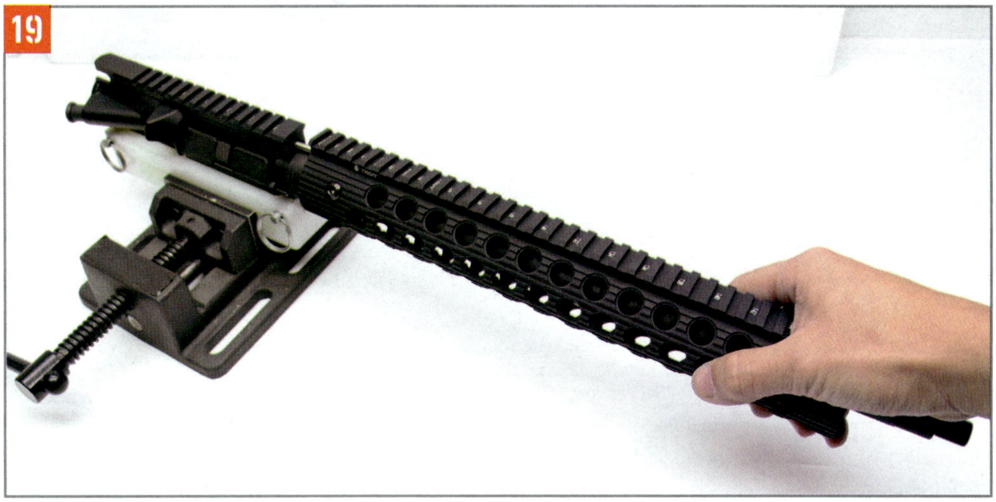

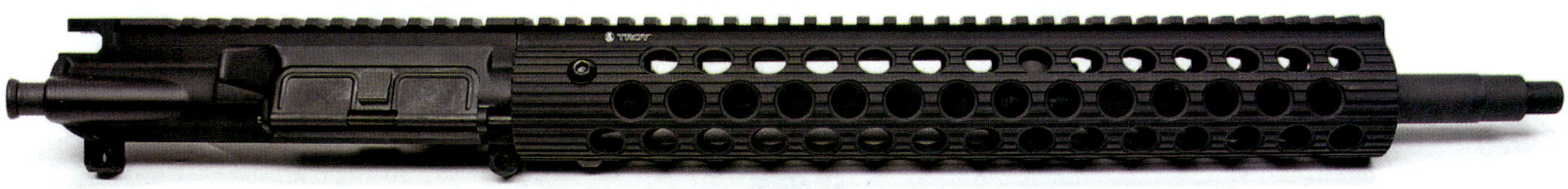

III. MUZZLE DEVICE TIMING

Most muzzle devices need to be timed, or aligned correctly, to get the most out of the device. We'll go through three common methods.

SHIMS AND PEEL WASHERS

PARTS

Surefire MB556K Muzzle Brake Adapter
152-000-045 $134.10

1. Some muzzle devices such as this brake come with a shim kit that allow for it to be timed correctly on the barrel. The shims vary in thickness and are added onto one another to get the brake to the correct orientation. Peel washers work in very much the same way. Instead of adding layers of shims, you actually peel off layers of the washers until you get the desired result. A little heat added to a peel washer makes it easier to peel.

2. Following the instructions, we tighten the brake after installing the proper shims.

3. This brake is properly timed.

1

2

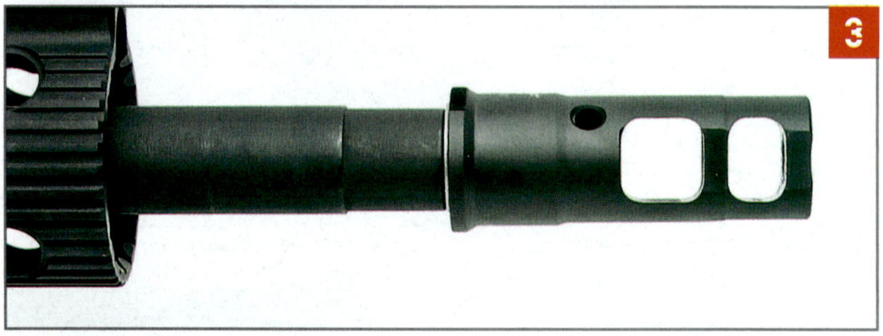
3

CRUSH WASHER

PARTS

DPMS A2 Flash Suppressor
231-000-198 $8.99
DPMS Crush Washer
231-000-175 $2.79

4. The most common way to attach a muzzle device is by using a crush washer. Here, we attach an A2 flash suppressor and crush washer to the barrel by hand.

5. The solid bottom of the A2 must face downward, but at hand tight, this one is over 180 degrees in the wrong orientation.

6. That's OK, crush washers can be turned well over 360 degrees, so all we need to do is use the armorer's wrench to turn it right side up.

7. This A2 flash suppressor is properly timed. Note that crush washers cannot be reused, so use a new one each time you reinstall a muzzle device.

4

5

6

7

NO WASHER

PARTS

Smith Enterprise G6A3 Vortex Flash Hider
851-000-091 $59.99

8. The simplest of them all are devices such as the Smith Enterprise Vortex series that do not require timing at all. In fact, they discourage it.

9. The flash hider is hand torqued and will actually tighten itself with use.

10. This device is properly installed.

8

9

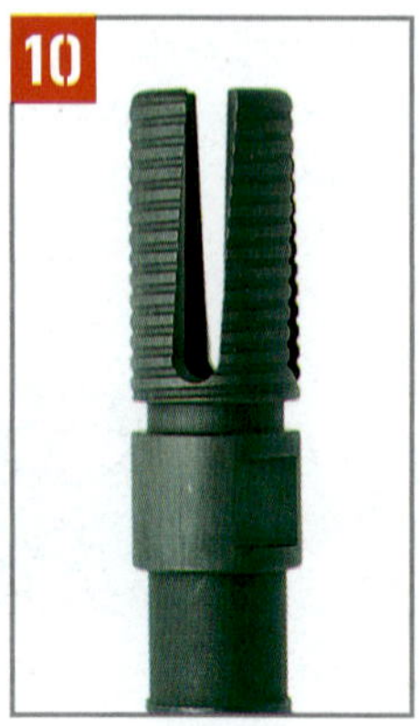
10

IV. INSTALLING BOLT CARRIER GROUP, CHARGING HANDLE, AND ACCESSORIES

PARTS

Les Baer Custom AR-15 Bolt Carrier Group
124-000-001 $229.99
Phase 5 Tactical Ambi Charging Handle Assembly
100-009-955 $79.95
Magpul MBUS Gen 2 Sight Set
100-006-941 $89.99

1. Everyone has his or her preferred lube; we use Break Free CLP in this example. Whichever lube you go with, it's important to know that AR-15s need to be lubricated, especially when new.
2. After lubing the charging handle, we continue lubing the bolt carrier group inside and out. A light coat will do for now.
3. Install the charging handle and bolt carrier group by inserting them into the upper receiver.
4. Following the instructions, we install the Magpul MBUS sights.
5. Our new upper is just about complete! All we need to do now is a final inspection.

V. FUNCTION AND SAFETY INSPECTION

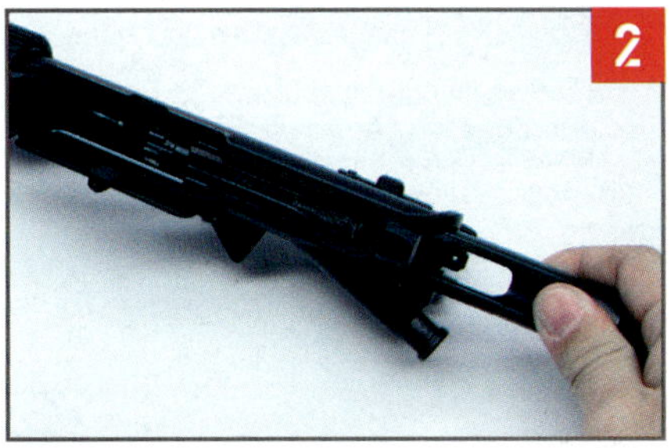

Now that the upper is assembled, we need to check that it is safe to fire and functions correctly.

GAS TUBE AND CARRIER KEY CLEARANCE

Let's first make sure that the gas tube and carrier key do not touch. If they make contact, it could cause the upper to shoot less accurately.

1. Remove the charging handle and bolt carrier group.
2. Replace just the bolt carrier group.
3. Gently push the bolt carrier group into battery. While doing so, see if you can feel or hear the carrier key make contact with the gas tube.
4. If it does, adjust the gas block until the gas tube does not make contact with the carrier key. All you should feel and hear is the bolt locking into the chamber.

CONFIRM ACTION

Now we want to check the action to make sure that the bolt cycles correctly.

5. With the charging handle and bolt carrier group installed in the upper, pull back on the charging handle to disengage the bolt from the chamber.
6. Cycle the bolt into the chamber by hand a couple of times, each time pulling the charging handle to disengage the bolt. Feel for any binding surfaces.
7. The action should be smooth and the bolt should lock up in the chamber securely. If you encounter binding or the action is not smooth, check your work or try adding lubrication and confirm the action again.

V. FUNCTION AND SAFETY INSPECTION (CONTINUED)

CHECKING HEAD SPACE

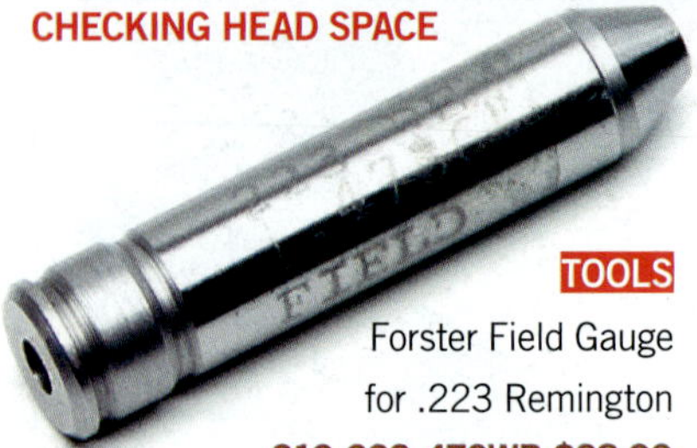

TOOLS
Forster Field Gauge for .223 Remington
319-223-470WB $28.99

The beauty of the AR-15 is that it is a modular system and made according to a strict set of specifications. This means that most parts made by one manufacturer will fit parts from any number of other manufacturers. Headspace is important to check in all firearms, but because of the AR-15's inherent design, most bolts and chambers will be within safety tolerances even when each are made by different makers, especially when the parts are new. However, headspace is a concern for AR-15s with heavily worn barrels and bolts.

The subject of headspace can be a lesson on its own, so we'll just say that if your rifle's headspace is not correct, it could lead to a catastrophic malfunction of your rifle leading to the destruction of your rifle or severe injury, even death. So spending the extra minute to check your upper's headspace is well worth the effort.

8. There are several types of gauges made to check headspace; here we use a Field Gauge. Pull back on the charging handle to extract the bolt from the chamber.

9. Insert the Field Gauge into the chamber.

10. Push the bolt carrier group to close on the Field Gauge. Do NOT reassemble the gun and cycle the action on the gauge — just push gently on the back of the BCG by hand.

11. If the bolt does not lock up, then your upper is safe to fire. If the bolt does lock up, then your rifle is not safe to fire. The bolt, chamber, or both might be worn to the point that the rifle is no longer safe.

9

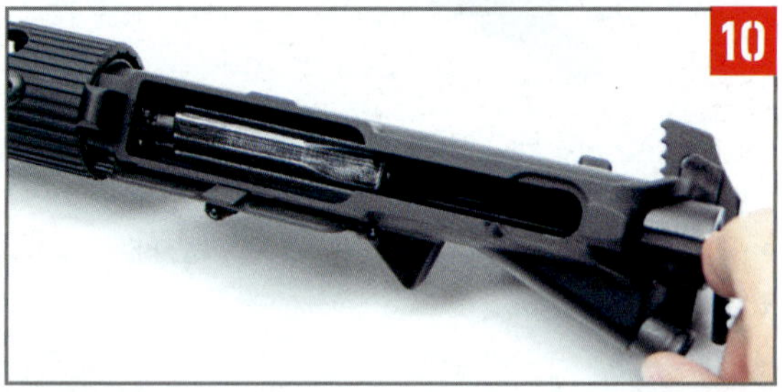
10

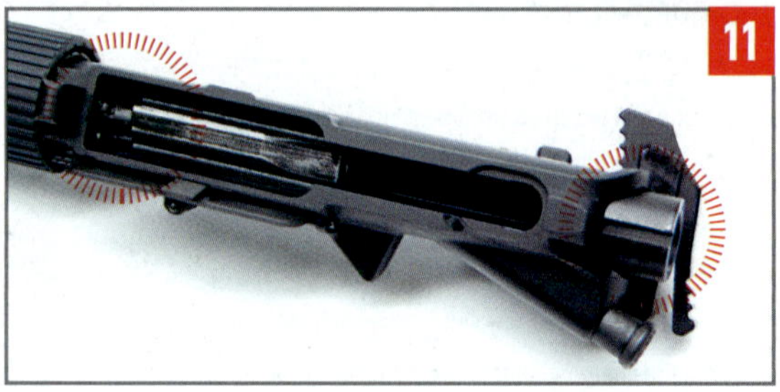
11

There you go; you now have an upper receiver assembly that you put together yourself. Attach it to a lower receiver and you're ready for the range. Remember to keep the upper lubricated, especially during its first couple hundred rounds. Also don't forget to double-check that any screws don't come loose after your first outing with it.

With this newfound knowledge, you can now build, maintain, and modify an AR-15 upper receiver to your heart's content. But please remember to follow all safety precautions and triple-check your work. Since we always love a dose of gun porn, share your creations with us and fellow readers on our Facebook page. R

www.facebook.com/recoil.gun.lifestyle

BROWNELLS
SELECTION. SERVICE. SATISFACTION.®
WWW.BROWNELLS.COM

COMPONENTS

BROWNELLS STOCK#	ITEM	PRICE*
100-004-633	DANIEL DEFENSE M4 STRIPPED BARREL	$176.95
430-000-424	HIGH STANDARD AR-15 BARREL NUT	$10.23
124-000-004	LES BAER AR-15 CARBINE GAS TUBE	$14.99
100-008-858	SYRAC ORDNANCE AR-15 ADJUSTABLE GAS BLOCK (WITH GAS TUBE ROLL PIN)	$99.95
100-009-696	TROY ALPHA RAIL, 13-INCH	$179.99
152-000-045	SUREFIRE MB556K MUZZLE BRAKE ADAPTER	$134.10
231-000-198	DPMS A2 FLASH SUPPRESSOR	$8.99
231-000-175	DPMS CRUSH WASHER	$2.79
851-000-091	SMITH ENTERPRISE G6A3 VORTEX FLASH HIDER	$59.99
124-000-001	LES BAER CUSTOM AR-15 BOLT CARRIER GROUP	$229.99
100-009-955	PHASE 5 TACTICAL AMBI CHARGING HANDLE ASSEMBLY	$79.95
100-006-941	MAGPUL MBUS GEN 2 SIGHT SET	$89.99
100-005-851	DANIEL DEFENSE AR-15/M16 A4 UPPER RECEIVER	$139.99

TOOLS

BROWNELLS STOCK#	ITEM	PRICE*
231-000-157	DPMS AR-15 Upper Receiver Block	$45.99
851-000-068	Smith Enterprise .223 Armorer's Wrench	$49.99
532-000-012	Loctite C5-A Antiseize Lubricant	$7.99
080-000-244	Brownells #3 Starter Roll Pin Punch	$6.99
080-000-493	Brownells Bench Block	$18.99
102-100-004	Break Free CLP	$6.99
319-223-470WB	Forster Field Gauge for .223 Remington	$28.99

*Brownells' pricing subject to change

AR-15 LOWER RECEIVER OVERHAUL

By RECOIL Staff

PRESENTED BY

BROWNELLS

SELECTION. SERVICE. SATISFACTION.®

WARNING!

When handling firearms, always observe safety rules and the precautions set forth in the firearm's owner's guide. Be certain that your firearm is unloaded and made safe before proceeding with this DIY.

The AR-15 is one of the most customizable rifles on the market. With multiple alternatives available for virtually every one of its components, its modification possibilities are nearly endless. However you decide to change up your AR-15, knowing how to actually affect these changes will free you up to personalize it to your heart's content.

Although work on firearms is commonly recommended or required to be done by trained armorers or gunsmiths, these mods on an AR-15 lower receiver are easy do-it-yourself (DIY) projects, as long as you are careful and pay attention to the details. If you are in doubt at any point during this DIY, stop and consult a professional. That being said, if you are up for the challenge and capable of working with your hands, read on and let's get this overhaul party started.

AR-15 LOWER RECEIVER OVERHAUL

We'll tackle the modifications in sections. Parts for each modification will be called out at the beginning of each section.

I. Trigger and Hammer Assembly
II. Grip and Ambidextrous Safety
III. A2 Butt Stock to Collapsible Butt stock & Extended Takedown Pins
IV. Trigger Guard

For this DIY, we sourced the featured parts and specialized tools from Brownells. You can use the part numbers, item descriptions, and prices listed in the spec box at the end of the article to easily find each part or tool on Brownells' website: www.brownells.com.

TOOLS

In order to do all of the listed work, you will need the following tools. Having the right tool for the right job makes all the difference.

1. Brownells Lower Receiver Vise Block
2. Brownells Bench Block
3. Brownells CAR-15/M4 Butt Stock Wrench
4. Brownells Pivot Pin Installation Tool
5. Brownells #3 ($\frac{3}{32}$) Starter Roll Pin Punch
6. Brownells #4 ($\frac{1}{8}$) Starter Roll Pin Punch
7. Vise
8. Large Flathead Screwdriver
9. Hammer
10. Curved Pick or Similar Apparatus

START

We begin with an Aero Precision lower receiver (P/N 953-000-001) with a Double Star Lower Parts Kit (P/N 100-010-669) and DPMS AR-15 A2 Butt Stock Assembly (P/N 231-115-004) already installed.

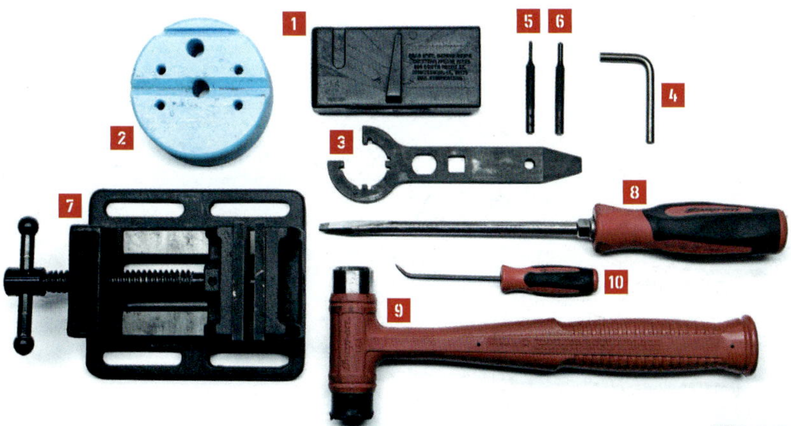

I. TRIGGER AND HAMMER ASSEMBLY

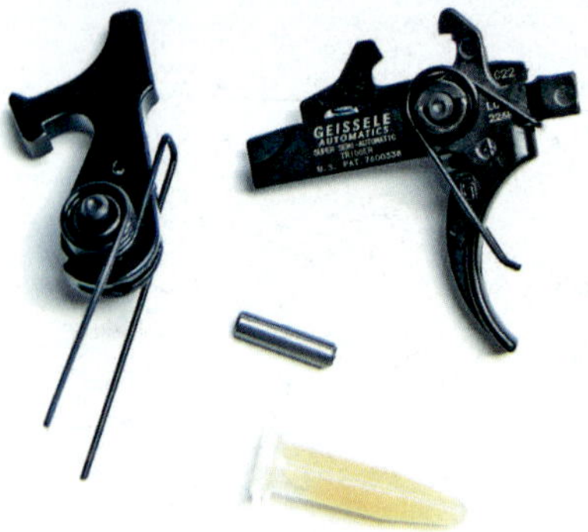

PARTS

Geissele Automatics .154" SSA Trigger

1. With the Lower Receiver Vice Block firmly installed into the vise, place your lower onto the block. Make sure the hammer is not cocked back.

2. Then, use a $\frac{3}{32}$ pin punch to tap out the hammer pin. The upper, forward pin is the hammer pin. It doesn't really matter if you go left to right or vice versa. But, be careful not to lose any parts, because the hammer is under spring tension and will jump when the pin is completely punched out. Here, we use a #3 Starter Roll Pin Punch, but since we're punching out a regular pin, any correct size punch will do.

3. Now do the same with the trigger pin.

4. The lower receiver's fire control pocket is now clear of the old trigger and hammer.

5. Be sure to read the instruction booklet that came with the trigger kit and follow the manufacturer's installation steps.

6. Place the trigger into the receiver with its spring legs pointed forward.

7. Install the trigger pin by hammering the pin in, while aligning the pin with the trigger, disconnector, and receiver holes. Geissele triggers come with a slave pin that keeps everything in line making for an easier install.

8. Make sure the hammer spring is oriented with the upper portion of the spring hooked up on the rear of the hammer, while its legs are pointed down.

9. With the legs of the spring pointed back toward the rear of the receiver, force the spring to bend while you place the hammer into position, so that its holes are aligned with the receiver's holes. The spring's legs should be pointed back, resting on either sides of the trigger pin you just installed.

10. The spring is going to fight you on this, so it might be easier for you to get the hammer pin started on one end before you hammer in the entire pin in. After it is started, it will be easier to align the hammer with the other receiver hole.

11. Now that the trigger and hammer are installed, it's time to do a function check. Cock the hammer back. The hammer should lock back and stay in position.

12. Make sure the safety selector works by putting it into Safe position. The trigger should not be able to fire.

13. Now switch it off Safe, and using your thumb, hold the hammer back and pull the trigger.

14. The hammer will release and fall forward, but use your thumb to ease it forward so that it doesn't slam into the bolt catch or receiver. Gently let it come to a rest.

15. Now with the trigger depressed and held bottomed out in firing position, pull the hammer back until it clicks and locks into place.

16. Slowly release the trigger until you hear and feel a loud click from the disconnector engaging. The hammer should now be in ready to fire position.

17. If all this checks out, you have a properly functioning fire control assembly. If it doesn't, retrace your steps to figure out where you might have gone astray. Common errors involve the trigger and hammer springs oriented incorrectly. Do not fire the gun if it does not pass this function test.

II. GRIP AND AMBIDEXTROUS SAFETY

PARTS

Magpul MOE+ Grip

Battle Arms Ambidextrous Safety Selector [BAD-ASS]

1. Remove the grip's screw through the opening at the bottom of the grip with a screwdriver.

2. A safety detent spring is located at the top right side of the grip. There is also a detent that is located just under the safety selector inside the lower receiver on the same side. Collect those parts and be careful not to lose them when you remove the grip.

3. When the grip, safety detent spring and safety detent are removed and the hammer is cocked back, the safety selector switch will easily come out the other side of the receiver.

4. Now we can reinstall the safety selector. As always, follow the instructions that the manufacturer includes with its product. Here, we are installing a Battle Arms Ambidextrous Safety Selector through the left side of the receiver. Again, the hammer should be cocked back during installation.

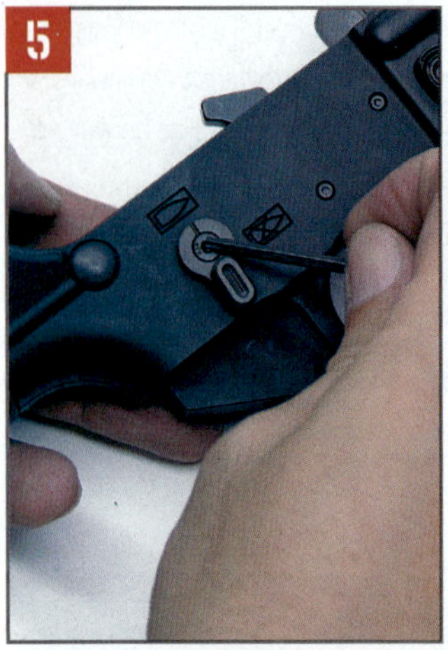

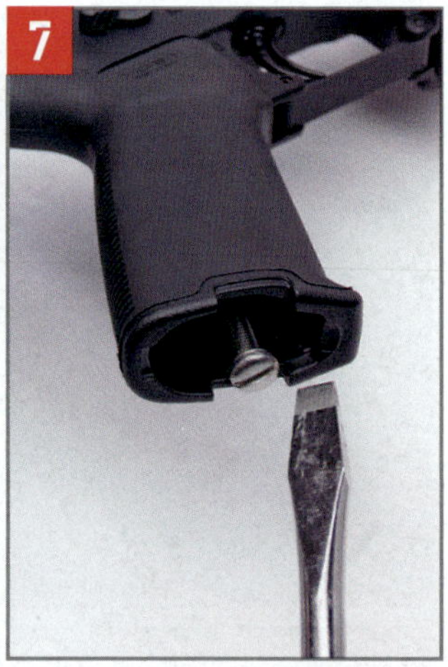

5. On the right side of the receiver, we screw in the right side selector arm.

6. After making sure that the safety selector operates correctly in Fire and Safe modes, we reinstall the safety detent, safety detent spring, and our new Magpul MOE+ Grip (in that order).

7. The screw goes in through the bottom just like the standard grip we are replacing.

8. Now that everything is buttoned up, perform another function test, looking for correct operation and feel. If the grip or safety selector doesn't feel right or operates correctly, retrace your steps and correct any mistakes.

III. A2 BUTT STOCK TO COLLAPSIBLE BUTT STOCK & EXTENDED TAKEDOWN PINS

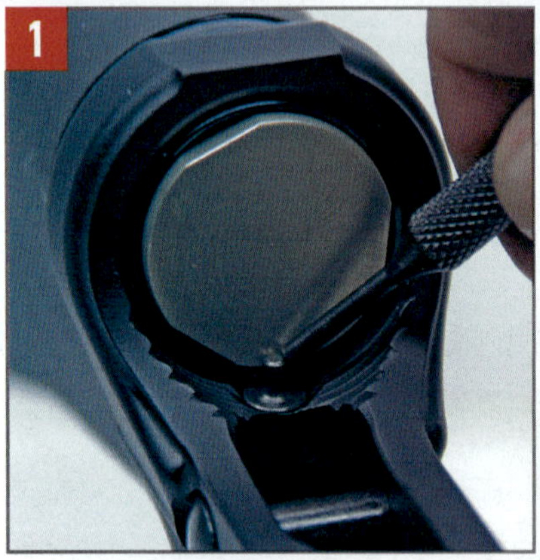

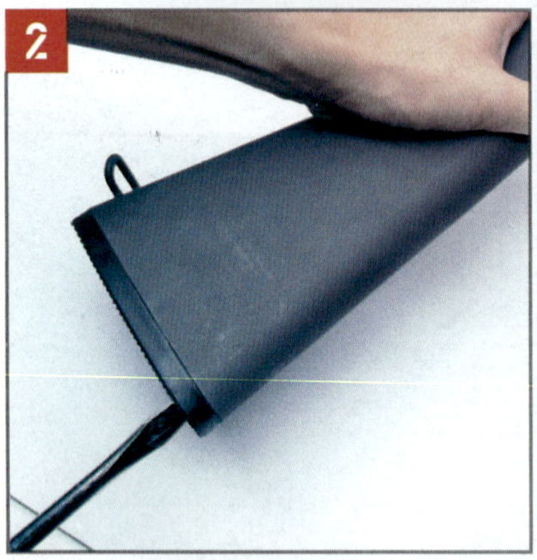

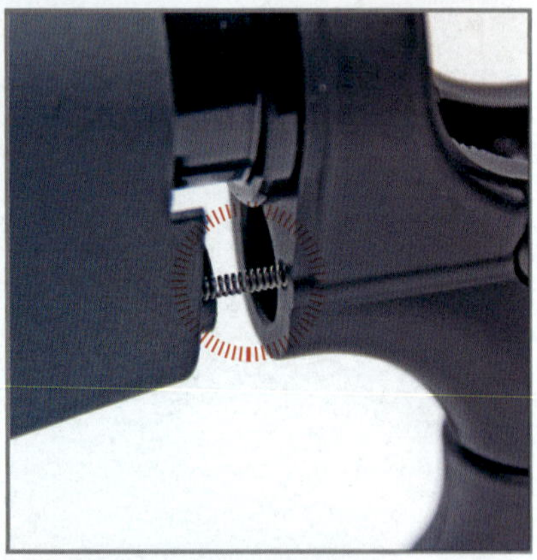

PARTS

Magpul ACS-L
VLTOR Mil-Spec Receiver Extension
Noveske Rifleworks QD Receiver End Plate
High Standard Extension Lock Nut
Colt H2 Heavy Buffer Assembly
DPMS Carbine Buffer Spring
YHM EZ Pull Pivot & Takedown Pins

1. Depress the buffer retaining pin and remove the buffer and action spring from the lower receiver.

2. Unscrew the screw found at the rear of the butt stock. Be careful, the rear takedown pin detent spring is located where the butt stock and receiver meet on the right side. As the butt stock is unscrewed, it will loosen and the spring will want to jump out. As the stock is slowly removed, grab the spring.

3. Following the rear takedown pin detent spring inside the receiver is a detent. Most of the time it's hiding in there and will need to be coaxed out. Make sure to collect it, so that you don't lose it, either.

4. Next, unscrew the receiver extension. Again, be careful because another spring will want to escape. Just in front of the open end of the extension are a buffer retainer pin and its spring just underneath it. The receiver extension holds these two parts in place, so unscrewing the tube will sending them flying, never to be seen again. Place a finger or tool on the buffer retainer as you unscrew the receiver extension. Once clear of the retainer, carefully collect the two parts and put them away; you'll need them during the reinstall.

5. With the butt stock assembly removed, the rear takedown pin can come out. After removing the stock pin, we replaced it with a YHM EZ Pull Takedown Pin.

6. Now let's prepare the buffer assembly for the collapsible stock for installation. The receiver extension we are installing is a five-position VLTOR Mil-Spec Receiver Extension.

7. Start by screwing the High Standard Extension Lock Nut, also commonly known as a castle nut, on the tube, with the wide notches facing toward the rear.

8. Follow that up with a receiver plate. Here, we are installing a Noveske Rifleworks QD Receiver End Plate. This end plate accepts Quick Detach sling swivels to attach your favorite sling. There's a protrusion at the 6 o'clock position on the plate that corresponds to a notch in the threads of the buffer tube. Make sure they match up.

9. Reinstall the takedown pin detent and its spring into the receiver.

10. Now screw the buffer tube into the rear of the receiver. When the front of the tube approaches the buffer retainer hole, reinstall the buffer retainer and its spring. Hold them in place until the buffer tube engages the retainer, keeping it from flying out.

11. At the same time, make sure the rear takedown detent pin spring is being pushed in by the rear receiver end plate nice and straight.

12. When all is in position, hand-tighten the castle nut, so that everything is snug.

13. Place the lower back on the lower receiver block. With the CAR-15/M4 Butt Stock Wrench, tighten the castle nut. We tighten it as much as we can by hand without getting all muscleman on it. Many prefer to also stake it in place for extra security.

14. Next, we installed the Magpul ACS-L butt stock, according to its included instruction manual.

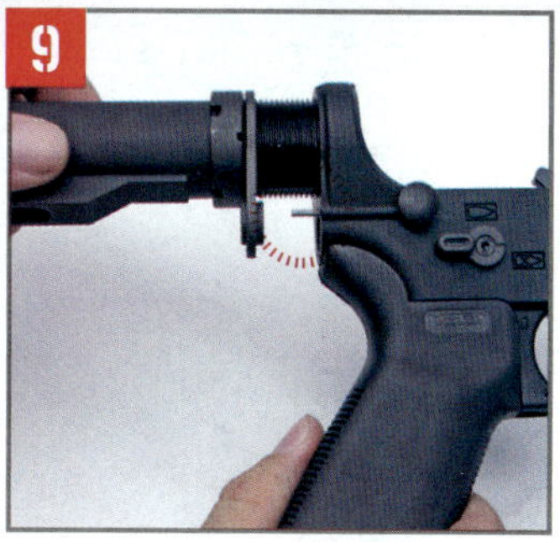

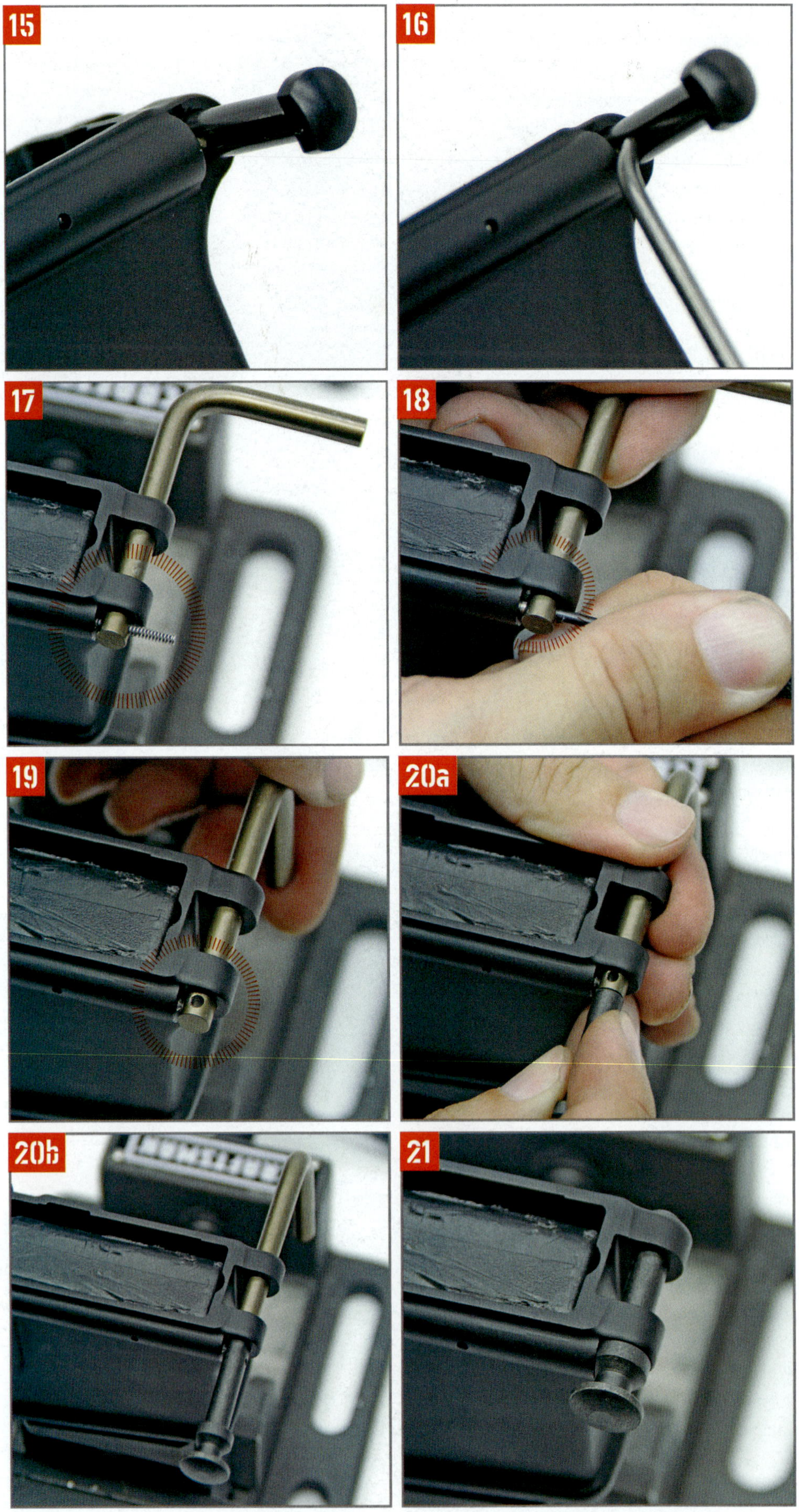

15. Now let's move on to the YHM EZ Pull Pivot Pin. In order to install the pivot pin, the old one needs to be removed. Begin with it pushed out of the receiver.

16. Using a thin yet strong pointed tool, you will need to push the detent that retains it back into the receiver so that you can free the pivot pin for removal. We use an angled pick, which makes things a bit easier, but a sturdy paper clip would do the job, too. Be patient, because the detent spring will want to battle against your efforts, but you will prevail.

17. Now that the old pin is out, let's put in the new pin. Using the Pivot Pin Installation Tool, slide the detent spring into the tool's hole.

18. Follow the spring with the detent by pushing it through the tool's hole with a #3 punch.

19. Carefully turn the tool so that it helps keep the detent and spring retained within the receiver.

20. Now, we can finally install the new pivot pin. With a good solid grasp of the pivot pin, apply a bit of force toward the lower receiver, while sliding it into the receiver front right to left, so that you use the pin to push the tool out. Be careful, because the detent will want to launch out where the pin meets the tool. Avoid this by applying firm rearward pressure, making sure there isn't a gap between the pin and the tool.

21. When it's in, it should be retained within the receiver, so that it can't be pulled completely out of the receiver.

IV. TRIGGER GUARD

PARTS

Magpul MOE Trigger Guard

1. If you've spent any time on the gun forums, you will probably have seen the horrifying sight of trigger guard ears that have broken off of a lower receiver, due to a bad trigger-guard install. Let's not do that. It's important to brace the ears when uninstalling or installing. Here, we use a Brownells Bench Block to brace that delicate area.

2. With a #4 roll pin punch, carefully hammer out the roll pin. If the roll pin doesn't want to budge, try heating the area with a torch and try again.

3. The front of a standard trigger guard is kept in by a spring-retained pin. Push it in and pull the trigger guard out.

4. To install the Magpul MOE Trigger Guard, first use the included setscrew to fasten the front of the trigger guard to the receiver.

5. Now, it's back to the block to brace the receiver's ears, while knocking the rear roll pin back in. Once the pin is in and flush on both sides, the install is complete.

If you've never worked on your AR-15 yourself, you might want to try it out. It's fun and rewarding, and it can save you a few bucks. Remember to always be careful. When in doubt, double- and triple-check your work before proceeding to the next step. Share your results with us on our Facebook page, www.facebook.com/recoil.gun.lifestyle; we always like to see before and after pictures of lower receiver overhauls. **R**

TOOLS

BROWNELLS STOCK #	ITEM	PRICE*
080-000-692	BROWNELLS AR-15/M16 LOWER RECEIVER VISE BLOCK	$49.99
080-000-492	BROWNELLS RIFLE BENCH BLOCK	$18.99
080-000-670	BROWNELLS CAR-15/M4 BUTT STOCK WRENCH	$79.99
080-216-011	BROWNELLS PIVOT PIN DETENT INSTALLATION TOOL	$15.99
080-000-244	BROWNELLS #3 (3/32") ROLL PIN STARTER PUNCH	$6.99
080-000-245	BROWNELLS #4 (1/8") ROLL PIN STARTER PUNCH	$5.99

UPGRADES

BROWNELLS STOCK #	ITEM	PRICE*
100-003-620	GEISSELE AUTOMATICS SSA TRIGGER	$186.00
100-006-786	MAGPUL MOE+ GRIP, BLACK	$22.99
100-005-841	BATTLE ARMS DEVELOPMENT AMBIDEXTROUS SAFETY SELECTOR [BAD-ASS]	$53.99
100-009-325	MAGPUL ACS-L CARBINE STOCK, BLACK	$95.99
100-003-750	VLTOR WEAPON SYSTEMS MIL-SPEC RECEIVER EXTENSION	$44.99
100-004-756	NOVESKE RIFLEWORKS QD RECEIVER END PLATE	$24.99
430-000-532	HIGH STANDARD EXTENSION LOCK NUT	$4.99
160-307-200	COLT H2 HEAVY BUFFER ASSEMBLY	$44.99
231-000-261	DPMS CARBINE BUFFER SPRING	$5.29
100-003-690	YHM EZ PULL PIVOT & TAKEDOWN PINS	$17.99
100-004-156	MAGPUL MOE TRIGGER GUARD	$8.99

*Brownells' pricing subject to change

VINTAGE RESTORATION

The 1911 May Seem Intimidating At First, But Follow This Step-By-Step Guide And You Can Build One From Scratch

By RECOIL Staff

Photography by Henry Z. De Kuyper

WARNING!
When handling firearms, always observe safety rules and the precautions set forth in the firearm's owner's guide. Be certain that your firearm is unloaded and made safe before proceeding with this DIY.

Compared to a Glock or M&P, the venerable 1911 is decidedly old-school when it comes to assembly. Looking at the designs side by side, it's easy to see where progress has been made in the last 100 years — with the help of modern materials and technologies, there's been a ton of effort made to simplify both the manufacturing process and human interface with the weapon, whether shooting or stripping it.

That said, for a century-old design, the 1911 is pretty straightforward once you get the hang of it. Don't try to rush the process. Do work through it slowly and methodically. If you follow these instructions, you can turn a small pile of parts into a functioning firearm in about 10 minutes. For most stages of reassembly, we find it best to group parts together in pairs. We have no idea if that was John Moses Browning's intention, but it seems to work pretty well.

Note: This process works for both an existing 1911 that you may be upgrading, or a bare 2011 frame, such as the one we're building here. Disassembly is simply the reverse of the steps shown, so don't be afraid to follow it in order to detail your gun every few thousand rounds, particularly if you're shooting lead bullets.

1. Trigger - STI standard
2. Trigger - 1911 Standard
3. Sear Spring
4. Mainspring Housing
5. Grip Safety
6. Manual Safety
7. Hammer Pin
8. Sear Pin
9. Slide Stop Plunger, Plunger Spring, and Manual Safety Detent
10. Mainspring Housing Pin
11. Hammer
12. Disconnector
13. Sear
14. Magazine Catch Retainer
15. Magazine Catch Retainer Spring
16. Mag Catch STI Standard

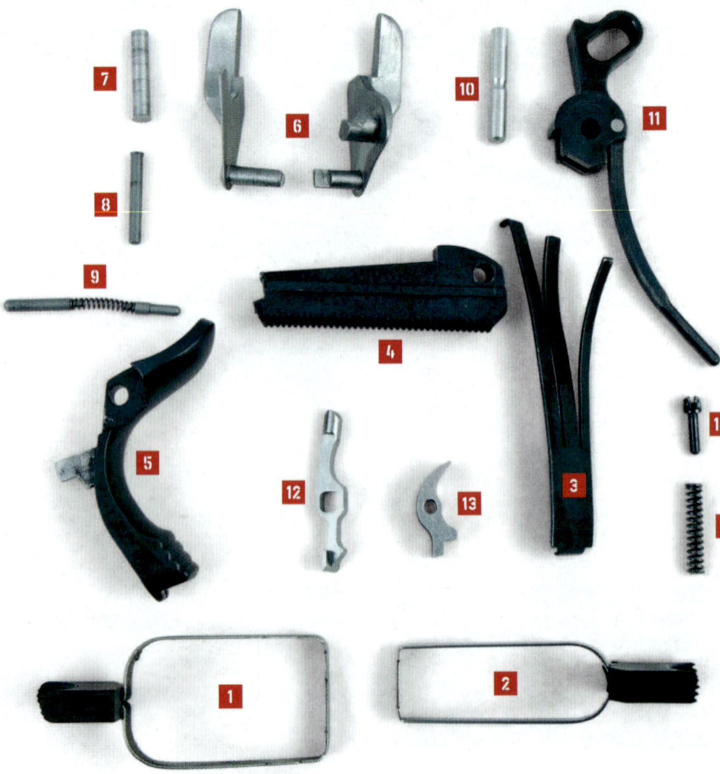

I: FIT THE TRIGGER AND MAG CATCH

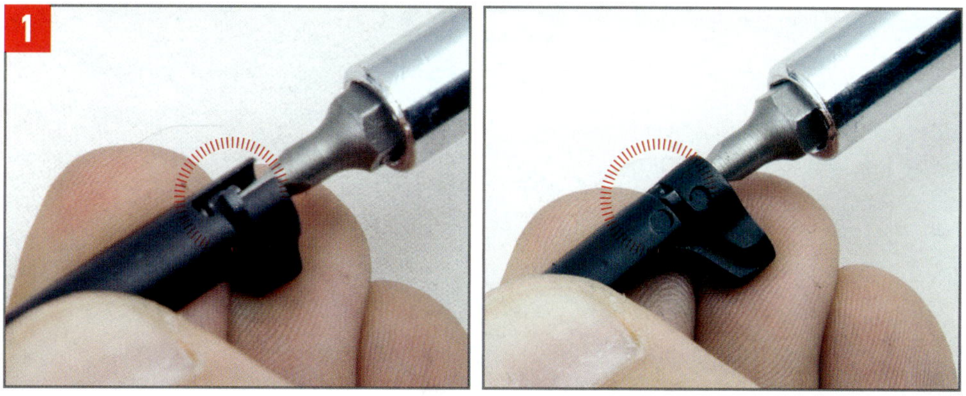

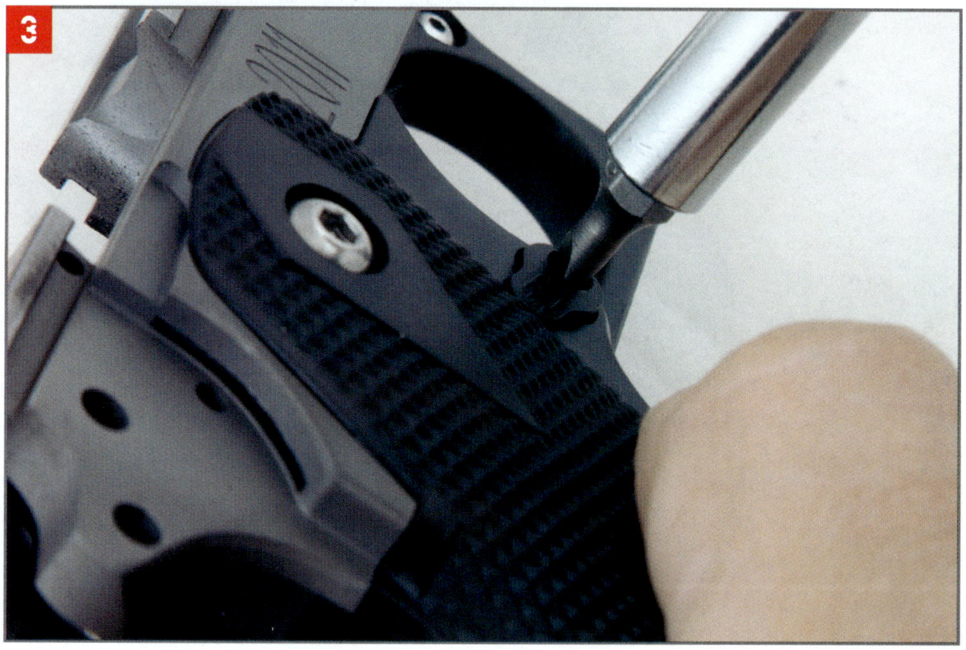

In a standard 1911, the trigger is retained by the magazine catch, so these two parts are typically grouped together. The mag catch is one of the easiest parts to screw up, as it's retained by a little shaft with an ear on it that is prone to shear if you get too ham-fisted with it. So don't. The trick is to press the magazine catch button in while gently applying torque to the retainer until it turns into the locking slot.

In the case of a 2011, the trigger is retained by the grip. To get to it, remove the upper grip screws and the sleeve at the front of the trigger guard, pop out the grip bushings, and separate the grip from the steel receiver.

1. Magazine catch is retained by the tiny ear shown here in its locked position. Make sure you don't apply excessive torque with the screwdriver.

2. Insert magazine catch from the right side.

3. Slightly depress the mag catch while applying torque. The mag catch retainer will then rotate into its notch in the frame.

TWEAK:

Using a Dremel and felt bob with polishing paste, buff the back of the trigger bow where it contacts the disconnector.

II: SEAR AND DISCONNECTOR

Assemble the sear and disconnector as shown in the photo. Slip the assembly into the frame so that the tip of the disconnector sticks up a little and the flat surface is in contact with the back of the trigger bow. Press in the sear pin from the left of the frame (you may have to wiggle the sear and disconnector in order to line up the holes).

1. Disconnector and sear shown in their proper orientation.

2. Disconnector and sear in place in the frame. Note flat surface in contact with back of trigger bow and sear pin about to be pressed home.

TWEAK:

Polish the tab at the bottom of the disconnector, where it touches the trigger bow, and the top, where it passes into the frame. If you want to get serious, polish the frame hole, too. This will eliminate almost all grittiness in the trigger pull.

1

2

III: HAMMER AND SEAR SPRING

Drop the hammer into the frame and press the hammer pin in from the left of the frame to secure it. Note that the sear pin that you installed in the previous step will want to drop out if you turn the frame over, so try to assemble everything without laying the left side down. Lift up the hammer strut and drop the sear spring into position with the little tab engaging in the frame notch.

1. Hammer is placed into frame and pin inserted from left of frame.

2. Lift hammer strut out of the way, then place the sear spring tab into the notch cut in the lower part of the frame.

TWEAK:

The spring has three legs. You can adjust tension on the trigger by slightly bending the left (sear) leg. Be extremely careful when doing this and do not go too light, as an unsafe condition may result. If you make any changes to the sear spring at all, make sure you religiously follow the pre-flight check at the end of this article. Test fire the pistol with only one round in the magazine until you have established that it functions as intended. Do not lighten the middle leg, as this can cause the hammer to follow and ruin your sear.

1

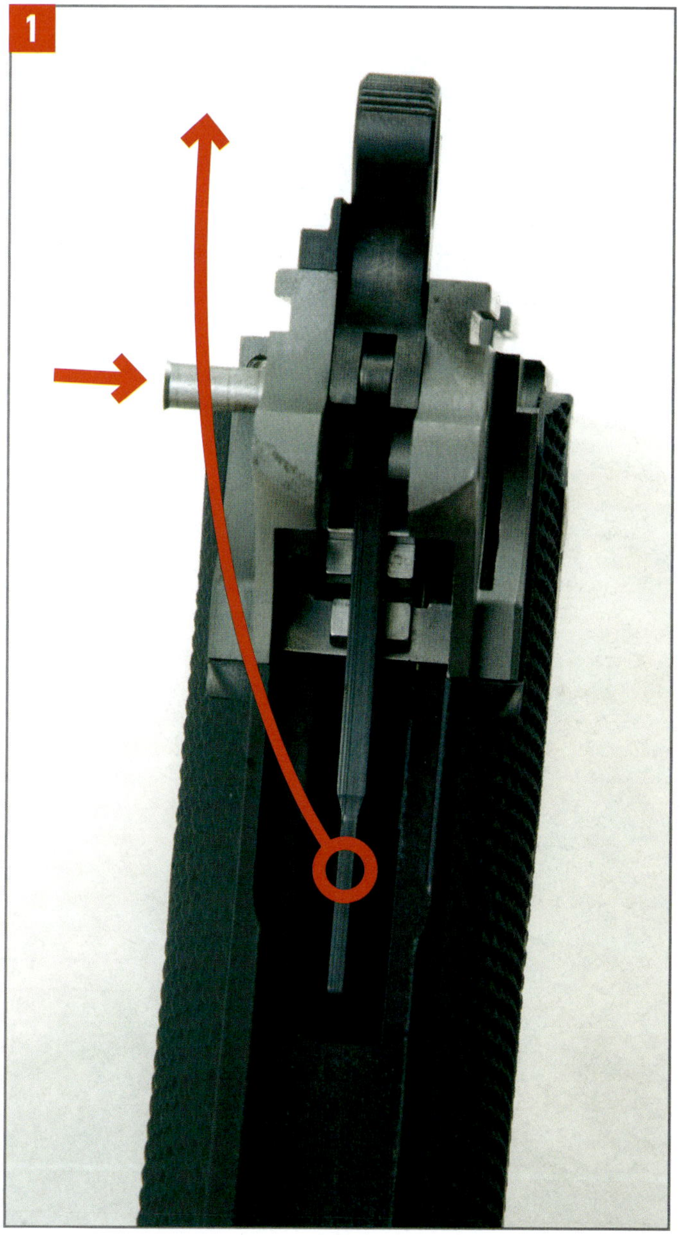

2

IV: MAINSPRING HOUSING AND GRIP SAFETY

Here's where things get a little fiddly. Slide the mainspring housing (MSH) into the slot in the back of the frame, up to where the end of the hammer strut just engages in the cup-like mainspring cap. Now drop the grip safety into position, pressing it against the sear spring so that the little tab at the bottom of the grip safety clears the top of the MSH, then push the MSH the last 1⁄16 inch or so until it bottoms out. The mainspring may become slightly compressed at this point, and you might feel like you need three hands as you need to keep tension on it as you press the MSH retaining pin through the frame. Stick with it.

1 & 2. Slide the mainspring housing into place and make sure the hammer strut is sitting in the little cup on top of the mainspring cap, as shown.

3. Drop the grip safety into place, then slide the mainspring housing home.

4. Press the mainspring housing pin through the frame.

1

3

2

4

V: MANUAL SAFETY AND DETENT

Insert the manual safety detent assembly into the plunger tube, small end first. Cock the hammer. (Note: do not dry fire the gun at this stage, as damage to the fire control components may result.) Press in the grip safety so that the hole through it lines up with the frame, then push the pin at the back of the manual safety through this hole. At this point, the safety won't go completely in, as the detent plunger is sticking out. Using a small punch, push the detent plunger into the plunger tube and press the manual safety home. Note for ambi safeties: if you have a 2011, the right side goes in first, followed by the left side. In a conventional 1911, this order is reversed — secure it by attaching the grip panels and you're done.

1. For a 2011 with an ambi safety, install the right side first.

2 & 3. Then insert the left side, depress the detent plunger with a small punch, and push the manual safety all the way in.

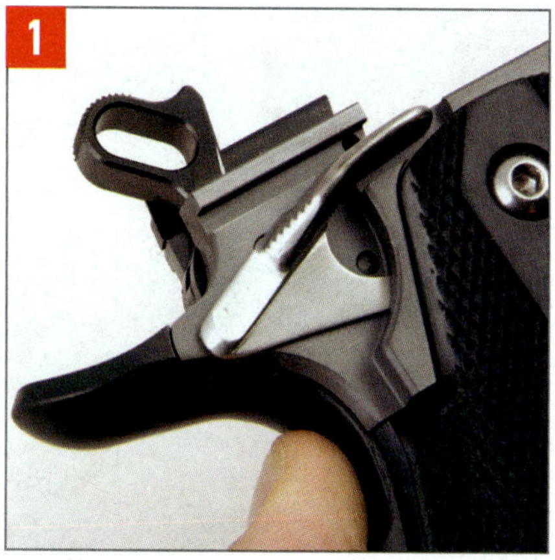

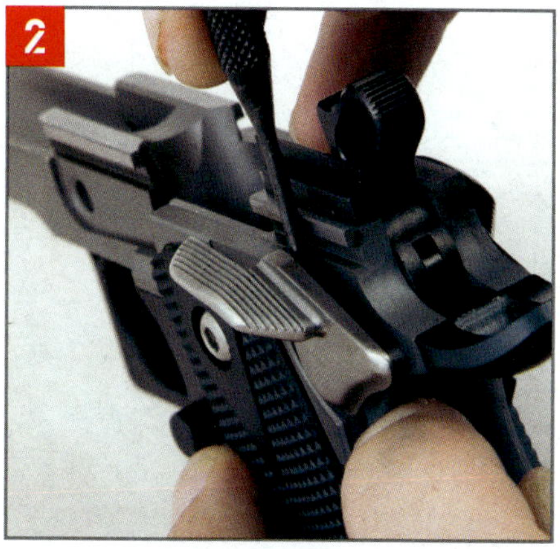

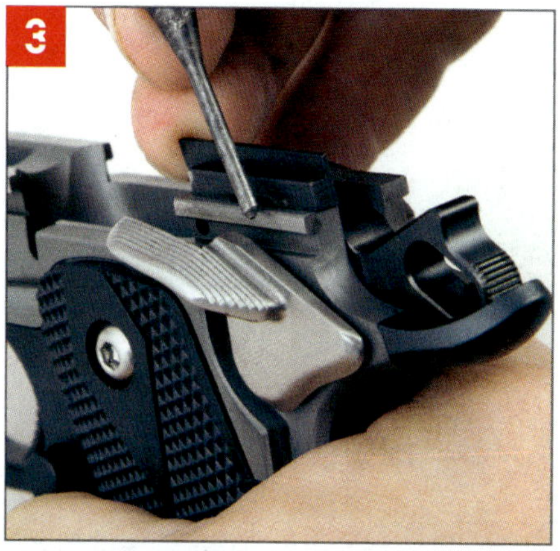

VI: SAFETY AND FUNCTION CHECK

You can use this pre-flight check when inspecting a new gun, or after you've done any work on your existing 1911. Again, if the slide is not installed, don't dry-fire the gun — catch the hammer with your free hand as it slips off the sear so that it doesn't make an expensive "click" sound as it slams against the frame, or worse, the half-cock notch. After clearing the weapon, leave the hammer cocked. Without activating the grip safety, and with the manual safety disengaged, attempt to activate the trigger. If the hammer doesn't fall, congratulations — your grip safety works. Now engage the manual safety and press in the grip safety. Once again, try the trigger. The manual safety should prevent the gun from firing.

If both safeties work as they're intended, with the pistol completely assembled and pointing in a safe direction, take the gun in a firing grip, disengage the thumb safety and press the trigger. The hammer should fall. Keep the trigger pinned to the rear and vigorously rack the slide. The hammer should stay cocked. If the gun passes these three tests, you can now take it to the range and test it with live ammo. **R**

BROWNELLS
SELECTION. SERVICE. SATISFACTION.®
WWW.BROWNELLS.COM

COMPONENTS

STOCK #	ITEM	PRICE*
100-011-124WB	HAMMER, SEAR AND DISCONNECTOR SET	$134.99
965-315-102WB	COMPLETE PIN SET, STAINLESS	$18.95
965-602-145WB	AMBIDEXTROUS SAFETY	$59.99
087-867-000WB	BEAVERTAIL SAFETY	$39.99
087-045-822WB	SEAR SPRING	$5.99
100-003-021WB	MAINSPRING HOUSING	$45
791-111-101WB	MAGAZINE CATCH	$31.99

TOOLS

STOCK #	ITEM	PRICE*
539-000-019WB	PISTOL PUNCH SET	$17.50
080-000-444WB	1911 SCREWDRIVER SET	$24.99

*Brownells' pricing subject to change

CH. 6

Suppressors

TITANIUM DREAMS

The JK Armament SBR PRO "Solvent Trap" and the Bad-Idea Oil Filter

By Dave Merrill

Just like nearly everything else in this issue, we went the DIY route with our silencers. Normally when you purchase a suppressor, it'll sit at your dealer for months — sometimes more than a year — before the federal government approves your paperwork. But it's faster if you make your own; in fact, we did just that in RECOIL Issue 46.

"Solvent trap" is the current underground moniker for parts and kits used to manufacture your own silencer with an approved Form 1. These are perfectly legal in most jurisdictions, provided you haven't drilled holes in anything just yet. Why would you want to fill out a Form 1 instead of a factory Form 4 transfer? Time, mostly. As we write these words, an online Form 1 takes an average of three weeks whereas even the fastest Form 4 transfer now sits between three and four months. Cost savings is another measure, as most solvent trap kits are available at a significant discount relative to factory silencers (invariably for a reason).

FILTERING "OIL"

We'll get the only-worth-it-for-memes alternative out of the way first: the oil filter adapter.

About a decade ago, we first laid eyes on oil filter adapters; they were originally made by Tom Cole of Cadiz Gun Works and called "solvent trap adapters." Cole tells us, believe it or not, that the original intention was to actually trap solvents. Later, they released their Econo-Can, a registered adapter that's transferred like a standard silencer.

But this type of filter being featured in the 2019 release of *Call of Duty: Modern Warfare* really brought them into the world of mainstream memes. While you can find sketchy adapters on Amazon and Wish, Cadiz Gun Works actually manufactured and legally registered oil filter adapters as silencers and even have an oil filter replacement service for when they wear out.

We teamed up with an SOT to legally register our own. Engraving actually took more work than we anticipated because our adapter didn't have enough surface area. A stainless steel washer was welded to the base for this purpose.

So, how effective is it? Setting aside that it makes most pistols a single-shot affair because there's no way to attach a booster, the ungainly size that blocks sights on both rifles and pistols and the massively increased weight, simply put: It's not all that quiet. An oil filter silencer essentially is a simple expansion chamber that works for a few shots before having to be replaced. And the larger the bore, the faster it loses any efficacy.

JK Armament has everything you need sans a vise and cordless drill to make an SBR-ready modular titanium silencer.

And oh yeah — you can't even legally swap out that oil filter unless you're a manufacturer.

In a world where silencers are OTC just like a receiver, rifle, or pistol, these oil filter silencers would make for a fun party trick, similar to taping a soda bottle to the end of your .22LR. But since legally manufacturing or purchasing adds on $200, and you can't even swap the filter yourself? Hard pass.

JK ARMAMENT SOLVENT TRAP

This brings us to the exact other end of the DIY spectrum, the JK Armament.

JK Armament is a new company, just releasing "solvent trap" kits for sale in 2020. Though the company is new, the people behind it aren't. The namesake of the company, Jake Kunsky, has over two decades of experience in industrial manufacturing and weapons design. Kunsky's company is the culmination of his experience developing products for NEMO Arms, Gemtech, and Maxim Defense. His background with silencer design and engineering gives him unique insight into what today's customers are looking for in a user-configurable and maintainable solvent trap.

DESIGN

JKA's first product was the JK 155 Modular Solvent Trap (MST) Kit made for either rifles or pistols. The rifle kit is nearly identical to the pistol, except the tube is one piece instead of two, it comes with a fixed barrel adapter, and the eight stackable baffles are optimized for the higher pressures of rifle rounds, so they can't be drilled so wide.

The name of the game is modularity, and it's certainly evident here. Each baffle section threads right

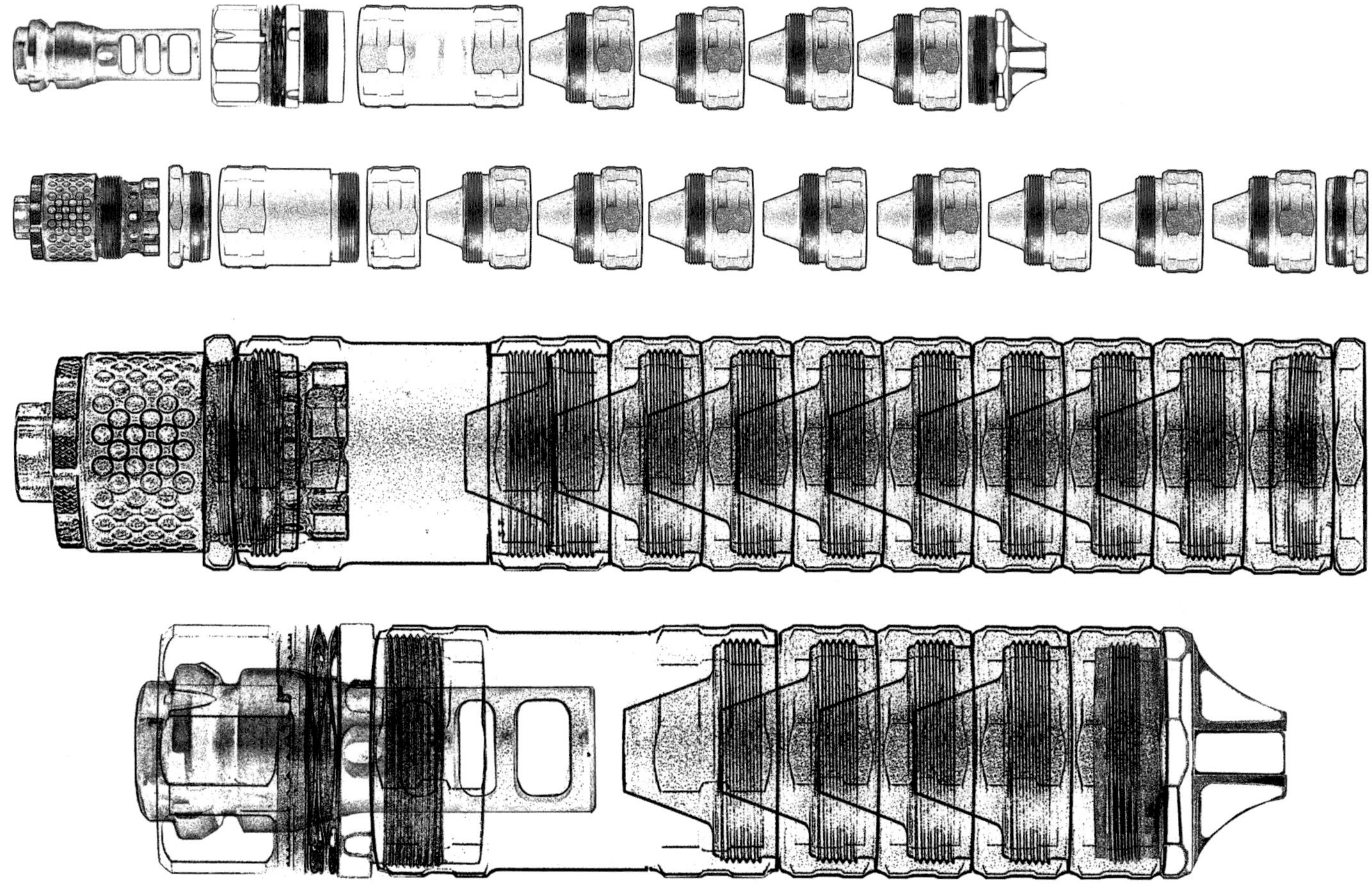

on top of another, closely mirroring what we've seen from the likes of Schultz & Larsen for the last decade, the Erector from Q LLC, the Dead Air Odessa, Rebel Silencers SOS, half the product line from Torrent Suppressors, and the new Sig Sauer Mod X.

No need for a tube, spacers, or math, nor do you have to worry about dubious construction, as you would with a cheap Chinesium kit.

However, these first kits were made from aluminum. Aluminum is just fine for rimfire, most pistol calibers, and subsonic 300BLK, but once you start rolling out higher pressures, the aluminum will quickly erode away.

Now, JKA does what we anticipated from the very beginning, making them out of titanium. Aside from being turned from titanium, the JK 115 SBR Pro kit comes with four baffles instead of eight, and a new flash hider endcap is available. If you want a larger kit with more baffles, you can simply add them prior to manufacturing.

The base tube and all baffles are threaded HUB 1.375x24 TPI, so mounts will never be an issue as nearly two dozen companies on the market currently make parts and pieces for the HUB. For pistols, a SilencerCo, Torrent, or Griffin Armament 1.375 booster will nest inside, and JKA makes an adapter for those with Gemtech, SDTA, or Liberty boosters.

PLANNING

Planning ahead will help you maximize the utility and versatility of your silencer by making it multi-caliber AF. Not only can you buy the kits, but you can buy individual and incomplete baffles at the same time. Want to make a silencer with 10 baffles? You can do that. Want to make your own bastard version of a SilencerCo Wizard Staff? Sure. Want to drill baffles with different aperture sizes in order to make this a modular silencer that's always caliber-specific? Oh yes, you can do that too.

You can figure whatever length, caliber, or muzzle attachment you want. Dead Air KeyMo? Sure thing. Pistol boosters? Also, yes.

There's some word going around that you can't register a silencer in a smaller diameter than the bore holes, but that's hogwash. There's no regulation or dictate about how large or small each aperture has to be for a given caliber, so long as you can put the registered caliber through it in any configuration — just no extras and no spares. Of course, all of this may be subject to change at the whim of the BATFE.

The Enfield Rifle Co Novus, as seen here in RECOIL Issue 47 also works on this principle with their nesting baffle stack; it's a 5.56 silencer that can be assembled to effectively suppress .30-caliber or .45ACP.

SOCOM Suppressor Firing Table		
Magazine #	**Rate of Fire**	**Rounds Fired**
1	(1) round per second	30
2	(2) rounds per second	30
3	(1) round per second	30
4	(3-5) round bursts	30
5	(1) round per second	30
6	(2) rounds per second	30
7	(1) round per second	30
8	Full-auto mag dump	30
	Total Rounds:	240

There's currently no reason why you couldn't make a JKA kit with 12 baffles with three different caliber apertures, but we'd recommend Cerakoting or otherwise marking each caliber so you don't inadvertently get a strike.

MANUFACTURING

After we planned out our silencer and received our approved stamps, it was time to hit the workshop. The required info was lasered onto the main tube, so we needed to drill the baffles. Kunsky actually advises against using a drill press, even if you have access to one.

JKA solved any centricity issues with a custom drilling jig. It's meant to be used with a hand drill and a vice — no need for a mill here; the undrilled baffles simply thread onto the jig. A caliber-specific bit available from JKA aligns with a bushing on the jig to ensure it hits center.

The same jig also works for endcaps, and since the jig itself is made of aluminum, your vice can squeeze it hard enough that the endcap won't rotate out when you drill.

With the jig at the ready, along with a cordless drill at a low-speed setting and some machine oil, it was fairly easy. If you drill too fast or don't use cutting oil, you can wreck one of these by chipping it if you're not careful. Take your time, and you'll be fine.

After each new section, we threaded the completed baffles on top of each other to confirm alignment. Make sure you clean the baffle well, or else an errant chip may mislead you.

After 3 SOCOM tables, the blast baffle (left) looks a bit sandblasted, but you can just rotate it away to increase longevity.

AT THE RANGE

The original JKA kit was aluminum, relegating it for us solely to pistol calibers and limited use with 300BLK. But with the titanium PRO kit, we knew we could press it much further. And press it further we did.

Using a select-fire 10.3-inch rifle, we decided not to run one or two, but three SOCOM Table II tests. Table II specifies eight fully loaded 30-round magazines with rates of fire ranging from one round a second to full-auto mag dumps. The entire table is performed with no rest between stages, but everything should come down to ambient temperature prior to repeating the table.

Effectively, this is a stress test for the silencer. This accelerated wear

Bits and jigs are available for 5.56, .30-caliber, and 9mm.

gives you a pretty good idea of how the silencer will perform long-term.

Heat is the bane of titanium. Heat erodes titanium, and you can see it happen in real-time when sparks shoot out of the end of your silencer. There was a time when no one would have seriously considered titanium for anything other than a precision rifle or rimfire silencer, but that was nearly 15 years ago.

All said and done, we fired 720 rounds with this test, and while there was noticeable wear on the blast baffle, the others were mostly fine. But, since this is a modular can, you can simply rotate your blast baffle to the end of the stack to easily increase the longevity of your silencer. And in the worst-case scenario? Have an SOT replace one of the baffles ($125 for a Ti cup, $45 for aluminum).

The flash hider endcap performed exactly as expected too, which didn't surprise us because it's definitely on rather long. For our next build, we'll probably trim them down a bit to reduce the overall length.

LOOSE ROUNDS

You'd be well-advised to add some anti-seize to the threads to aid in disassembly and reconfiguring your can. If you carefully plan your build in advance, you'll have a handy silencer that can effectively eat all sorts of ammunition without a compromise like an oversized bore.

When we're dealing with a new company and a new product, there are typically some warts here and there. Kunsky's experience definitely shows in this design, bringing a lot of advanced features to a home builder that would otherwise require expensive equipment or regaling yourself with eBay parts.

We can't wait to see what comes from JKA next.

DIY SUPPRESSOR

Here's How to Legally Make Your Own Can, Save Yourself a Year of Wait Time, and a Few Hundred Bucks

By Iain Harrison

Photos by Kenda Lenseigne

Making your own firearm from a partially completed receiver is something we've covered extensively in the past, but there's a perfectly legal route to achieving the same ends with a suppressor, which in many ways is more effective and attractive for the average law-abiding citizen. In order to transfer a factory-made can from your local dealer, you'll have to fill out an ATF Form 4, pay a $200 tax, submit fingerprints and a passport photo, and then wait. And wait.

After around 11 months, you might get the chance to go pick up the property you paid for so long ago, or you might just have to wait a bit longer. If this doesn't sound quite so appealing, then you could always go the DIY route. There's no getting around the paperwork and tax stamp, but you end up with a workable solution in weeks, rather than months. We covered the Form 1 E-File process in Issue 44, and the article is currently on RECOILweb.com, so if you haven't already, you may want to familiarize yourself with it. There's one departure from the procedure outline in the article, and that's the bit which deals with describing the manufacturer of the item.

On the drop-down menu, select FMI (for Form 1 manufactured), and you're on your way. You'll also have to describe the length and caliber of the can you're going to make. Tip: Some people get wrapped around the axle when it comes to fingerprints. There's absolutely no reason to make an appointment and pay a third party to fingerprint you, when you're perfectly capable of smearing ink on your own digits. Order a fingerprint kit from Amazon, and do it yourself in the comfort of your own home.

Once your Form 1 has been approved, which usually takes around three weeks, you can then buy a tube, spacers, baffles, and endcaps from the many online vendors that exist on the fringes of the interwebs. Due to the nature of NFA law, these will be described in rather coy terms, and you may wind up purchasing "barrel shrouds," "solvent traps," "oil filter kits," or "storage cups," all of which are largely useless for their advertised purpose, but give the vendors a fig leaf of deniability. Yes, it's all a bunch of bullsh*t, but it's the system we're stuck with.

Once your components arrive, you can then set to work engraving the tube to meet the legal requirements of the National Firearms Act (see RECOIL Issue 44). You could go get this done on a laser engraver and make it look all professional-like, or you could just bust out the Dremel. We did the latter, as it's going to be wrapped in a suppressor cover anyway. With your tube engraved, you can then drill holes in the baffles and endcap, screw everything together, and head to the range with your shiny new can. Enjoy! R

Our DIY suppressor consists of a grade 5 titanium tube and titanium end caps. The first two baffles in the stack are 17-4 stainless steel, and hardened with a propane torch, as these take the brunt of propellant gasses — the rest are titanium to keep weight to a minimum. Note the small cuts on the side of three baffles, designed to induce more turbulence and hence slow gas flow further; these were done with the ubiquitous Dremel and finished with a Swiss file.

» SOURCE

ATF E-Forms: https://eforms.atf.gov/

Form 1 Suppressor forum: http://form1suppressor.boards.net/

Parts Vendors: https: //superprecisionconcepts.com/

https://sdtacticalarms.com/

CH. 7

Striking Implements

Transfer your blade design onto the steel using the layout fluid and a metal scribe for the most workable design sketch.

KNIFEMAKING

DIY Bladesmithing Your Own Knife at Home

By Chad McBroom

There's no doubt that every OFFGRID reader understands the importance of a well-constructed knife. It is arguably the most valuable tool in any survival kit. Whether out of love of the blade or an innate desire for self-sufficiency, many have considered putting their own hands to the forge only to be dissuaded by the expensive equipment and technical skills employed by modern-day bladesmiths.

In this step-by-step instructional article, we'll guide you through the knife-making process from start to finish, using a stock removal method and basic hand tools.

STEP 1: SELECT A STEEL

The foundation of the knifemaking process is steel, so consider your options carefully. Choosing a steel that has the right balance between hardness, durability (toughness), and corrosion resistance can be challenging, especially when you're working with limited resources and technology. Many of the steels you might find at the local salvage yard or hardware store are mild steels unsuitable for anything more than a prison shank, while the many "super steels" used in high-end cutlery must undergo hardening processes that require specialized equipment and extreme precision. It's best to stick with basic, high-carbon steels like 1095 or 1075. Some tool steels like O1 are also relatively easy to work with.

Whenever possible, you should purchase your steel from a reputable dealer that specializes in cutlery steels. This is the best way to make sure you're getting correct and uniform materials. Plus, you can purchase the steel in the width and thickness you wish to work with, which will save you a lot of time and effort.

Mystery steel (any steel with unknown properties) should always be a last resort. In the event of a global meltdown or zombie apocalypse, a car leaf spring would be a fair gamble, especially in older models where 5160 or similar spring steels were commonly used, but with modern cars it's still a crapshoot. Plus, repurposed steel, if not already flat, will require additional cutting, bending, twisting, and/or grinding to make it flat and useable.

STEP 2: LAY OUT THE DESIGN

Once you have acquired your steel, the next step is to decide on the blade and handle design and then transfer that design onto the metal.

The easiest way to do this is to sketch your design on a piece of cardstock or cardboard with a pencil. Once you're happy with your design, cut out the design and then trace it onto your steel. You can use a steel layout fluid like Dykem to coat the surface and trace around the template with a metal scribe, or you can simply use a black marker on the bare steel surface.

STEP 3: SHAPE YOUR BLANK

Shaping the knife blank with minimal tools will be the most tedious step in the process. This can be done with nothing more than a hacksaw and a file, but it will take time and patience. If you can introduce an angle grinder or metal bandsaw into the mix, it will be much faster with less tendonitis.

The process is the same regardless of the tools you're able to work with. Use a saw or grinder to cut along the lines of your design. Once you get to tight spots like the finger grooves, make a V-shaped cut to remove as much material as possible.

Once you've removed as much excess metal as possible, use a file to straighten your lines, round your corners, and remove everything else that isn't a knife. It's helpful to have a few different shapes and sizes of files so you can fit into those smaller grooves and odd-shaped crevices.

Don't forget to drill holes for your handle pins or cord wrapping during this step of the process. It's much easier to drill your holes prior to hardening the steel.

An angle grinder is an inexpensive power tool that can help you make quick work of shaping your knife blank.

STEP 4: SMOOTH & SURFACE

Once you have finished shaping your knife blank, you'll need to smooth everything out and make sure the surface of the knife is even. This can be done using anything from a power sander to a homemade sanding block. The objective here is to make everything nice and even so you can grind your bevels.

Assuming you're working with minimal tools, a sanding block made out of mild steel with a piece of leather glued to the bottom works very well. Simply cut a sheet of sandpaper a little wider than your knife and fold it over the leather side of the bock near the end of the strip. Pinch the sandpaper between you thumbs and index fingers to hold it in place. As the grit wears down, you can slide a fresh portion of sandpaper into place and continue until the entire strip is worn out.

It's recommended to start with 80-grit sandpaper for this step. Adding some WD40 onto the sandpaper helps extend the paper's life and makes hand-sanding much easier and cleaner. You can use a bench vise to hold the knife while working on the outer edges. Once you begin working on the flat surfaces, you can place a long, flat board inside your vise so it extends out from the workbench, then clamp the knife onto the board to hold it in place. The back side of the jig we'll be discussing in the next section makes a great working surface.

Use a file to smooth and straighten your cut lines. Different sizes and shapes of files to fit different areas will make this process much easier.

STEP 5: GRIND YOUR BEVELS

Grinding the bevels into a knife is the most difficult task every knifemaker faces, whether they're using a $4,000 knifemaking grinder or the most primitive tools. This author discovered a highly effective way to grind perfectly uniform bevels using a simple homemade jig.

Jig Construction

To build the jig, you'll need three 2-foot-long ¾-inch by 3½-inch boards, wood glue, four or five eye bolts of different lengths, two nuts for the eye bolts, and a long screw with tight threading to use as a stop

The homemade jig shown opposite page and above is cheap to build and, if used properly, will produce clean, uniform bevels.

screw. You'll also need a table vise or other method to hold the jib in place when you start using it.

Using one of the boards as a base, glue and clamp the other two boards on top to form a "T" shape. Drill holes the size of your eye bolts down the center of the jig and holes the size of your stop screw down the center of the cross board. Use 1-inch spacing between the holes on both sections.

Using the Jig

To use the jig, begin by attaching a two-foot metal rod to the back of a file with hose clamps. Insert a stop screw into one of the pre-drilled holes on the cross portion of the jig. The purpose of this screw is to create a consistent stopping point as you file your primary bevel.

Next, determine where you want to set your plunge line (where the grind stops) and orient the knife so the desired plunge line lines up with the stop screw. Once you have the proper alignment, clamp the knife to the board.

Screw an eye bolt into one of the pre-drilled holes on the jig to set your grinding angle, and use the two nuts to lock the bolt in place. The angle should be set steep enough to achieve about a 30-degree bevel. The longer the bolt and the closer it is to the knife, the greater the angle.

Using the eye bolt to guide the file rod, grind down the bevel until you reach the midline of the edge, then repeat the process on the other side. Switch back to the original side and decrease the grind angle. File until the bevel reaches the desired height and blends with the original bevel, then repeat on the other side. You may have to make slight adjustments to your angle along the way until you reach the desired effect.

STEP 6: HEAT TREATING

This is a multi-step process used to change the molecular alignment of steel particles to create a hardened edge that isn't too brittle. The first step in this process is to heat the steel to a point where it temporarily loses its magnetic properties. The heating process can take place in a coal or gas forge, if available, but you can also make do with a propane or butane torch.

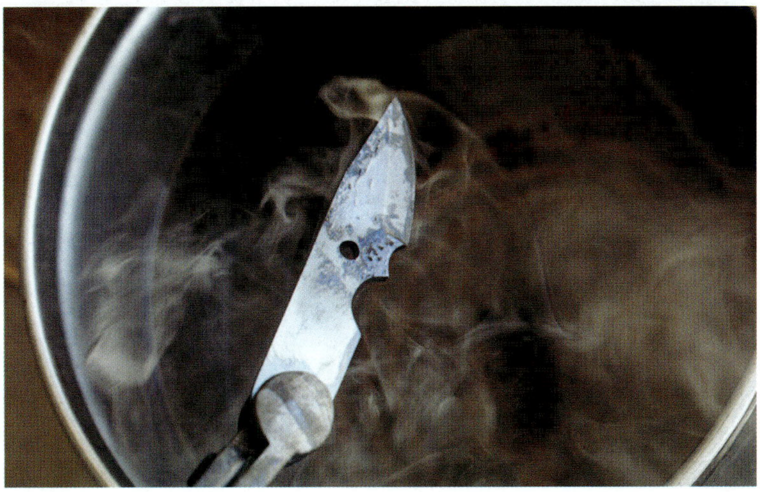

Right: Heating the blade to critical temperature and quenching it in oil to harden the steel.

Heat the blade from the cutting edge up to at least the midline of the primary bevel. The heat should be applied until the steel reaches a cherry red color that is even throughout. The heating point can be confirmed by touching a magnet against the heated portion to check for magnetism. If the heated portion of the steel doesn't pull toward the magnet, then the critical temperature has been reached and the blade is ready to be cooled using a quenching solution. After performing a magnet check, be sure to apply additional heat to regain the temperature you just lost prior to quenching.

Dip the blade into the quench as rapidly as possible after removing the heat source. Be sure to move the blade back and forth so it cools evenly. Remove the blade and check the hardness with a file to make sure the quench was effective. The file should skate off the edge if the steel is hard.

Right: Applying a cord-wrap handle to the finished knife.

Below: Using a metal block with a leather surface and 80-grit sandpaper to surface the knife.

The purpose of the quench is to cool the steel rapidly and force the rearrangement of atomic positions to produce martensite, which is an exceptionally hard phase of steel. Martensite is very brittle, so it's necessary to modify the mechanical properties by heat treating at a lower temperature range for an extended period of time, a process known as tempering.

Once the metal cools to the touch, place it in the oven for one hour at 400 degrees (temperature can vary depending on the steel). Remove the knife and allow it to cool, then repeat the process for one additional cycle.

STEP 7: RESURFACE

After heat treating the blade, you will need to remove the fire scale and carbon deposits from the surface. This is basically a repeat of Step 4, so there's no need to describe the process in detail. The amount of resurfacing you do is based on personal preference and how rough or polished you want the final product to be.

STEP 8: ADD HANDLES/ WRAP

Since handle-making and wrapping can itself be an art form, we won't go into detail here. A simple paracord wrap is the easiest method to start with. Wrap the cord around the top of the handle once, then cross and twist the cord each time the ends meet until the entire tang is covered. When you reach the end, feed the cord through the lanyard hole on the bottom and tie a knot on the other side. Cut and melt the ends of the cord to secure the knot.

STEP 9: SHARPEN

The final step is to sharpen your blade. Depending on the thickness of your primary bevel, it may take some time to form a secondary bevel resulting in a cutting edge. Sharpening can be done using stones or mechanical sharpeners. The more consistent the angle, the finer the edge will be. For a polished, razor-sharp edge, work in stages and finish by running the edge backwards along a leather strop. It's possible to improvise a strop from an old leather belt — look for one with a natural, suede-like inner surface — treated with automotive polishing compound or toothpaste.

Sharpening is the most satisfying portion of the knife-making process. Shown below is a Japanese water stone.

CONCLUSION

The art and science of metalworking can take a lifetime to master, but it doesn't have to be a difficult arena to enter. No matter if you're motivated by the survivalist's code of self-reliance or a desire for artistic expression, the information presented here — though it only scratches the surface — is enough to get you started without breaking the bank.

CASTING DECORATIVE BRASS KNUCKLES FROM BULLET CASINGS

An Incentive to Clean Up After a Day at the Range

By Patrick Diedrich

After spending hours at the range and (hopefully) cleaning up as much of your spent casings as you are able, why waste all that valuable metal when you can put it to good use? Sure, reloading spent casings is always a great option, but there are other ways to make use of brass besides using them as décor for your belt or guitar strap.

Compared to other metals, brass has a reasonably low melting point, lower even than silver or gold, making it the perfect candidate to turn into liquid and cast into something that's sure to impress. It's very important that all of the casings you collect for this project are brass, not another kind of alloy. So, grab a bucket, fill it with spent casings, and let's get started casting some brass knuckles!

Starting with a template will help fine-tune the design and prevent casting a size too large or too small.

Below: Not having reloading tools doesn't mean brass has to be thrown out.

DESIGN

To get started, you will want to flesh out a design on paper, or better yet, cardstock. Use a notecard or the cardboard from a cereal box or case of beer. Just sketch out a brass knuckle design approximately the same size as you would have the finished project be. If you are artis-

tically challenged, there's a plethora of brass knuckle outlines on the internet that you can use for inspiration or copy-paste-print. The design I am using is modeled after the UK military issued brass knuckles circa WWII.

When you're satisfied with the design, carefully cut the design from the paper, including all of finger holes. This template will be used as a stencil for the following steps. I also suggest reading this article all the way through before starting, making note of any materials you will need to acquire. Careful preparation will reduce the risk of heating and pouring molten metal.

Next, you'll need to find a piece of Styrofoam that matches your desired dimensions and use the template to stencil your design and cut away the excess foam. I used some scrap home insulation foam, but any stiff Styrofoam will work. Don't use the spongey foam used in mattress toppers or pillows. The result should be foam knuckles as close in shape and quality as you would like the finished brass knuckles to be. The fewer the defects in your foam knuckles, the better the final product will be.

PREPPING THE CAST

Plaster of Paris, which can be found in most art sections of

Right: Plaster will produce a smoother casting and reduce the amount of filing and polishing that needs to be done.

Middle: Ensure the plaster is completely dry before casting to prevent catastrophic mold failure.

Bottom left: The cast will take the form of the foam, including any engravings or embellishments you carve into it.

Bottom right: Sand will help distribute heat as the brass cools and acts as an additional safety precaution if the plaster cracks while pouring.

department stores, should be well mixed to a consistency slightly more runny than peanut butter. The smoother the mixture, the better. Any lumps or inconsistencies could be a headache later on. Place the knuckles on a piece of aluminum foil with the most aesthetically pleasing surface facing up and cover any exposed foam thoroughly with the plaster. Any exposed foam should be covered with ¼ to ½ inch of plaster. Let the plaster of Paris dry completely overnight or until you are sure there's no moisture left.

As the plaster is drying, you can prep the next steps. You will need a container wide and deep enough to hold your Styrofoam knuckles and at least 3 inches of sand in every direction. If you do some internet sleuthing on metal casting, you'll find recommendations for products like professional casting sand and tutorials on how to make your own casting sand. Buying those products is not necessary for the sake of this project and a normal bag of sandbox sand will work. When you are sure the plaster is dry, you can carefully embed the plastered foam knuckles into the top of the sand, still leaving the top exposed. Be very careful not to crack or break the plaster.

Having a dedicated smelter is ideal, but unnecessary. With patience and practice, the same effect can be done with a brake drum forge and a graphite crucible.

FIRE UP THE FURNACE

To heat melt the brass, you'll need some kind of refractory to get the temperatures high enough. There are countless tutorials online on how to do this, and you can invest as much or as little time and money as you prefer. Because I do some hobby blacksmithing, I just use my forge (it is not ideal, but it will get the job done). Lastly, you'll need something to use as a crucible, the container you'll melt the casings in. A graphite crucible is ideal and inexpensive, but there are ways you can improvise. The key is to ensure your crucible's melting point is much higher than that of the brass. A steel crucible will work in a pinch, but if your refractory gets too hot, the steel will melt, and you'll lose all your melted brass.

Now it's finally time to melt those spent casings, but before you heat them up, carefully inspect the bot-

you'll need. This is also the time to make sure your plastered foam knuckles is set properly in the sand and nearby, so you won't have far to move the crucible.

Brass is an alloy of copper and zinc, and as it melts, the zinc may burn and create smoke which, if inhaled, could cause flu-like symptoms. Also, if you melt it for too long, you'll create a layer of yellow-looking dross, the impurities which coagulate and collect as the metal burns. You may end up trying several times before you get the hang of it, but the rewards are worth the efforts! When the brass is melted, carefully pick up the crucible with the tongs and pour it evenly over

Top: Always wear appropriate PPE while working with molten metal.

Left: Casings may take time to melt. Be patient and add more as necessary.

Right: A crucible tong makes pouring the liquid brass more efficient and is much safer than improvising with something else.

Bottom: It will take a while for the brass to cool. Don't break the mold until you're certain it has cooled enough to handle.

toms and ensure the blasting caps have been fired. The last thing you want is for a blasting cap to pop while it is sitting a puddle of molten brass. Get your refractory fired up, fill your crucible with the casings, and carefully place it in the heat with tongs long enough to keep you from getting burned. As the brass heats up, you'll see the casings liquefy. Estimating the volume of what is needed with any degree of accuracy is difficult, so it's wise to melt more brass than you think

the exposed foam. The foam will burn away and be replaced with the brass. Fumes from melting brass and Styrofoam are toxic and these steps should be done in a well-ventilated area, and any precautions you wish to take to safeguard your respiration should be taken. I also suggest wearing safety glasses, welding gloves, and a long-sleeved shirt and pants made of non-synthetic fiber.

FINISHING TOUCHES

It will be tempting to check your results as soon as possible but resist the urge! Even after the brass solidifies, it will be hot enough to inflict serious burns for a while afterwards. To find out how hot a cast piece of metal was, I dropped it in a bucket of cool water 10 minutes after the liquid metal solidified. The water began to boil instantly! Play it safe and let it cool several hours before plucking it from the sand. When you are certain it's cool enough to inspect, clean or break off any plaster that's on and admire your work. If there are any flaws or deformities, they can be corrected with files or grinders, and if it's to your liking, you can make it stand out with some sanding, buffing, or polishing. The best part about this project, is if you're not happy with the results, you can always prep another piece of Styrofoam, re-melt the brass, and re-attempt it as many times as you wish.

As you can see, my casting didn't quite fill out as perfectly as I would have liked, and that's perfectly OK. Because I'm using a forge instead of a refractory, the molten brass cooled too quickly and didn't fill as much of the mold as I would have liked. On top of that, there were some impurities (dross) that I was unable to remove while it was in the crucible. But even though the last knuckle is not entirely enclosed, I would like to think of it as what Bob Ross would term "a happy accident," and utilize the gap as a bottle opener! Next time I will improve my method and make something even more impressive. This project is not exclusive to brass knuckles, this same casting concept can be used to make anything you carve out of the Styrofoam. So don't throw away the casings of the ammunition you spent all that money on, put it to good use and create something! R

DISCLAIMER: Check local laws before attempting to make your own smoke grenades.

CH. 8

Bonus Projects

SMOKE AND SPICE

Making Your Own Smoke Grenades

By Forrest Cooper

Photos by Samantha Lauraina

From theatrics to tactics, smoke has many uses other than perfecting roasted meat and getting in our eyes around a campfire. Whether to obscure movement or to signal allies, the use of smoke long predates firearms, harkening back to the days of large-scale maneuver warfare. In the 21st century, as both technology and industry have permanently changed the appearance of war and conflict, so too have combat units in the military adapted to smaller unit structures. While the concept of a whole army in conflict once dominated the scope of war, we now live in the age of small unit tactics.

And that both translates to — and is enabled by — the increased abilities of the individual. Infrastructure has adapted to accommodate small teams of highly capable individuals, providing vast resources to a smaller and smaller roster of combatants, further augmented by more extensive training as well as technology such as drones, global positioning systems, and night vision. Yet at the same time, operators are still deploying age-old tools like knives, axes, and hammers to accomplish their missions. As long as fire has been in the hands of man, smoke has had its place in war, be it the symbol of danger on the horizon or a sacrifice rising up to please the gods.

The modern military has two primary purposes for smoke: as a signal for medical support, supply drops, or target identification, and as a way to obscure troop movements. Video games frequently capitalize on both of these by imitating the use of smoke to call for reinforcements, or the completion of a task or mission. While not too far off from reality in this aspect, the use of smoke in high-paced first-person shooters leans toward the fictional, evoking the illusive disappearing act of a ninja. Outside of an organized team, smoke grenades belong more in fiction and film than reality.

That being the case, they add a certain panache to any range day, theatrical event, or training operation. And while the military supplies their operators with those iconic canisters, making a smoking compound really isn't as difficult as safety-worshipping fearmongers would imagine. The trick is matching the tool to the job and stoking the fires of one's imagination to make it work.

Legally speaking, smoke grenades, or smoke bombs, should fall under the category of pyrotechnics, not destructive devices, unlike other home-made improvisations such as Molotov cocktails. That being the case, it falls on each individual to know their state and local laws. Playing too much with fire will likely get you burned once or twice, and playing too much with smoke will certainly catch the attention of neighbors, if not law enforcement. A friendly word of advice: Courtesy and situational awareness will go a long way.

INGREDIENTS AND TOOLS REQUIRED

The easiest part of making a smoke grenade is finding out what you need. Common table sugar and potassium nitrate, generally known as saltpeter, constitute the smoke-making compound. In rural areas, you can likely buy both in the same store, but for urban dwellers, there's the internet. Potassium nitrate is cheap, used for various gardening tasks like removing stumps, but that doesn't mean every form of stump remover is made of the right stuff. Check product labels before buying. As for kitchen tools that you use to create your concoction, assume that they'll all be forfeited for future use. Thankfully, cooking utensils abound in most big-box stores. A medium-sized pan and a wooden spoon or spatula are all that you'll need. If given options, choose a smaller spoon.

Once you've sourced the base ingredients, you must choose how to carry and deploy the smoke, and how to ignite it. Luckily, various forms of fuses can be purchased online, and with the popularity of fireworks for films and festivals, there's plenty to be learned from those who spend their time imitating their idols instead of becoming them. If options are available, choose the slower-burning varieties in order to better ignite the compound. Our first attempt ended up comically resembling cartoon dynamite, with results that were much more show than bang. We used thick cardboard tubes designed for fireworks as the body, with cardboard plugs supported by heavy-duty tape to create a good seal. When constructing a smoke canister, limiting the airflow arrests the compound from burning too fast, producing more smoke along the way.

Most of your work will be done around a stove top. Throughout our trials, it became so apparent that an open flame should be

DIY SMOKE BOMBS ITEMS LIST
Cannonfuse.com
Slow Cannon Fuse: **$6**
1¼-inch Paper Plugs: **$3**
1¼-inch Tubes: **$3**
Sugar: **$8**
Potassium Nitrate: **$20**

-or-
Stump Remover: **$8**
Pastels: **$15 per big stick**
Capsaicin: **$20**
Fry Pan Set: **$17**
Utensil Set: **$7**
Electrical Tape: **$3**

Using a pastel marker sugar as base, some color could be applied to the smoke grenade itself.

For a few minutes, the sugar would show no signs of melting. Adding capsaicin to the mixture turned quickly, clearing out the kitchen and one's sinuses.

Color-coding the smoke grenades is easy with basic craft supplies.

avoided at all costs that we suggest purchasing a budget hot plate in the event that your home is equipped with gas burners. To spice things up, depending on the intended use, you can add oil-based pastels or capsaicin extract. More on these later.

CONSTRUCTION

Preparation is key to success here, so begin by pre-building the tubes. If you recycle, empty aluminum cans of the thicker soup and vegetable variety are most of the way there, the only challenge being the open end. If opting for fireworks cardboard tubes, insert the bottom plugs and wrap the whole canister in tape. Stage the tubes near the stove top or cooking surface, along with pre-cut fuses. Then, begin heating the sugar.

Our tests confirm that a 2-to-3 ratio of sugar to potassium nitrate makes the ideal smoke sauce. Making smaller batches takes more time, but they're so much easier to control that we quickly stopped using a 10-inch pan in favor of a more wieldy 8-inch one. After dumping ½ cup of sugar into the pan and setting the burner to medium-high, we added ¾ cup of potassium nitrate and stirred it into the mixture. For about 4 minutes it felt like nothing was happening, but soon the sugar began to melt and clump together. With the first batch, we completely caramelized the mixture, but we got the hang of it after that. Go for a sticky, slightly grainy texture, like sandy dough.

Once the smoke compound is complete, spoon it into your tubes, filling them to about ¾ inch from the top. That space accommodates a top plug if you use one, as well as providing the oxygen needed for the mixture to start burning. Packing it down tightly, use a skewer or similar disposable pointy thing to stab a channel in the center for the wick. Then proceed to cap the newly created smoke bomb, keeping some of the wick exposed.

Let it cool and test them out, someplace that you know you won't attract too much attention, especially considering 2020's massive spike in arson.

ADDING FLAVOR

Achieving different colors comes at the cost of buying the right materials. At a local art store, chonky pastels are sold for the artistic-minded. While the military may use different colors to mark different situations such as medical or troops in the open, for us, it's all for appearances. Don't expect to nail perfect color combinations the first time. To add a particular color to a smoke bomb, melt the pastel in the base of the pan before adding the dry ingredients and follow the same procedure described above.

While it's technically not illegal to own pepper spray and the like, making an eye-watering, cough-inducing smoke grenade may attract the attention of those whose attention you don't want to attract. To turn a smoke grenade into a dispenser of spicy air, add capsaicin extract to the smoke mixture. The extract is an oil, so it's supposed to vaporize as the compound burns. Just a little flaring up in the pan was enough to clear out our kitchen, so this is a perfectly reasonable time to wear your damn mask.

The legality of tear gas, as it's listed in various penal codes, is dubious at best, but we couldn't find consistent prohibitions on making or possessing peppered smoke. In many cases, it appeared that difference between legal and illegal would come down to chemical compounds, primarily ones that belong in a lab. Since a smoke bomb self-

can't hold onto them. Always have a fire extinguishing source on hand.

As to adding color, again we were met with mixed results. Blue color added minimal visible difference — easy to identify on a sunny day, but unnoticeable at dusk. Red added more optical flare, but as the bloom dissipated, the color blended into the expected gray of uncolored smoke. If you're looking for vibrant colors, consider buying commercial options. Sadly — and also thankfully — the capsaicin tended to burn off in our tests, adding a little agitation to our spicy air cans but not nearly as much as expected.

Finally, regarding cost, making smokes is analogous to cooking food. Large batches are cost effective, but don't justify making less than 10 servings. A medium-sized batch will easily produce upward of 20 small canisters, with the key limiting component being the throwable body. Soup cans make an excellent, albeit bulky option.

If you've ever wanted to pop smoke, this is a cool weekend project. Be safe and have fun.

identifies as a pyrotechnic apparatus, not a destructive device, and adding spice to one wouldn't make it a biological weapon, you can give it a try. But double check your local laws, and don't be surprised if it's not looked on lightly if used for a prank, even by the law.

TESTING AND EVALUATION

While our testing produced varied results, marking the tubes allowed us to narrow down the variables and make adjustments for our second batch. First and foremost, the consistency of the smoke mixture determined if the bomb would merely bloom or turn itself into a sugar rocket. The tubes filled with the more caramelized mixture tended to treat the smoking agent like a propellant, launching themselves in a cartoon-like random pattern. Conversely, the less melted and subsequently less bonded compound burnt itself out quickly to the tune of 30 to 40 seconds, producing less smoke along the way. One tube was more of a cannon, launching a glob of sticky, burning anger. Thankfully, nothing was lost but a 6-inch circle of lawn grass. Take this as a warning, though — home-made smokes shouldn't be deployed without taking proper safety considerations. They're not like Roman candles; you

Of the tests performed, the capsaicin-infused smoke grenades created more disappointment than crowd-dispersing clouds.

The sugar to potassium nitrate ratio determined whether the cardboard tube would become a smoke grenade, a sugar rocket, or a cannon launching burning, molten sugar. So always point in a safe direction.

TO ROLL 'EM OR NOT

The Economics of Reloading

By John Childs

Over just the last dozen years, ammunition prices have been as wild as a bucking bronco. They've ranged from dirt cheap (pre-Obama era) to paying a-dollar-a-shot for military surplus rifle rounds after Sandy Hook. Lately, prices have come down and appear to continue declining even as availability is on the upswing. In light of the changing price and availability dynamic, now's a good time to look at the value proposition of rolling your own ammo.

There are more reasons to reload than saving money alone, so we'll look at three common scenarios that bring shooters to the reloading bench: economy, performance, and necessity. We'll cover the benefits of making your own 9mm FMJ rounds, 6.5 Creedmoor match ammo, and then something more obscure, 300 Ultra Mag hunting ammo.

Before we get too far along, let's look at the cost of getting into a reloading setup. Currently you can find basic, single-stage reloading kits from companies such as Hornady, Redding, Lee, and RCBS for around $300. These include most of the tools needed to craft your own ammo, but you'll have to add caliber-specific dies and perhaps a few other things. Regardless of what direction you go, figure on spending about $400 to get the tools needed to get started. Of course, this is before any of the consumable (bullets, primers, etc.) or reusable (brass cases) components are purchased. To determine if reloading will be economically beneficial, you'll have to invest and recoup this amount before you'll see any savings that comes from reloading ... but be warned, very few individuals start down this rabbit hole and end up saving money, because it's a long dirty burrow you're climbing down.

THE CASE FOR 9MM PRACTICE AMMO

9mm Luger pistol ammo is ubiquitous, and its pricing is down to almost pre-Obama era levels, $10 to $15 for a box of 50 cartridges. Bought in bulk, 9mm factory training rounds can be had for as little as $0.18 a pop, though you're scraping the bottom of the Spam can at that price.

The fairest way to compare the cost of factory loaded ammo versus the home-spun stuff is to begin with all-new componentry.

A quick note on powder: There are 7,000 grains of powder per pound, so to get the cost per round, divide 7,000 by the grains used in each cartridge and then divide the cost of the powder by how many cartridges can be loaded per pound for a total per-round powder cost.

All told, it's significantly more expensive than factory ammo. But a key benefit of reloading (and the reason it's called "reloading," not "loading") is that brass is reusable. So the cost of the next box of reloaded ammo drops considerably to about $0.18 per round, or $9 per box. Of course, these calculations assume buying components in lots of 100. If you buy components in bulk the pricing will drop, sometimes substantially.

If we're just looking at 9mm as our measure, it's hard to make reloading worth it from a strictly financial standpoint. If purchasing components in bulk, reusing brass a few times, and using powder loads that aren't hard on the brass, it can pay for itself. But we haven't factored in the value of the time spent sourcing, prepping, and loading our ammo, and that's a considerable investment in itself. Another thought is availability. If you have the components on hand, if ammo becomes scarce, you'll be sitting on easy street for some time.

9MM COMPONENT COSTS	
CASES	100 pieces Starline brass = $22 / $.22 each
BULLETS	100 Winchester 115 FMJ = $11.50 / $.12 each
PRIMERS	100 CCI = $3.59 / $.04 each
POWDER	1 lb. Unique (6 grains/round) = $25 / $.02 each
TOTAL	$.40 per loaded round / $20 per box

Dies and bullets to load 6.5 Creedmoor Match ammo

LOADING FOR PERFORMANCE

Another thing to consider when contemplating loading your own ammo is the benefit of tailoring a load to suit a particular rifle and creating ammo that's more accurate than we can purchase over the counter. This alone makes reloading for performance worth the effort. Let's look at the economics of reloading the darling of the intermediate range rifle scene, 6.5 Creedmoor. Whether it's match or expanding hunting rounds, 6.5CM is everywhere these days. Its popularity has led to an extremely competitive retail market for the round, and that's great for consumers ... but what about reloaders?

6.5 CREEDMOOR COMPONENT COSTS	
CASES	100 pieces Hornady brass = $35 / $.70 each
BULLETS	100 Hornady 140 grain ELD-M= $36 / $.36 each
PRIMERS	100 Federal Gold Medal Match = $4.09 / $.04 each
POWDER	1 lb. Hogdon H4350 (41.5 grains/round) = $33 / $.17 each
TOTAL	$1.27 per round / $25.40 box

We calculate a total cost of $1.27 for each roll-your-own cartridge compared to about $1.50 for a round of factory-loaded Hornady ELD-M ammo. Aside from a significant savings on match day, let alone cheaper practice days, hand-loaded ammo comes with the ability to customize the load's bullet seating depth, powder charge, and componentry for maximum accuracy and reliability in your rifle.

While this example shows significant savings, it requires a considerable amount of time to develop a load that competes with factory offerings, unless you're just lucky enough to hit the ball out of the ballpark on your first time up at bat. Generally speaking, load development takes anywhere from 40 to 200 rounds to wrangle the most out of your rifle, especially if you're attempting to wring out every single ounce of performance, like bug-hole groups, extreme spreads in the single digits, and lightning-fast muzzle velocities. The only way those ends

are achieved is from many hours at the loading and shooting benches. The time can add up over the course of a few weeks.

A quick note on wringing out low extreme spreads and standard deviation values (geek-speak for more consistent velocities) on reloads compared to factory ammo. It's not hard to get better numbers than you'd get from factory ammo, especially if you're using top flight componentry and weight-sorting both brass and bullets. But some factory ammo is cheap enough and consistent enough to make the margin between factory and reloaded ammo pretty thin. So, do some homework and make sure you factor in the cost of your time when deciding if reloading is worth it.

300 REMINGTON ULTRA MAGNUM	
CASES	100 pieces Nosler brass = $75 / $1.50 each
BULLETS	50 Nosler 180 grain Accubond = $35 / $.70 each
PRIMERS	100 Federal Gold Medal Match = $4.39 / $.04 each
POWDER	1 lb. Hogdon H1000 (94 grains per round) $35 / $.47 each
TOTAL	$2.71 per round / $54.20 box

COMPELLED BY SCARCITY OR OBSCURITY

Let's look at 300 Remington Ultra Mag, using Nosler Trophy Grade ammo loaded with its 180-grain Accubond bullet. It costs around $80 for a box of 20 cartridges, or $4 a pop, compared to $2.71 for DIY.

Bottom: Dies, bullets, primers and dies to load 300 Remington Ultra Mag Ammo

So, it would appear reloading obscure hunting cartridges is a tremendous value and a fine way to justify a few hundred bucks in reloading equipment ... until we realize how few 300 RUM rounds the typical hunter uses in a year. Figure three to five rounds to check zero, add a few more for troubleshooting the occasional malfunction, then add anywhere from one to three rounds to take a big game animal ... that's only 10 rounds a year for a big-game hunter. But, even at 10 rounds a year, we're still saving money and maximizing accuracy by tailoring the load to our rifle. We just need to accept it'll take while to recoup the cost of our reloading equipment.

BRASS TACKS

In each of the examples above, we calculated the per-round cost using virgin brass. But reloads are exactly that. Collecting and reusing the brass from your last shooting excursion provide considerable savings since you don't have to buy it each time. Brass cases are reusable anywhere from six to 15 times, depending on the pressure of the individual loads and the care taken in handling the brass itself. In the 6.5 Creedmoor example, the price per box using reloaded brass drops from $18.40 to $11.40, a $20.59 savings over factory ammo. The savings on 300 RUM using recycled brass is even greater, dropping from $39.20 per box to $17.20, saving $68.59 over purchasing factory ammo.

Judging by economics alone, reloading ammunition can be a worthwhile endeavor depending on the cartridge you're loading. But knowledge is where it really pays

Dies, bullets, primers and dies to load 300 Remington Ultra Mag Ammo

Reloading aisle at a local retailer with a plethora of choices of bullets, powder, and primers for the reloader.

off. Developing and loading your own ammo is the absolute best way to unlock the mysteries of accuracy and reliability, ultimately elevating every shooter's ballistic abilities. You'll never watch enough videos or read enough articles to match what you'll learn about guns after just one cycle of load development.

That being said, while you'll learn a lot, by the time you've invested in all the gear it can be a financial wash. For 9mm, it's not necessarily any cheaper, and as shown if time is added into the equation, the reloader ends up upside down. Loading your own 6.5 Creedmoor can save you a few bucks while maximizing the cartridge's performance. And, finally, once you step into obscure cartridges with premium projectiles, the finances will eventually favor you, but what you really gain is the ability to shoot without the hassle of tracking down and/or waiting for tough-to-find ammo to appear in stock at your retailer.

Finally, reloading ammo is a hobby by itself for a lot of people. It'll extend the pleasure of concentration from just the time on the range, to the loving (or maddening) task of building rounds that stack bullets neatly in one hole. Conversely, if detail-oriented and fairly monotonous hobbies aren't your idea of fun, today's factory ammo is better than it's ever been. Whether looking for savings or performance, you can't make a bad choice. **R**

» SOURCES

Gun Digest – Reloading: Is it Still Worthwhile by Phil Massaro: gundigest.com/gear-ammo/reloading/reloading-is-it-still-worthwhile

Cartridges of the World – 16th Edition by Frank C. Barnes

Endcap at a Sportsman's Warehouse of plinking pistol ammo. Notice the aggressive pricing!

DIY TRUCK DRAWERS

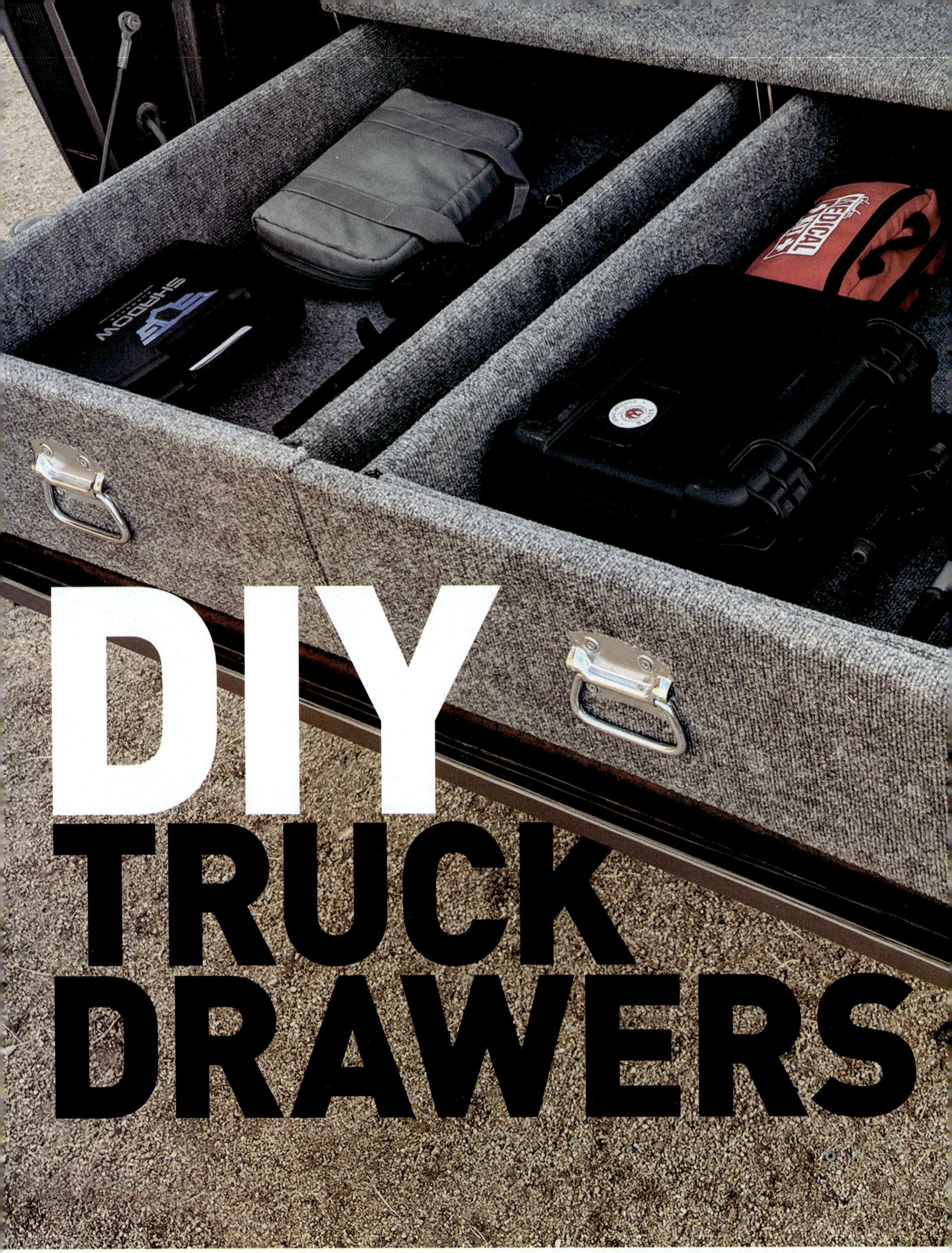

Need Storage Space for Your Guns and Gear, But Don't Want to Pay Out the Ass? We Took a Weekend to Put Together This Storage Solution for Less Than $250

By Iain Harrison

Now, you could just go out and buy a set of sliding drawers. Nothing wrong with that, and there are plenty of good companies who'll gladly take your cash for an off-the-shelf product that'll fit the bill. When you have a vehicle with other modifications, however, finding something tha'll work with say, a rollcage or existing hardware like toolboxes or saddle tanks, then things start to get a bit spendy, as you're now deep into the realm of the non-standard accessory. Or you could be like us, and just feel like paying in time rather than money is the way to go, and wish that Heinlein had added, "make a truck box," to his list of stuff a man should be able to do.

We put this together over the course of a weekend, while taking plenty of breaks for important stuff like hitting the range and gym, as well as the inevitable Home Depot runs to pick up stuff we forgot. We're giving you a complete list of hardware, as well as schematics for the critical items, so you don't have that excuse.

GRAB NAPKIN, START SCRIBBLING

This isn't the be-all and end-all of designs. You can make it your own by adding refinements, such as locks, limit stops, flush-fit handles, and so on — or you could simplify it by omitting carpet. We framed out the whole bed and scribed around the wheel wells, but there's no reason to go that route if you don't need to keep the overall height to a minimum. We did, due to the presence of a slew ring and crew-served weapon mount up top, but we realize that's not something that comes standard on every Kia.

With that in mind, the basic design shown here will fit between the wheel arches of any F-150 or Silverado. Go ahead and make it fit your needs.

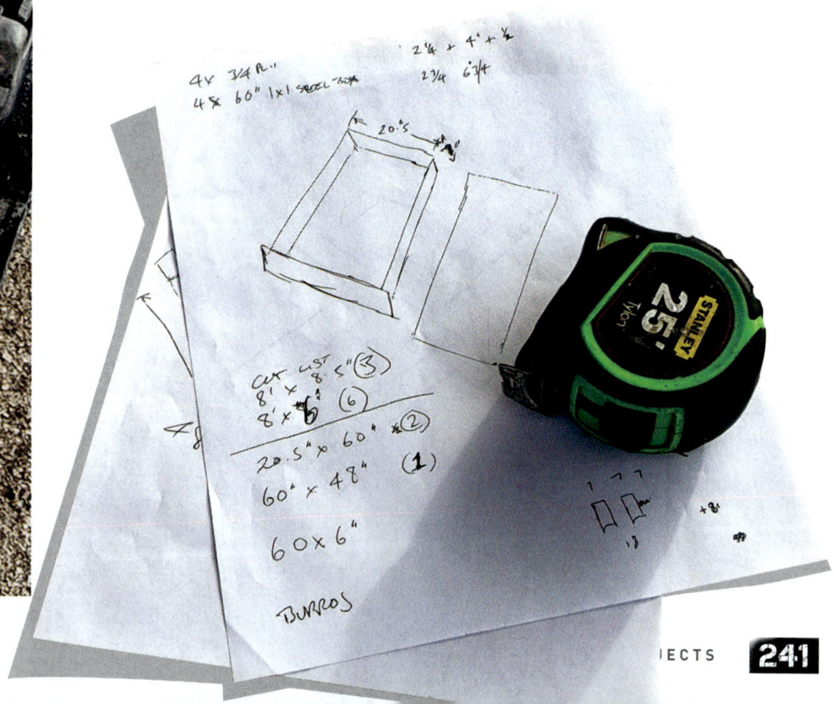

MATERIALS LIST

The total amount of each item will be dependent on how long your truck bed is. Our Ballistic Bronco has a 6-foot bed; we had to shorten up the box due to the aforementioned cage, but you'll need:

- ¾-inch exterior grade plywood (we used Radiata pine and wound up using three sheets with a fair amount of scrap)
- Skateboard bearings. A pack of 50 from Amazon will run you around $15 and leave you with a handful of spares
- 8mm x 60mm bolts — these act as axles for the bearings
- ¼-inch washers — budget five per bearing used
- 8mm nuts
- Thread locker
- 1-inch square tube — four pieces, each 4 inches shorter than the overall length of your drawers
- 1.25-inch long cabinet screws
- Polyurethane adhesive, such as Gorilla Glue
- Drawer handles
- Carpet offcuts
- Stain

TOOLS

- Circular saw
- Combo square
- Tape measure
- Utility knife
- Shears
- Kreg jig
- Cordless drill
- Clamps
- Staple gun

NICE TO HAVE

- Jigsaw
- Router with 1/8-inch roundover bit
- Impact driver (not really necessary)

METHOD

1. Start by framing out the base. We used 6-inch strips of plywood and laid them out across the bed to define the truck box outline. If you plan on being a basic bitch, then the simplest method would be to use one complete sheet as the truck box base, but this is both heavy and unnecessary from a structural standpoint. You can easily skeletonize the base, as the joints are in compression and only need to carry loads from the sidewalls, so this the perfect opportunity to bust out the Kreg jig. Glue and screw everything together, and run a couple of sheetmetal screws through the crossmember closest to the cab into the truck bed. Don't worry. It's a truck, FFS.

1a

1b

1c

1d

2. Attach the side and back walls. Make sure you use either a clamp block or speed square to ensure you don't have any problems later on down the line. More glue and cabinet screws.

3. Once the walls are in place, it's time to frame up the boxes. Size each one to fit its respective place in the base assembly, and lay out to leave a ⅛-inch gap between the side walls and the drawer sliders, as well as a ½-inch gap top and bottom.

4. Prepare the sliders by pilot drilling a hole through both sides, every 8 inches. Open up the hole on the outside face so that a ¾-inch screw head can pass through, then degrease and prime with auto primer. Once dry, use your combo square to ensure they're parallel to the bottom of the box base and attach with glue and screws.

5. Make a drill jig from a scrap of plywood to help with getting your bolt holes in the right place. Lay this out according to the diagram below. When the drawers are open, the set of bottom bearings closest to you act as a pivot point and transfer force onto the upper set of bearings, so these need to be closer together in order to handle the load.

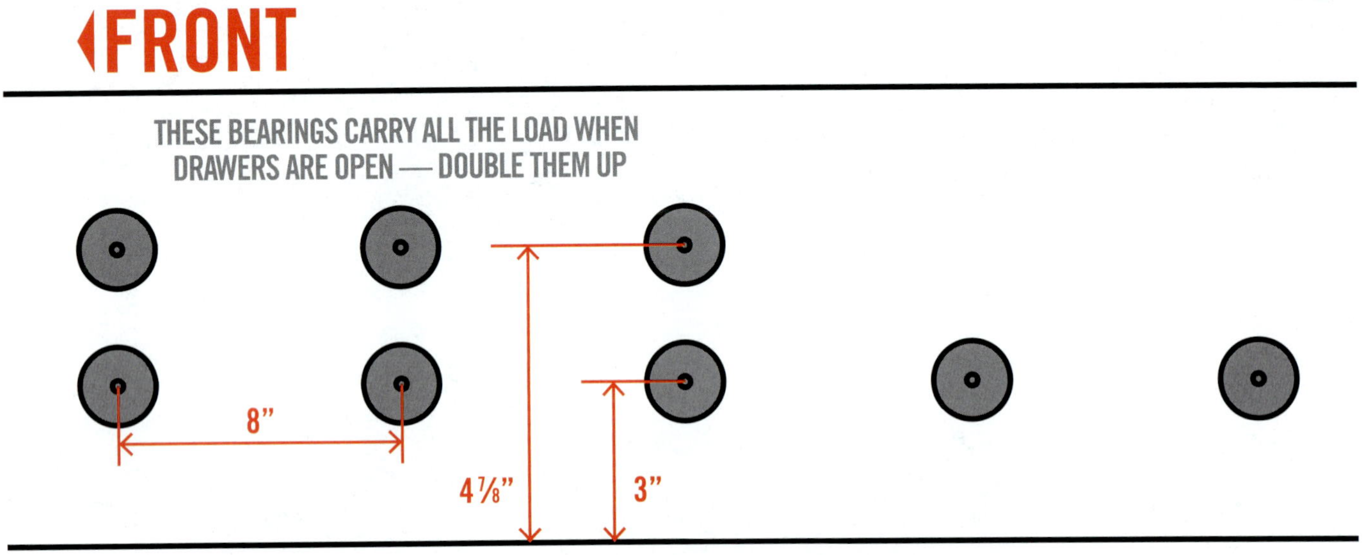

6. Drill your bearing axle holes, then assemble your bearings. Use stacked washers to shift the bearing away from the sidewalls and center it up on the drawer sliders. Use thread locker, as you don't want to be fishing inside the drawer assembly six months down the line in order to retighten a nut that's worked loose.

7. Screw down the lid.

8. Slide the drawers in and test for function. At this point, you should have a basic but serviceable way to cart your stuff. But we can do better. Using a few outdoor carpet offcuts, line the drawers to keep your guns from banging around. If you want, carpet the top of the lid also.

If you really want to go overboard, then make drawer fronts and carpet those before attaching them to the drawers, but if you're at the stage in your carpentry skills where you want to make drawer fronts, you don't need us to tell you how to do that.

Once you're happy with the fruits of your labor, kick back, crack open a cold one, and text your buddies pics to establish your dominance of the DIY truck box universe. **R**

8a

8b

8c

8d

PHOTO COURTESY OF NICHOLAS IOCCO

LONG RANGE GOALS

Want to Build a Gun Range? You've Got Your Work Cut Out for You

by Amy Bushatz

It's not that their neighbors hate shooting or guns. They all "like to shoot," and many of them "have no problem with guns" — or so they say. They aren't *those* people. It's just that they don't want to live near a shooting range.

That, at least, is the not-so-neighborly attitude many range owners encounter when they try to open their outdoor shooting complexes: opposition from the "not in my backyard" crowd.

The folks trying to keep ranges away have the ultimate American super weapon at the ready. They don't have to use the typical political showdown over gun rights, state weapons rules, regulations, or background checks to get their way. They have the power of bureaucracy.

You'd be hard pressed to find someone who has felt this force more acutely than former Marine sniper Alex Hartmann. He stumbled into the gun business after leaving active duty and found success with Ridgeline Outfitters, a company that caters to marksmen who don't mind a trek to New Hampshire to score gear and modern instruction.

With plans in his pocket for a state-of-the-art shooting park, he signed a purchase agreement for the perfect 200-plus acre parcel in Winchester, New Hampshire, with the sale contingent upon zoning approval. He did everything you're supposed to do when you're a savvy entrepreneur trying to open a business that might face an uphill battle for approval. He hired lawyers, did his homework, and poured thousands of dollars into noise studies and other preparations for facing the area's zoning board.

Alex Hartman of Ridgeline Training emerged from a zoning board meeting in Winchester, New Hampshire, to find his truck had been keyed, presumably by a neighbor opposing the construction of his proposed shooting center.
PHOTO BY ROB CURTIS

"My vision for it was this destination shooting facility that was open to the public," he says. "You're never going to make everyone happy. You could want to build a church somewhere and people are going to bitch about it."

But Hartmann made one fatal mistake: He underestimated the enemy. Perhaps he just didn't know about the six years' worth of zoning appeals his would-be neighbors brought upon one businessman before the Dunkin' Donuts was able to break ground in early September of this year. Or maybe he hadn't heard about the local neighborhood association that launched a red-tape zoning war with an asphalt company that's still ongoing.

Whatever the reason, Hartmann couldn't have imagined what would happen next. Right before the zoning committee was scheduled to make a decision in the spring of 2017, that he believed would likely be in his favor, he received a notice in the mail from his land seller: The deal was off and the land had already been sold to a group of undisclosed alternate buyers.

The buyers turned out to be a few in a group of anti-range neighbors organized under the name "Winchester Citizens Against Ridgeline." And they meant business. Members of that group, such as local resident Sander Lee, had poured thousands of dollars, much of it raised from donations, into their own sound studies to fight Ridgeline. They saw Hartmann's project as an encroachment on their own rights as landowners.

Lee insists that the neighbors like guns. They just didn't like Ridgeline.

"We live in a very rural setting that we live in by choice. We have 5 acres because we personally wanted the beauty and quiet that came from living in a rural setting," said Lee, who personally raised over $16,000 for the Ridgeline project. but said he wasn't one of those who bought the property. "We're not anti-gun, we're anti-location."

Hartmann thought he knew the neighbors' fighting style well. He had already faced them at zoning committee hearings and through their social media and online anti-Ridgeline campaign. But he was blindsided by the notice that the sale was off and the land no longer available. That was the end of the Ridgeline project in Winchester. With no land to build on, the zoning application was dead.

Hartmann is far from alone in his zoning battle. In Vermont, PRS enthusiast Nicholas Iocco has found that death by red tape is, perhaps ironically, not even a little stealthy. His troubles started in 2016 with a notice in the mail from his small town of Whiting that he was guilty of a zoning violation.

What Iocco wanted to do on his rural farmland was simple enough. With over 90 acres of terrain right outside his back door and permission from his neighbor to shoot over a portion of their adjacent 500 acres, Iocco knows his land is perfect for long-range competitions. The shooting space that he calls home is also hard to come by elsewhere in New England, making it the perfect place to host region-wide matches that can attract the cream of the PRS crop.

After registering Green Mountain Rifle Challenge LLC with the state, Iocco purchased 40 AR500 steel targets, a liability insurance policy, and a bunch of swag with his company name and logo. All told he dropped about $16,000 and settled into a swing of hosting several competitions a month with a $50 buy-in, as well as renting out his land for Vermont state police training, he said.

But then his neighbors started to complain about shooting noise and scared animals. They said the business was unsafe, concerns that Iocco said are unfounded.

Not long after the local zoning officials got involved, the state of Vermont also sent notice. If he wanted to operate his land as a business, they said, he would need commercial status, which would cause both his property taxes and the property taxes of his neighbor to skyrocket.

"The neighbors are all like 'Oh I'm a member of the NRA, I like the shooting, I own a bunch of guns' and all this stuff," Iocco said. "They can complain all they want, that's their right, but they make it seem like it's the end of the world."

Rather than deal with the taxman, Iocco abandoned the idea of making his range a profit-turning business, and now just wants to charge enough to cover his expenses. He says what he wants to do is operate a hobby. He argues that businesses make money, they don't just break even, but hobbies are a different story.

Even that requires city permission, lest he face potential fines

PHOTO BY MATT STAGLIANO

Nicholas Iocco hoped to host precision shooting matches on his property in Whiting, Vermont, but his venture, Green Mountain Rifle Challenge LLC, became entwined in a net of red tape constructed by the local zoning board, neighbors, and the state.

PHOTOS COURTESY OF NICHOLAS IOCCO

Iocco's own 90-plus acres combined with 500 more from an agreeable neighbor were to make up the Green Mountain Rifle Challenge shooting area. He envisioned a range that would've hosted precision rifle matches and given the Vermont State Police a place to train.

and problems later. At the time of this writing he was still waiting on a verdict — and still shooting and inviting friends over to do the same.

Up in Alaska, Chase Eckert is trying to avoid the battle Hartman and Iocco have faced. If Vermont and the Live-Free-or-Die State are gun friendly, Alaska is paradise. Guns aren't just a way of life on the Last Frontier, they're a daily safety necessity, even near the state's largest city. Locals generally agreed that a teen mauled to death by a bear during a popular trail race near Anchorage this year may have lived had he been packing.

But bureaucracy can still kill you even in Alaska, a place renowned for its lack of regulation. You don't need permission to open a pig farm in a residential neighborhood, but you'll have to cough up $1,000 for a special-use permit if you want to exceed 70 decibels on a commercial shooting range in the Matunsuska-Susitna Borough.

That assumes, of course, all goes well and your permit is granted. After you apply, the Borough will put your plans through a battery of tests to determine exactly what kind of nuisance your new business is likely to be. As in other municipalities in the Lower 48, the public gets a chance to file their comments and speak at a public hearing. After those steps, a seven-member board makes a determination.

In most circles in Alaska, even a hint of being anti-gun is political suicide. But that's especially true in the reddest of the red Matanuska-Susitna Valley. If neighbors in any part of the Anchorage bowl region are going to be pro-shooting range, it's there.

Eckert has billed his planned outdoor range just off the busy Glenn Highway as Livefire AK, a "shooting theme park" featuring machine gun rentals. Eckert grew up shooting rabbits on the property, and recently started renting it out as shooting space for a local tactical shop. In the summer he plans to use the range for a mix of instructor and rental lanes. In the winter it'll serve as a much-needed outdoor class space.

PHOTO BY AMY BUSHATZ

Chase Eckert and his brother, Hayden Leavitt, plan to open their shooting range, LiveFire AK, in Palmer, Alaska. They recently started construction on the property, which is currently home to a gravel pit.

But like Iocco and Hartmann, Eckert made a key miscalculation. He thought his neighbors would be at least OK with the improvements he's planning to the currently vacant property, which has been in his family for over 40 years. It's recently become host to vandals and discarded needles. Developing the property will keep those people out, he says.

He was wrong. Even before the project started, his closest neighbor, whose bedroom sits only a few hundred yards from the planned instructor lanes, began contacting the Borough and notifying the local paper of the upcoming project. A second neighbor is rumored to have asked other nearby property holders to sign a petition against the project, although he has denied doing so.

"We're going to proceed as best we can to be a good neighbor," Eckert said. "Frankly, we can do whatever we want on our land."

Officials with the National Rifle Association, who advise future range owners on best practices and how to navigate zoning boards and applications, say walking into the neighborhood approval process without enough preparation is a common mistake.

"Traditionally the people who are building ranges are ... passionate about shooting, but they may not necessarily be business people," said Elizabeth Bush, who manages recreational programs and ranges for the NRA. "We try to educate people so they go into the situation with as much information as possible, in a professional way."

Eckert thinks his odds with the zoning board are pretty good, and he's willing to take a risk. He started construction on his new range without his permit in hand, a gamble that he hopes will pay off in the long term. The permit is only required if he ends up going over the 70 decibel mark, and he's confident his planned noise abatement techniques will keep him under. Even if it takes a war, he said he plans to continue to work with the city to make his range a reality. He won't give in easily, he said.

That stubbornness isn't just an Alaskan trait. In the battle against bureaucracy there's one thing the zoning people and neighbors would be wise to remember: Shooters are persistent.

Iocco plans to continue to use his land for PRS whether his neighbors like it or not, and regardless of what the zoning board says — a right which Vermont protects, he says.

"If nothing else, Vermont does protect landowners and friends who shoot on their own property. Even if we can't have liability insurance, even if that has to all go away, even if I don't charge anything and allow them to come shoot, under Vermont law they would still be protected," he said. "Right now I couldn't care less if I could make money doing it."

And Hartmann won't let getting chased off of his original site plan keep him down, either. After the surprise sale he sued the seller for breach of contract, although he's barred by a nondisclosure agreement from revealing any details of the suit or its outcome. He said he's got his eyes on a new location for the Ridgeline park, declining to discuss any details.

But that's probably smart. After all, you can't let the opposition know what you're up to. **R**

DIY FUN WITH PVC

How to Make a Target Stand on a Budget

By C.L.'s Garage

WARNING!

Similar to handling firearms, take proper safety precautions and prepare the right tools when working on this DIY in your garage. Once on the range, follow all firearm safety rules.

You might be asking, "A target stand made of plastic? Are you serious?" It's true that PVC is not the most durable material for shooting. There are a variety of materials that are clearly more stable and last far longer. And there are many places to purchase these high-quality target stands, such as local firearm stores or online at www.letargets.com or www.brownells.com. Heck, you can do a Google search of "target stand" to find pages and pages of options.

Many can argue that PVC is not the best option for a target stand. But few can say that they can make a target stand for cheaper using less time, while having just as much fun. The parts for our do-it-yourself project cost less than $9 for the smaller size and under $13 for the bigger size.

Aside from the affordability factor, a PVC stand has other benefits, too:

LIGHTWEIGHT: Our stand weighs only a couple of pounds, even with the target backer.

REPAIRABLE: Yes, pieces will get shot out, but the PVC can be economically replaced — even on the spot at the range.

PORTABLE: The stand quickly disassembles and fits inside the trunk of most cars.

CUSTOMIZABLE: PVC stands are like life-sized Legos. You can remove, add, adjust, or resize at any time — you can even make a monster-sized one — so long as you have the additional parts prepared.

Your local hardware store should carry everything needed to build a PVC target stand.

The most important aspect of making a good PVC stand is cutting the pipes to the length needed. A PVC cutter might be a bit costly at $18, but it's one of the fastest ways to cut a plastic pipe. There are also PVC saws and hacksaw blades for cheaper. As long as you can cut the PVC pipe cleanly without hurting yourself, any of those tools will be fine.

To be fancy, you could use a staple gun, hole puncher, or drill to make a hole for a zip tie to go through. Also you will probably need something to cut the material of the backer to the proper size if it's not already sized and ready to go.

TOOLS

You may already have some or all the tools needed to measure, mark, and cut the PVC pipes in your garage. Below are our recommended tools:

1 PVC Cutter

2 Measuring tape

3 Marker

4 Masking tape

DIRECTIONS

Let us start by saying there's no single right way to make a PVC target stand. Being able to custom build your way is the best part of choosing plastic. So treat this story as a starting point, a general guideline that shows you what tools make the job easier and what materials are recommended.

MATERIALS

PVC

Plastic pipes are sold in varying thickness, but ¾-inch should do the job, ½-inch may be too thin and anything bigger than an inch might be overkill and overpriced. PVC pipes also come in various lengths, but we recommend the ¾-inch-by-10-foot-long pipes, which cost about $2.18 per 10-foot pipe. To build one stand, you need three pipes at these dimensions. Also, you will need a minimum of six 90-degree elbows (about 39 cents each) and four tees (about 40 cents).

BACKER

You will need a backer to affix your paper targets, bull's-eyes, or colored paper plates. You can use just about anything — from old cardboard and pizza boxes to foam core sold at craft stores — as long as your targets can be stapled, taped, or secured to it. Obviously, the backer will be filled with bullet holes after one use, but will need to be replaced only when you can no longer secure your target to the backer; knowing this might help decide on the chosen material.

ZIP TIES

Lastly, to secure the target backer to the stand, you will need at least four zip/cable ties that are at least 7½ inches long. These cost about 11 cents per zip tie, depending on the package you buy.

1 For the larger PVC stand, the pipes should be cut down as follows: two 41-inch tubes, four 31-inch tubes, and six 15-inch tubes.

2 Next, build the base. Connect each tee with one 15-inch tube on each side. Then take each 31-inch pipe and connect an elbow at each end.

3 Connect the four pieces just made to form the base of your target stand.

4 Next, it's time to assemble the vertical frame that'll hold the targets. Grab one 31-inch pipe and connect an elbow to each end. On the remaining 31-inch tube, connect a tee to each end.

5 Connect the two 41-inch pipes to the two pieces just made. The remaining 15-inch tubes should fit in the open holes in the tees at the bottom; these will act as legs to connect to the base. Now connect the vertical frame to the base.

6 There are numerous ways to secure the backer to the target stand; one of the most convenient methods is to use zip ties. You can use a drill, hole puncher, or even a pointy tool to make a hole in each corner of the backer where you'll attach the zip ties. To strengthen the holes on the backer, you can first apply masking or duct tape so the zip ties don't rip through.

7 Next, you'll need to secure the zip ties to the PVC frame. If you want to get fancy, you can drill holes in the 90-degree elbows so the zip ties have a dedicated spot to attach to.

8 Now you should have one completed PVC target stand ready to go. Once you're at the range, you can tape or staple your target to the backer.

9 You can follow the same procedure to build our smaller PVC target stand. For its base, you'll need two tees, four 90-degree elbows, four 10-inch tubes, and two 20-inch pipes. For its vertical frame, you'll need two tees, two 90-degree elbows, two 20-inch pipes, and four 24-inch pipes.

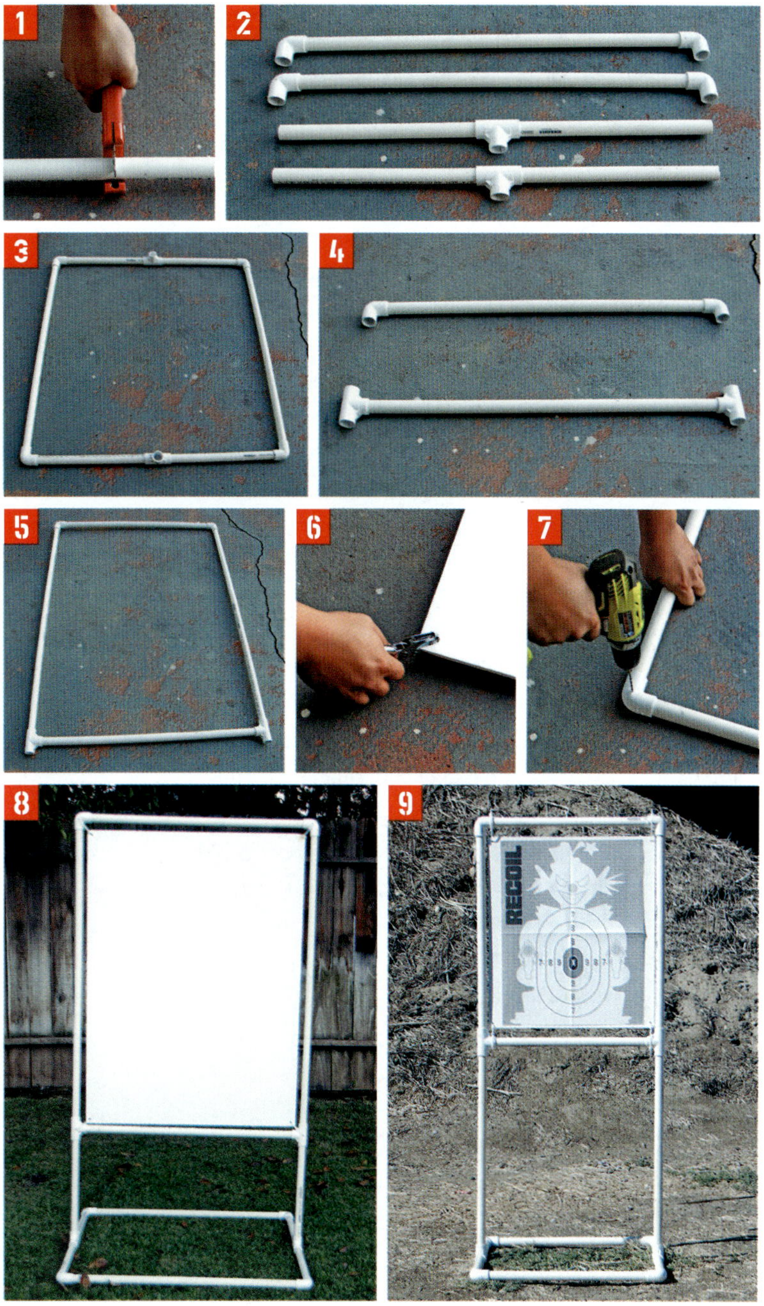

Preplanning can help produce one target stand within 10 minutes, even less if you have a helper (AKA sucker). Know that the elbow and tee joints don't ever change, so if you want to resize your stand after it's done, all you have to do is use different pipe lengths.

AT THE RANGE

Just like ammo, it's always good to have extra supplies when using your PVC target stand at the range. Here's a list of suggested gear to bring with you:

- Staple gun and extra staples: This will prop up targets much faster and more effectively than other means
- Masking or duct tape: It can be used for quick repairs or to secure targets to the backer
- An anchor: You will need something heavy to keep the target stand from blowing over, such as sand bags or tent stakes. (If you forgot to bring an anchor, you can use rocks found at the range.)
- Extra PVC: Your stand will inevitably get shot up, but most of the time it'll last the day (unless you're a worse shot than you think) and you can repair it when you get home. But extra PVC pieces mean you won't have to end the day early if you have a complete break in the plastic.

If you happen to have a rifle sighted in with a scope, you can always finish your day by testing your sniping skills: try to shoot the zip ties without hitting the PVC. So, be safe and be legal, and mostly just have fun building the PVC target stand and using it. **R**

WORKING WITH KYDEX

Learn to create any custom holster, pouch, or sheath for your gear

By Ray Tirona

Kydex was originally a low-cost alternative to custom-leather holsters. Several companies were established to make custom holsters that were handmade. In general, many Kydex holster manufacturers do great work, but at times, their designs will not fit all of their client's needs. Promotion by popular firearm instructors can drive up the waiting times and costs of custom-made holsters as well.

Kydex is a thermoplastic acrylic-polyvinyl chloride alloy that is manufactured in sheets exclusively by Kydex, LLC. Originally developed for aviation interiors, this strong, malleable material became the industry's darling for knife sheaths and firearm holsters. Because Kydex is a thermoplastic, heat is used to form the material, and if done properly, will allow a secure fit for the intended piece of gear.

This is where I started my quest to make my own holsters out of Kydex. Due to my law-enforcement background, I have been carrying a firearm for over 20 years and have four large drawers full of custom-leather and Kydex rigs. What I found was that the latest and so-called greatest designs will not fit everyone's unique lifestyle. In my opinion, the holster has to be made to fit the individual—just as you should have a new suit custom-fitted instead of just picking one up off the rack at a department store.

Because making a holster for a gun is one of the more challenging things to do with Kydex, we use it as an example in the following steps. Learn to do this, and you should be able to make any pouch or holster you'll ever need. The great thing about learning how to form Kydex is that you can finally make that holster for your one-of-a-kind handmade knife, trusty flashlight, mobile phone, or even that ninja katana you bought at the gun show.

WARNING!

When handling firearms, always observe safety rules and the precautions set forth in the firearm's owner's guide. Practice all steps on scrap materials before proceeding. Practice makes perfect. Perform this modification at your own risk.

TOOLS

1. Kydex 1/4-inch rivet press with rivet flaring dies
2. Kydex molding press, usually made of steel with high-density foam pads
3. Rivets (1/4-inch) for .080- or .093-gauge Kydex sheets
4. Electric drill
5. Kydex 1/4-inch drill bit
6. Chicago screws (1/4-inch, black)
7. Dremel rotary tool with various attachments
8. Sharp blade
9. Heat gun
10. Old toaster oven (shown on page 40)

MATERIALS

1. Kydex sheets (usually 1 square foot) in your choice of color and thickness (I use either .080 or .093 gauge)
2. Loctite Threadlocker (not shown)

01 Draw out or visualize the design that you intend to make. I like to draw it out, and it becomes my paper road map. What type of holster do you want? One with a straight drop, or an FBI cant? Where are you going to wear it? Inside, or outside, the waist—or maybe on the thigh? What firearm do you intend to use it with? What belt-loop size? Select your Kydex sheet for style, color, and thickness. You can draw a faint outline of the weapon on the sheet so that you can visualize how much you will need to make the right and left sides. When you have found the proper size, cut the excess material off.

02 Preheat your toaster oven (use an old one, do not use a contemporary one for preparing meals) to approximately 350 degrees. If you don't have a toaster oven, you can use a heat gun. Heat the Kydex until it is flexible, sort of like the thick Fruit Roll-Ups. I like to use a toaster oven, because you can heat up both sides equally; you can't do that with a heat gun.

03 When the Kydex is in a malleable state, place one side of the sheet in the molding press and the other half over the handgun, with the smooth surface toward the weapon. Make sure when you close the press that both pieces of Kydex don't shift off of the mold and weapon. Hold the press together for approximately five minutes.

04 When the holster cools enough to touch, inspect your work. At this point, you may also want to decide where the lock-up point of the holster will be, either the ejection port or the trigger guard, or a combination of both.

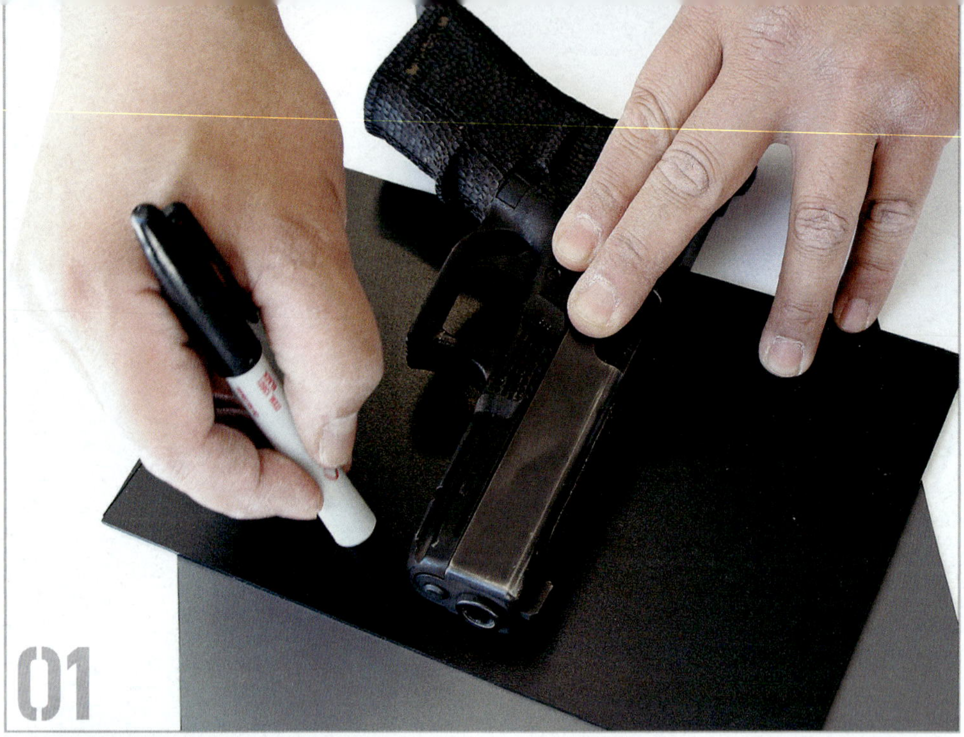

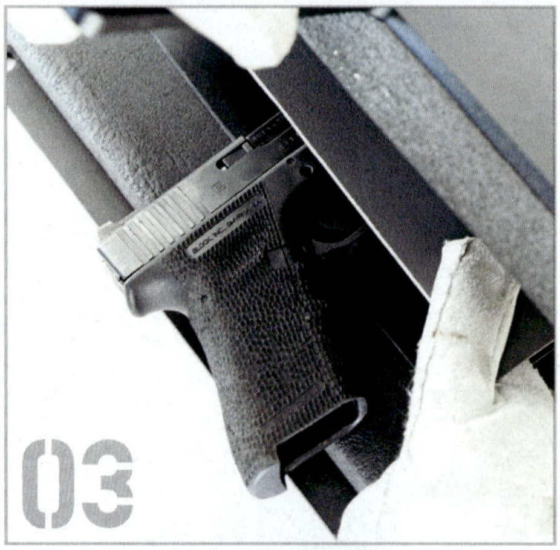

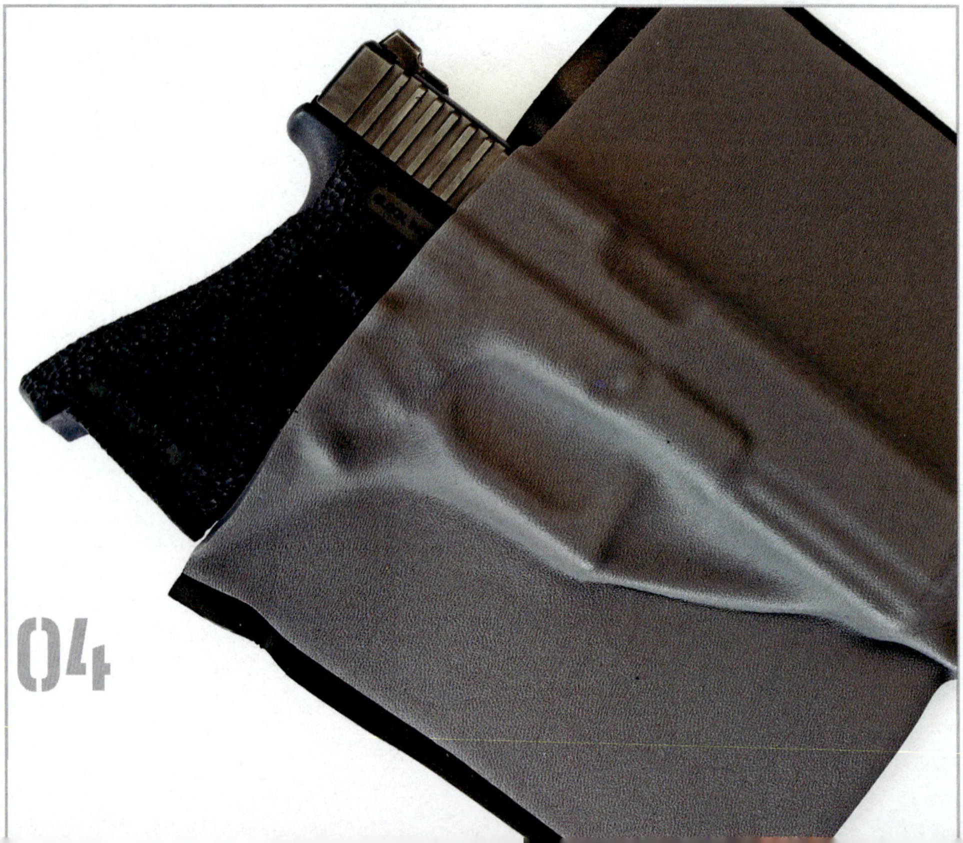

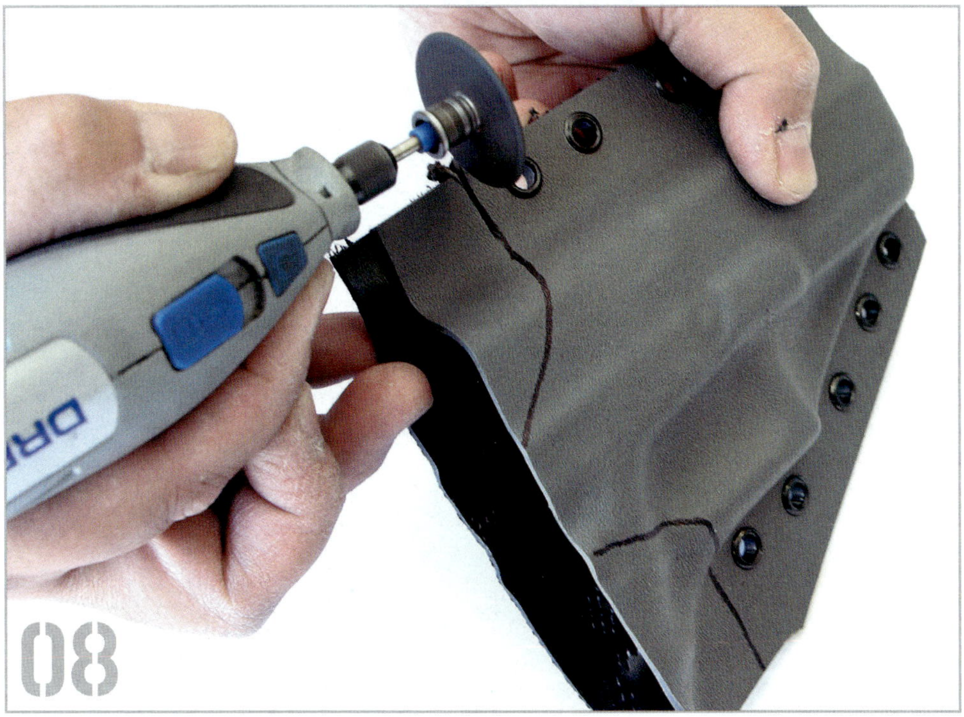

05 You can use the heat gun to soften the ejection port or the trigger guard, then sandwich the Kydex halves together and pull the handgun in and out of the mold until you reach the desired tension.

06 Now it's time to rivet the two halves together. Use the 1/4-inch drill bit to drill holes for the rivets. You only need to drill two holes at this point, so you can continue to check the shape and not get ahead of yourself.

07 Rivet each side to secure the base shape. At this point, you can use the heat gun to adjust the base holster retention. You can do this by lightly heating up the ejection port/trigger guard area again until you achieve the desired results. You can make other adjustments after the other rivets are in place. (Note: Now you can add the rest of the holes and rivets. There's no need to go overboard with the rivets; they are there to hold the two halves together and for mounting points for your desired belt attachments.)

08 Now, with both sides riveted and the tension adjusted, you can adjust the overall shape of the holster with the Dremel rotary tool. You can use the cut-off wheel or sanding drum attachments for this purpose. It's up to you how you want the holster to look. Angular looks are in vogue now, but I prefer a more organic shape. Be unique: it's your rig!

09 With Step 8 completed, the next step is to radius bevel the edges. This will make the holster smoother, preventing damage to your clothing and allowing a more comfortable carry. It will also give the holster a professional look. A good felt wheel or fine Cratex tip on a Dremel works great!

10 Next comes the belt attachments or loops. It's very important to use a belt that is designed to support a pistol's weight and spare magazines. Do not use a cheap, flimsy belt. If you decide to use a cheap belt, the holster will shift and become uncomfortable. The constant motion of adjusting the holster is easy to spot to the trained eye. You can make your own belt loops, or choose prefab ones from the many available in various catalogs. This is where proper planning is necessary to complete your design.

11 With the belt loop interface completed, it's time to test out your new holster with your chosen belt. This will show if you need to make some adjustment on the retention of the holster. Just be aware that if you overheat the rough outer surface to adjust the weapon retention, it may make that area shiny, so take care in this procedure. **R**

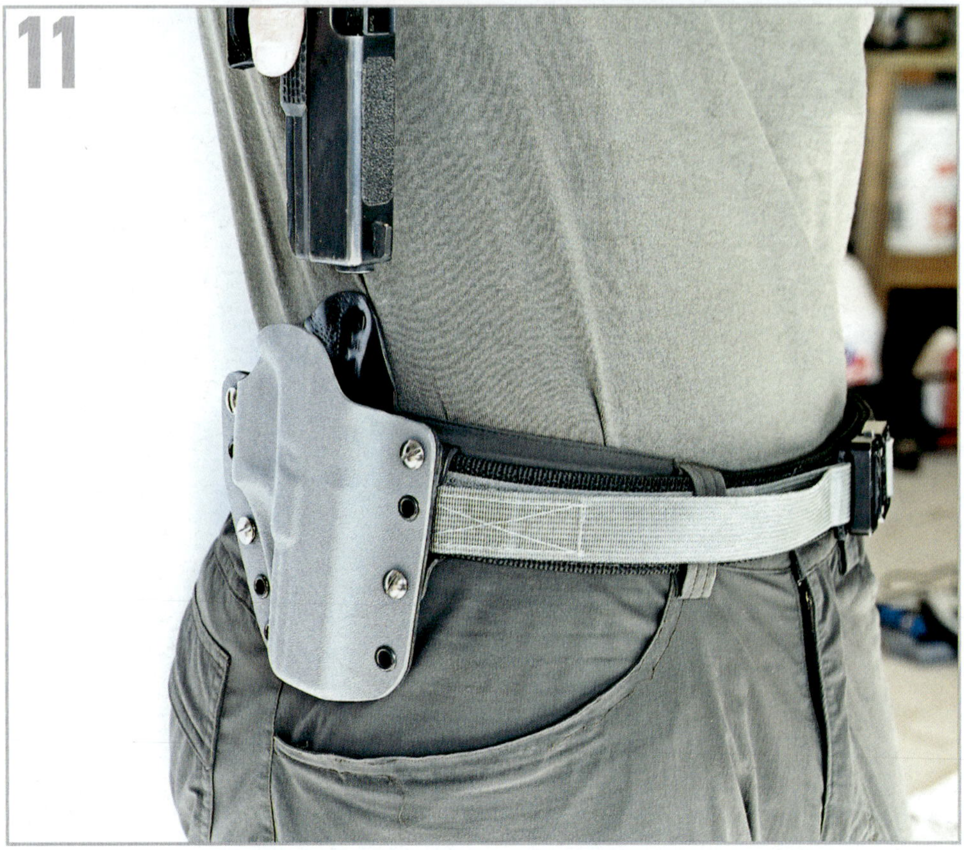

KNIFE KITS

Kydex Gun Holster and Knife Sheath Kits
URL: www.knifekits.com

DON'T WANT TO DIY?

Are the specialized tools, heat forming, riveting, and rotary sanding a bit more than you want to take on? Let a professional do it for you! The following companies are just some of the several out there that make great Kydex products:

Blade-Tech Industries
www.blade-tech.com

DSG Arms
www.dsgarms.com

G-Code
www.tacticalholsters.com

Griffon Industries
www.griffonindustries.com

JMC Custom Holsters
www.jmcholsters.com

Praetor Defense
www.praetordefense.com

Raven Concealment Systems
www.ravenconcealment.com

Safariland
www.safariland.com

Other items to consider making a Kydex holster for.

BENDING KYDEX

PHOTO BY ANTHONY KANE

Making Your Own Holster Isn't Easy, But Learning From an Expert Will Get You Started

Story by Jon Hauptman With Forward By David Reeder

Photos by Jon Hauptman

Many of us would no more attempt to customize our own gun or build our own holster than we would attempt to perform open heart surgery. Several RECOIL *staffers are included in that number — it's just important and potentially too expensive to leave to the sort of bumblefuqquery we're capable of.*

Such is not the case with everyone. Their hands are more capable, their talents more suited for building, and their efforts less likely to end in unmitigated disaster. Would-be craftsmen looking to indulge their inner do-it-yourselfer often have questions, or they should. So where should the hopeful holster handcraftsman go then for advice on DIY Kydex bending?

Jon Hauptman of PHLster, says we. We asked him what advice he'd give a new DIY bender, and here's what he told us:

Every day, I get at least one email or message from someone who's getting ready to make their first holster. In fact, as I write this, I'm holding two conversations with prospective or budding Kydex benders. Invariably, people want to jump into the deep end, head first. "Which vacuum former should I buy? How do I make an integrated holster/magazine carrier? Which clips should I use for my holster?"

As gently as I can, to avoid shattering and trampling their enthusiasm, I try to encourage temperance. There are so many questions to ask before getting started. Learning to make good holsters is a time- and resource-consuming undertaking.

Today, the technical details of the step-by-step Kydex holster forming and construction process are readily available. If it's not on my YouTube channel, it's on Kydexpro.com, where hundreds of experts share their knowledge and experience. If you need troubleshooting or advice, you have access to some great resources and helpful people. There's no reason to struggle with such basic problems as obtaining consistent well-defined molds or inconsistent edge finishing.

When asked about the most common mistakes that beginner Kydex benders make, some common technical errors came immediately to mind. From not working the hot material fast enough, to not doing any real preparation to their mold, to not using enough material, the real mistakes actually have very little to do with the technical process and more with simple fixes and easily learned lessons. If you're thinking about getting started down the path of making your own holsters, before you start setting up your garage workshop, listen up. Here are some dos and don'ts that'll be a big help in the short and long term.

Every drawing you make is a rehearsal. It's like free practice.

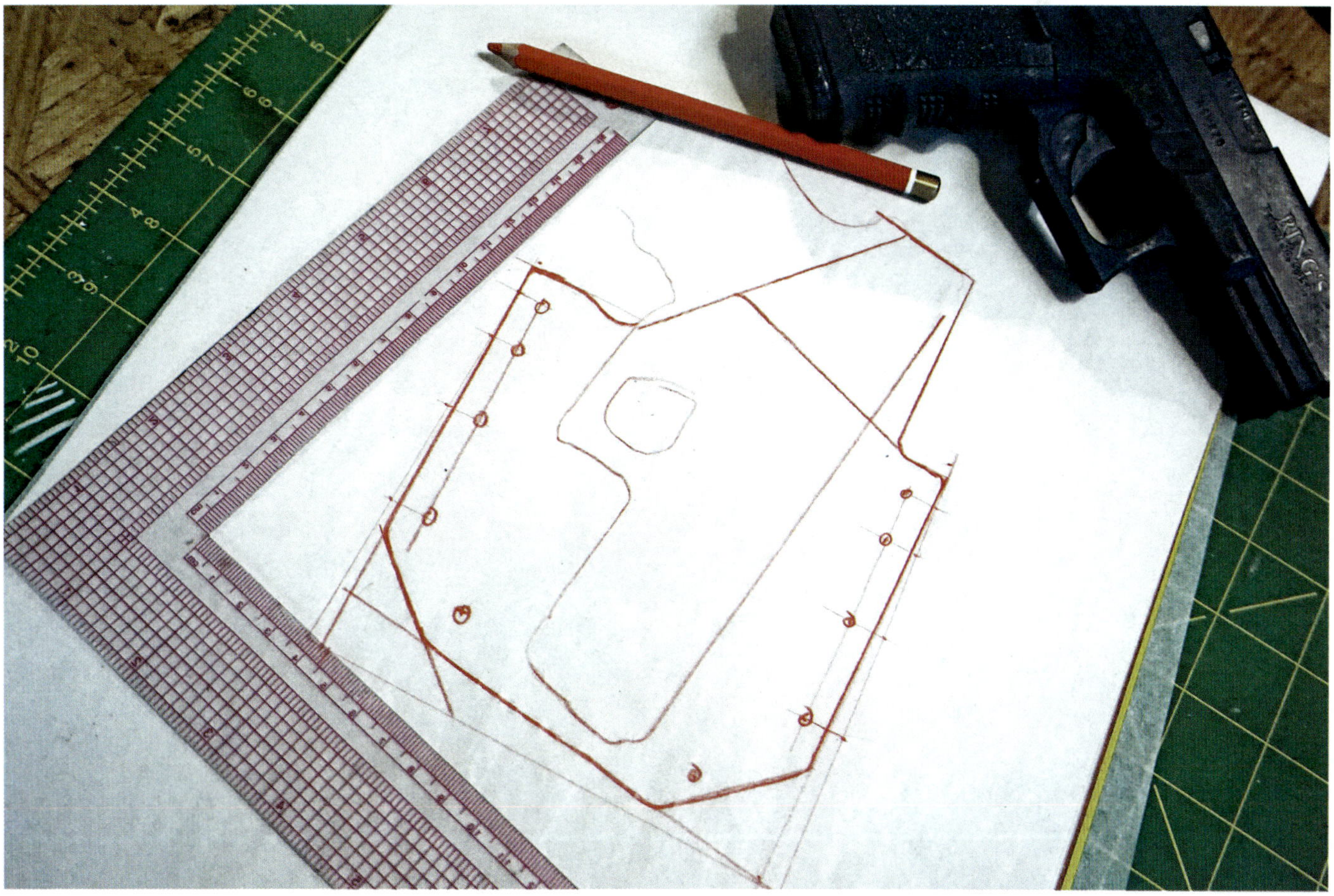

1. Don't start making holsters.

This doesn't mean, "don't pursue making holsters." It means: "Don't try to eat the whole elephant." Your first project should not be a holster. In fact, your first projects should have nothing to do with hot Kydex. Go to the art supply store and buy a pad of cheap newsprint paper, an L square ruler, some pencils, and start drawing. Fill your pad with drawings for holsters. Trace your (empty) pistol on each page and draw. Measure it and mark it the same way you would as if you were measuring and marking out a Kydex holster. Every drawing you make is a rehearsal. It's like free practice.

When the time comes to commit to drilling and cutting the real deal, you'll have done it dozens of times in practice. This is your dry-fire, so don't skip it. I've run several one-on-one weeklong boot camps here at PHLster for thermoforming students. Students draw every single night, including the evening of their arrival and the improvements over the course of the week speak for themselves. Since drawing is an activity virtually free of consequence and expense, it's a great place to do some exploration and learning.

While you're in the drawing phase, practice some other fundamentals. Practice drilling evenly spaced holes in rows like the rivets down the side of a holster. You don't want to go through all the work of making a nice mold only to throw it out or have to live with crooked rivets. Practice cutting curved lines with a scroll saw and then sanding and buffing those curves.

If you buy two 24x48-inch sheets of Kydex, break one sheet down into 6x8-inch pieces and drill a line of straight holes and cut, sand, and buff a complex curved line in each one. By the time you're done with that sheet, you won't be making disappointing mistakes on your early holsters. This isn't *Fight Club*, and nobody is throwing you into the ring on day one. You're building a skill and a craft for which you have the rest of your life to enjoy, so invest in some solid practice up front. You can also practice press skills.

Much learning is impeded by the frustration that accompanies a feeling of wasted time and material. People get so focused on making a holster and having something to show for their efforts, whether it's good or bad, that a level of stress or anxiety is introduced into the learning process. If you're having trouble working the press fast enough or getting a good, clean wrap on a one-piece IWB-style holster, you can practice this. Remember that pad of cheap newsprint? Tear out a few pages and use them to simulate a piece of hot Kydex. You can wrap those pages over the mold and work the press at your own speed to see where you're getting pinches, folds, and horns along the sight channel of your form.

When we've encountered forming errors during the boot camp, we have students stop using Kydex and switch to paper. We'll go through it step by step, when the clock isn't ticking on the material cooling, and really explore where errors occur in the forming process. If you work the press slowly, you'll be able to see the errors begin to take shape before the foam starts to obstruct your view of what's happening inside the press. This is as close to seeing inside the press as you can get, and it's very illuminating as to how the material behaves in the press.

Practice drilling evenly spaced holes in rows like the rivets down the side of a holster, cutting curved lines with a scroll saw and then sanding and buffing those curves.

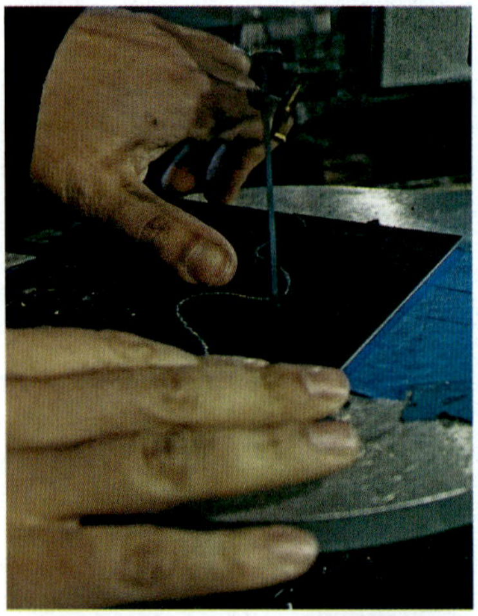

2. Don't cheap out on tools and materials.

Tin snips and Home Depot nuts and screws aren't going to get you anywhere near the holsters which have inspired you to start down this path. As far as benchtop tools go, you'll need a scroll saw, belt sander, and buffing wheel, at minimum. None of these are especially expensive, and you don't even need high-end ones to get started.

When I first began to get really busy, all I had was some cheap Skil brand belt sander. I think it took me a year of constant use and lack of maintenance to make hundreds of holsters before I finally killed that thing. As a beginner, you don't necessarily need the *best* tool for the job, but you'll only frustrate yourself by using the wrong tools for the job.

And, unlike when I started, the best supplies for making professional looking holsters are really easy to get. Companies like DIY Holster and Index Fasteners have the screws, hardware, rivets, belt loops, and belt clips that the pros use. Don't dig through that coffee can of nuts and bolts in your garage or toolbox. When the time comes, order up the good stuff to get the results you want.

3. Don't worry about making money right away.

When you're standing in front of your new work bench with your heater, thermometer, sander, buffer, saw, and the boxes of Kydex and hardware, you'll probably be thinking about how many holster sales it'll take to recoup this initial investment. Remember that you've signed up for a lifetime of enjoying a craft. And, like any craft, be prepared to spend your own time and money getting good at it.

Remember that you've signed up to learn something new and to enjoy yourself; by that metric, your investment is paying for itself from day one. Don't expect your customers to subsidize your learning process by asking them to give you money for your novice workmanship on a potentially life-saving device. Make the holsters for yourself or for free. When they're good, people will start offering you money for them, trust me.

Like any craft, be prepared to **spend your own time and money** getting good at it.

4. Don't be afraid to buy other people's holsters.

There are enough people looking at cool holsters on Instagram and copying what they see from pictures or buying cheaper holsters from gun show benders and thinking, "Hey, I can do this!"

I know they're higher-dollar items, but do yourself a favor ... buy the cool holsters that you admire. You will learn a lot from using them and studying them. These makers are your predecessors and, with the right eyes toward their work, each holster serves as a mentor.

If you pick up a Raven Concealment Phantom and know that every facet of the design is deliberate and considered, you'll find it endlessly full of lessons and clues about how to make good holsters. If their work has inspired you to start making holsters, then they deserve your business. Put together a collection of holsters that you admire and learn from using them. Pictures on the Internet won't tell you about how the holster feels and carries.

PHILLYEDC

If you're someone who learns more easily by watching than just reading you need to check out PhillyEDC, which is PHLster's YouTube channel. On it, Jon Hauptman and others do everything from demonstrating basic holster-making skills to giving advice on ancillary skills like buffing, compound use, and how to use decorative overlays. If you really want to get balls deep into this, watch the 17 video "Holster Clinic" series. If you're just interested in a good overview, take a look at the hour-long DIY Kydex Holster How-To. It must be pretty good, as nearly three-quarters of a million other people have watched it. If that isn't what you need, just look through what he has archived there. With over 200 videos available you should be able to do everything but earn your degree in Kydex bending.

5. Don't be afraid to ask someone to teach you.

There are holster makers everywhere. There's a pretty good chance that there's someone relatively local to you. Contact them. Every busy holster maker needs some kind of help at their shop. With the right attitude, you could land yourself a job or an apprenticeship with them. However, be decent enough to not take this experience and their generosity and turn around and open up on them. I have plenty of friends who've been burned by employees who show up, work for a few weeks, and then open their own shop. That's inexcusably disrespectful. Tell them your intentions up front. Maybe there's a place in their shop for you and maybe that's better than striking out on your own.

6. If you've done 4 and/or 5, make sure to give credit!

Always give credit to those who have inspired you. It's said that imitation is the sincerest form of flattery, but to acknowledgement is even more sincere. Too many people rehash some existing holster format, slap a different tacti-cool name on it, and advertise it as "our newest design."

Everyone who knows holsters knows where your ideas came from. Giving credit is important to your own design and developmental process. When something inspires you, deconstruct that. What specifics about the design and construction are inspiring? Can you trace where those elements came from? Now you're examining the origins.

If I were to tell you that the PHLster Skeleton and ACCESS holsters have their origins in the MIC holster, the Raven Vanguard 2, and in the Summer Special, that would give you a lot to explore and synthesize instead of just copying what I've done. Every time you look back on history with new eyes or different questions, you can find different answers in things that seem familiar. Keeping track of that history is important. It gives you a resource that you can re-explore and rediscover during your design process and it places your work in that lineage as well.

This era of Kydex formers won't have names with lasting legacies like Milt Sparks, Bruce Nelson, or Bill Rogers, if makers aren't honest and appreciative of their origins.

SPENDING YOUR DUCATS

Some of you will read through this whole thing and decide, "OK, I'm out. I'm not going through that whole process. I'm just gonna buy one." There's nothing wrong with that, or even with buying one until you get it right. If this is the case, and if Hauptman's expertise has you hankering for one of his rigs specifically, go to PHLster.com. There you'll have a couple of options. One is to remain there on that website to purchase something from the custom shop, including one-offs and limited editions and, of course, also to order something unique for your own needs. The other option is to take the link through to Snake Hound Machine, which has recently become PHLster's distribution partner. There you can buy the Skeleton Holster, Flatpack Tq Carrier (you're carrying some sort of med gear, right?), or any of PHLster's other minimalist gun carrying necessities. You can follow PHLster on Instagram to watch for specials (@phlster) or watch them on Facebook for updates: Facebook.com/PHLster/.

GHILLIE SUIT

A USMC Scout Sniper Gives Us the Lowdown on Building Your Own

By Pete Palma III
Photography by Henry Z. De Kuyper

The idea behind a ghillie suit is to provide a foundation to which you attach natural vegetation. Doing so will break up your outline, which is easily recognizable to the human eye. But you don't want to look like a bush — if you look into a field that has two trees, your eyes are drawn to the two trees. You want to look like the rest of the area, i.e. the field.

For example, when training for my second Iraq deployment, I took too long to get to my final firing position, and I was worried about running out of time. So I just crawled out from behind the bush that was concealing me, hastily set up my rifle, and fired. My objective was to get shots on target without being seen by my instructor. He was unable to locate me even after my second shot. But when the walker in the field revealed my location, the instructor could clearly see where I was. He had

Almost everything you need for a ghillie suit can be found at your local craft store.

been focused so intently on the two bushes bracketing me that he didn't even notice my poor position and camouflage.

A sniper platoon is one of the few places where being good at stalking is considered a good thing. However, since I was a machine-gunner rather than a sniper when I built my first ghillie, I knew absolutely nothing about the art. So I proceeded to build my suit like the ones in the movies or those you see in the big retail catalogs. It was a ghillie monster and stuck out like an officer's hot daughter.

I did just about everything wrong. When I first showed up to sniper school, where everyone has been in platoons for years, not only was I laughed at, but as penance I was made to dance around in my furry-legged ghillie bottoms in front of everyone. It was rough. Here's how to avoid making the mistakes I did.

CLOTHES

First, you'll need to pick a uniform to start with — this can be either a shirt or jacket and pants, or a one-piece like a flight suit. The jacket and pants give you more flexibility, but a coverall-type makes it easier to crawl backwards out of position as there's no way it can be dragged over your head. Tan is better than green, because it is always easier to make light things darker than to make dark things lighter (this is the theme of camouflage in general).

It should be a size larger than you usually wear because when you start sewing on pads and netting, the suit will "shrink" and you will want the extra room when you are stalking. In the movies and in old war stories from Vietnam, you always hear about snipers crawling on their bellies for days. So I initially put most of my padding on my chest and thighs. Don't that.

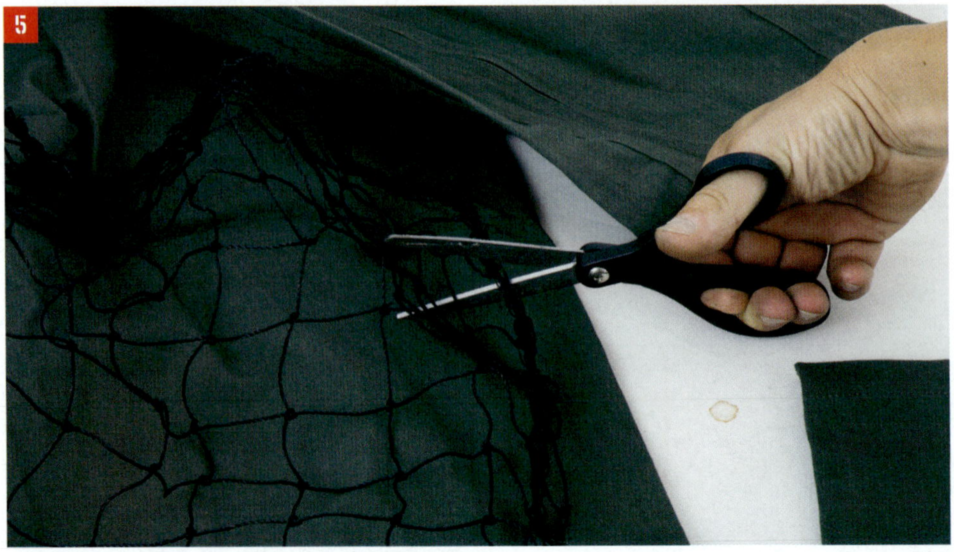

1-4 Dye your suit to match your environment with water-based dyes.

5 Lay out netting on base garment and cut to size.

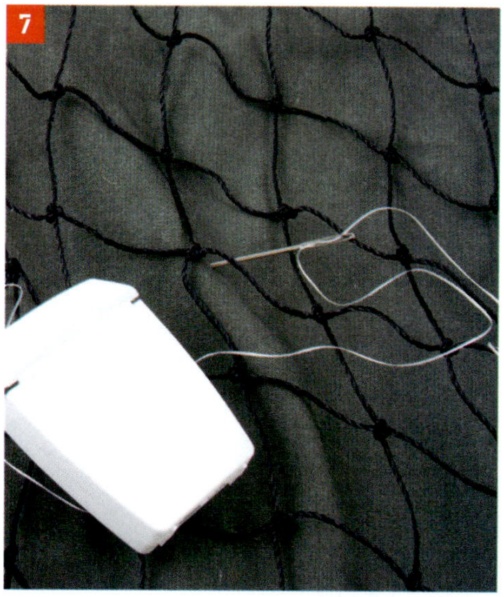

6 Attach netting with fabric adhesive.

7 Or sew at the knots using dental floss.

8 Cut jute into 12-18-inch-long chunks.

9 Form a loop and thread through netting.

10-11 Pass loose ends though loop and pull tight.

There will be times where you will be skull dragging, but the majority of your time will be spent on your hands and knees either in a high or low crawl. This provides for relatively good cover, but also allows you to move quickly. If your jacket buttons in the front, you'll need to sew a piece of cloth over it so that it doesn't open constantly, but I'd focus the padding on the elbows and knees. If you don't have experience running a sewing machine, buy some decent wine and flowers and get someone who knows what they're doing to sew it for you. Otherwise you'll be wrestling with it for hours. Ask me how I know this.

NETTING

Lay out netting on the suit and cut it to size. The netting placed on the back should have openings of around 1 to 1.5 inches. This will be large enough to attach jute and shove in vegetation. It can be attached quickly with Shoe Goo, or you can sew each knot with dental floss. The netting should cover the back, shoulders, triceps, ass, thighs, and calves.

Worthwhile additions to the suit include a piece of 550 cord on the cuffs as thumb loops to keep the sleeves from riding up your arms and annoying the crap out of you, as well as possibly cutting out the back of the jacket and inletting mesh to help keep you cool.

JUTE

Once you have attached your netting, you can then tie on jute as a base layer, dyed to match your area of operations. Do *not* go overboard with the jute, as the majority of your camo should come from natural vegetation. Find vegetation from low on the ground — things that look like nothing special, such as grass or weeds.

VEIL

The veil is arguably the most important part of the suit. It usually starts as a boonie cover with the bill cut off to avoid casting a shadow on your face, an unnatural dark area that is a dead giveaway. Netting should be placed on the veil in much the same way as on the ghillie top, and mesh can be placed in the top for cooling.

There is more to sniping and stalking than the ghille suit. I have known people who, for a bet, have done perfect stalks buck-ass naked, simply using proper movement and concealment. There is no one way to make a ghillie. If you lined up 100 snipers, you'd see 100 different ghillies and every sniper would immediately recognize his own suit. Follow the steps above, use good principles of camouflage and concealment, and make your suit your own. It is a matter of pride to create and modify your ghillie. And it will evolve with you — as you refine your skills, you will refine your ghillie. **R**

12 Add different colors to break up outline...

13 ... but don't go overboard.

14-15 Untwist strands to get more volume.

16-19 The ghillie's real job is to carry natural vegetation from the area you're working in.

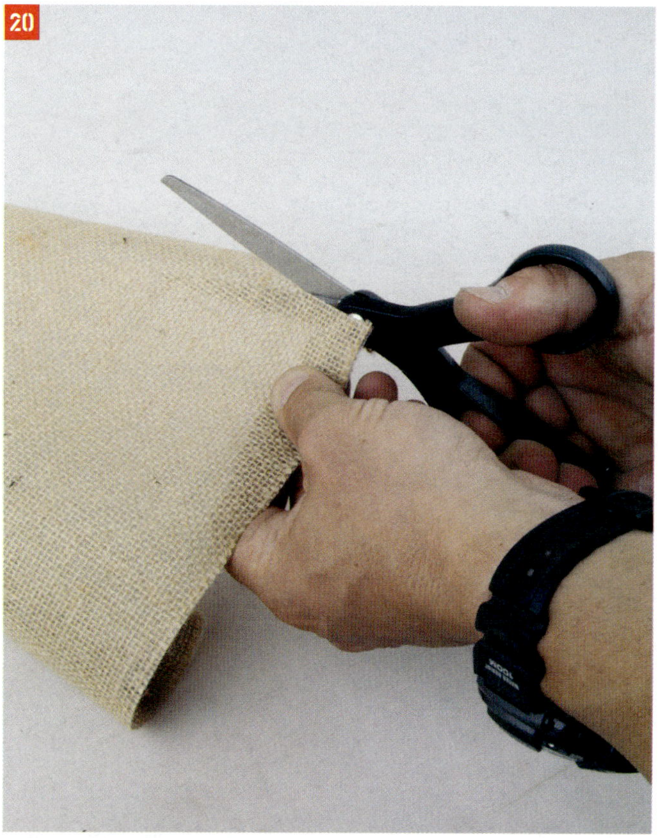

20-22 Don't have jute cord? Cut strips from sandbags and tie them to the netting instead.

RECOILTV
DOWNLOAD THE FREE APP TODAY!
RECOILTV
Navigate between a variety of channels
RELEVANT. HONEST. UNFILTERED.
RECOILTV IS THE PREMIER AUTHORITY FOR THE FIREARMS NEWS AND MEDIA YOU WANT AND DESERVE.
WWW.RECOIL.TV
RECOILTV is a 2A friendly streaming platform created to entertain and inform you on the latest guns and gear associated with the firearms lifestyle.
BROUGHT TO YOU BY:
BIG DADDY UNLIMITED
REVOLUTIONIZING THE ONLINE GUN STORE
PRESENTED BY:
BCM
EOTECH
SIGHT
norma
DEDICATED SINCE 1902
SIONYX
RECOILTV is available on:
Download on the App Store
Download on Apple TV
GET IT ON Google Play
Roku
fire tv
Chromecast